NTRES

Biology *of* Floral Scent

Biology *of* Floral Scent

Edited by

Natalia Dudareva
Eran Pichersky

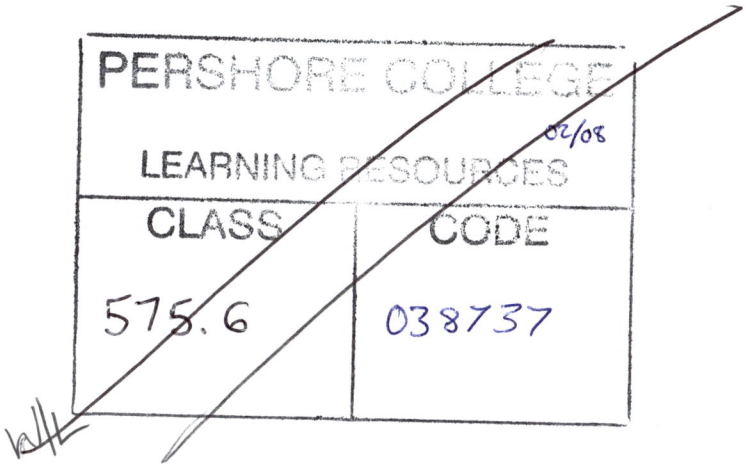

CRC Taylor & Francis
Taylor & Francis Group

Boca Raton   London   New York

A CRC title, part of the Taylor & Francis imprint, a member of the
Taylor & Francis Group, the academic division of T&F Informa plc.

Cover image by Liza Pichersky

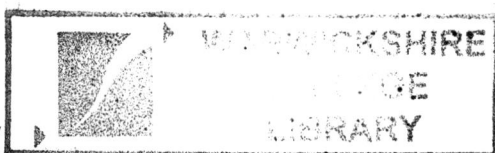

Pershore College

Published in 2006 by
CRC Press
Taylor & Francis Group
6000 Broken Sound Parkway NW, Suite 300
Boca Raton, FL 33487-2742

International Standard Book Number-10: 0-8493-2283-9 (Hardcover)
International Standard Book Number-13: 978-0-8493-2283-9 (Hardcover)
Library of Congress Card Number 2005058202

Library of Congress Cataloging-in-Publication Data

Dudareva, N.A. (Natal'ia Arnol'dovna)
  Biology of floral scent / Natalia Dudareva and Eran Pichersky.
       p. cm.
  Includes bibliographical references (p.).
  ISBN 0-8493-2283-9
  1. Flowers--Morphogenesis--Molecular aspects. 2. Flowers--Odor. 3. Plant osomophors. 4. Pollination by insects. I Pichersky, Eran. II. Title.

QK653.D83 2006
575.6--dc22   2005058202

---

informa

Taylor & Francis Group
is the Academic Division of Informa plc.

Visit the Taylor & Francis Web site at
http://www.taylorandfrancis.com

and the CRC Press Web site at
http://www.crcpress.com

Preface

The sense of smell is the most basic and universal sense. Even bacteria have mechanisms to detect the presence of chemicals in their environment. The scents that emanate from flowers have been noticed by humans since antiquity; a fact that has been documented in ancient texts. In 3000 B.C., when the Egyptians were learning to write and make bricks, they were already making primitive perfumes and using them for religious rituals. Humans' admiration for the fragrances of flowers has made these volatile substances into many commercial products. Volatiles are heavily used in the perfume, cosmetics, and fragrance industries, which are continually researching new and unusual volatile compounds and scents. Consumers are also constantly searching for new scented ornamental crops. However, the biosynthesis of floral volatiles and the roles of floral scents in plants are topics that have only recently begun to receive serious scientific attention. While we can certainly detect scent molecules in the air, the sheer number of such scents, and their complexity, confound us. Our olfactory sense is simply not good enough to separate the components and identify each one with any certainty. The consequences of our inability to clearly and objectively measure smells with our nose mean that, in the absence of appropriate instrumentation, scientific research in this area is greatly impeded.

Recent advances in practical methodologies and affordable instrumentation to collect, separate, and identify volatile compounds have allowed floral scent research to become a standard scientific research topic accessible to many investigators, which in turn has resulted in many exciting new discoveries. Thus the fourteen chapters of this book summarize and represent the progress in our current understanding of the major areas of investigation into floral scent: the techniques used to study it, how the various scent compounds are made, where they are made and how they are emitted from the flower, the effect of floral scent on the various ecological interactions between insects and flowers, and finally, how researchers are using the newly identified scent genes to genetically engineer flowers that will produce new scents. We realize that there is much more to be learned in this area and we hope that this book will stimulate new research to advance our understanding of floral scent biology.

Editors

Eran Pichersky is the Michael M. Martin collegiate professor in the Department of Molecular, Cellular, and Developmental Biology, University of Michigan, Ann Arbor. Professor Pichersky received his bachelor of science degree from the School of Natural Resources at the University of California, Berkeley, and his Ph.D. from the University of California at Davis. He was a postdoctoral researcher at Rockefeller University from 1984 to 1987, when he moved to the University of Michigan. In 2000 he was a senior Fulbright fellow as well as a senior Alexander von Humboldt fellow while visiting the Max Planck Institute for Chemical Ecology in Jena, Germany.

Professor Pichersky has trained more than 20 graduate students and postdoctoral fellows, and has authored more than 150 scientific papers. His research over the years has involved the biosynthesis of scent volatiles in the flowers of the California annual plant *Clarkia breweri* and in the model plant *Arabidopsis thaliana*. His group has also been studying the volatile compounds that are stored in special glands on the leaves of tomato and basil and are released when the plant is injured by insect herbivores. Such volatiles act as deterrents against the herbivores and also help to attract predators of the herbivores.

Natalia Dudareva is a professor at Purdue University in West Lafayette, Indiana. She received her B.Sc. and M.Sc. in biology and biochemistry at the Novosibirsk State University, Russia, and Ph.D. in molecular biology at the Institute of Biochemistry, Kiev, Ukraine, in 1982. From 1982 to 1991 she worked as a senior scientist in the Institute of Cytology and Genetics of the USSR Academy of Sciences in Novosibirsk and her research focused on the structural organization and transcription of the plant mitochondrial genome. Dudareva then completed her postdoctoral training at the Institut de Biologie Moléculaire des Plantes, Strasbourg, France (1991–1993), and in the Department of Biological Science, Windsor University, Windsor, Ontario, Canada (1993–1995), with emphasis on isolation and characterization of pollen-specific genes in sunflower.

As a postdoctoral research fellow in the laboratory of Professor Eran Pichersky at the University of Michigan, Ann Arbor, she became interested in plant secondary metabolism and biosynthesis of plant volatile compounds. Using *Antirrhinum majus* and *Petunia hybrida* as model systems, she continued the investigation of the regulation of floral volatiles' production at Purdue University, where she became an assistant professor in 1997 and an associate professor in 2001. Dudareva's laboratory is now combining the power of biochemical and genetic engineering approaches with metabolic modeling to gain new insights into the metabolic network leading to volatile secondary metabolites and to obtain a comprehensive understanding of the regulation of their production and emission in *planta*. In 2005 she received Purdue's 2005 Agriculture Research Award for her contributions to the understanding of the biochemistry of floral scent compounds.

Contributors

Susanna Andersson
Max Planck Institute of Chemical
  Ecology
Beutenberg Campus
Jena, Germany

Manfred Ayasse
Department of Experimental Ecology
University of Ulm
Ulm, Germany

Harro J. Bouwmeester
Business Unit Bioscience
Plant Research International
Wageningen, The Netherlands

Diana Buss
Department of Biological Sciences
University of Rostock
Rostock, Germany

Mikael A. Carlsson
Division of Chemical Ecology
Department of Crop Science
Swedish University of Agricultural
  Sciences
Alnarp, Sweden

Kevin C. Daly
Department of Biology
West Virginia University
Morgantown, West Virginia

Heidi E.M. Dobson
Department of Biology
Whitman College
Walla Walla, Washington

Natalia Dudareva
Department of Horticulture and
  Landscape Architecture
Purdue University
West Lafayette, Indiana

Uta Effmert
Department of Biological
  Sciences
University of Rostock
Rostock, Germany

Jonathan Gershenzon
Max Planck Institute for Chemical
  Ecology
Jena, Germany

Bill S. Hansson
Division of Chemical Ecology
Department of Crop Science
Swedish University of Agricultural
  Sciences
Alnarp, Sweden

Reinhard Jetter
Department of Botany and Department
  of Chemistry
University of British Columbia
Vancouver, British Columbia,
  Canada

Jette T. Knudsen
Ecological Institute
Lund University
Lund, Sweden

Efraim Lewinsohn
Department of Vegetable Crops
Newe Ya'ar Research Center
Agricultural Research
  Organization
Ramat Yishay, Israel

Joost Lücker
Biotechnology Laboratory
University of British Columbia
Vancouver, British Columbia,
  Canada

Eran Pichersky
Department of Molecular, Cellular, and
  Developmental Biology
University of Michigan
Ann Arbor, Michigan

Birgit Piechulla
Department of Biological Sciences
University of Rostock
Rostock, Germany

Robert A. Raguso
Department of Biological Sciences
University of South Carolina
Columbia, South Carolina

Diana Rohrbeck
Department of Biological Sciences
University of Rostock
Rostock, Germany

Ursula S.R. Röse
Institut Phytosphäre
Forschungszentrum Jülich
Jülich, Germany

Brian H. Smith
Department of Entomology
Ohio State University
Columbus, Ohio

Dorothea Tholl
Department of Biological Sciences
Virginia Tech University
Blacksburg, Virginia

Alexander Vainstein
Faculty of Agricultural, Food, and
  Environmental Quality Sciences
Hebrew University of Jerusalem
Rehovot, Israel

Linus H.W. van der Plas
Laboratory of Plant Physiology
Wageningen University
Wageningen, The Netherlands

Harrie A. Verhoeven
Business Unit Bioscience
Plant Research International
Wageningen, The Netherlands

David Weiss
Faculty of Agricultural, Food, and
  Environmental Quality Sciences
Hebrew University of Jerusalem
Rehovot, Israel

Geraldine A. Wright
Mathematical Biosciences Institute
Columbus, Ohio

Contents

Section III   Cell Biology and Physiology of Floral Scent

Section IV   Plant-Insect Interactions and Pollination Ecology

Chapter 13

Section V Commercial Aspects of Floral Scent

Chapter 14

Section I

Chemistry of Floral Scent

.

1 Detection and Identification of Floral Scent Compounds

Dorothea Tholl and Ursula S.R. Röse

CONTENTS

1.1 INTRODUCTION

Any treatment of the subject of floral scent must begin with a description of how its components are detected and identified. Such investigation is often referred to as "headspace" analysis, a term derived from the beer industry, where the analysis of the volatiles in the "head" of the beer was first developed. Floral headspace analyses were developed more than 30 years ago and have since greatly improved as analytical methods have become sensitive enough to collect and analyze volatiles by gas chromatography (GC) and gas chromatography-mass spectrometry (GC-MS) directly. Several previous reviews have discussed and compared different headspace techniques.[1-7] In this chapter we discuss several practical approaches to floral scent analysis and their advantages and limitations.

The first step in choosing a sample technique should always focus on the biology of the plant system and the purpose of the floral scent analysis. A first consideration is whether volatiles need to be collected in the field or whether the volatiles can be collected in the laboratory without affecting the composition of the blend. While some collection methods are transportable and can easily be taken to the field, other more sophisticated methods may require a complicated setup suitable only for the laboratory. A second consideration should focus on whether the floral scent of the investigated plant is already known and identified, and is only being confirmed (e.g., repetitive insect behavioral experiments), or whether the volatile blend is unknown and needs to be identified completely. Total characterization of a volatile blend often requires additional analytical steps and therefore necessitates more source material for compound identification. Also, if several flowering species are screened for the presence of only one or two compounds, the most appropriate technique may differ from the one necessary for complete identification of a complex volatile blend. Flowers may emit large amounts of volatiles that can easily be detected, even by the human nose, or they may appear rather odorless. Depending on the expected detect-ability of the floral scent, one may have to choose different types of collection methods that vary in their sensitivity. Flowers of *Arabidopsis thaliana*, for example, which release only very small amounts of volatiles (see Chapter 4),[8] require different collection techniques than flowers that release large amounts of volatiles, such as *Mirabilis jalapa*.[9]

Another important consideration is the developmental stage of the flower and timing when volatiles are collected. Some flowers, such as the orchid *Ophrys sphegodes*, are known to change their odor emission after pollination has occurred.[10,11] The volatile profile emitted by the flowers may also vary depending on the time of day,[9,12-14] as some flowers are mainly pollinated by moths and emit volatiles at night to attract their pollinators, while others are pollinated by insects that are mainly active during the day.[15] While some collection techniques allow for a very high time resolution of the volatile emissions, other techniques require several hours to collect sufficient material for further analysis. Depending on the time intervals during which volatiles should be sampled, some collection methods allow for an easy automated setup for 24 h collections, while others are most appropriate for taking a "snapshot" of the current volatile release.

In the following sections we present a selection of collection methods that range from low-tech, inexpensive, quick sampling methods to high-tech methods that require a complicated laboratory setup and can automatically collect samples in short time intervals over several days. We also present an overview of detection and identification methods for volatiles, including GC-MS, enantioselective GC, and multidimensional GC, and discuss the latest developments in ultrafast volatile analysis techniques, such as zNose™ and proton-transfer reaction mass spectrometry (PTR-MS).

1.2 FLORAL VOLATILE SAMPLING TECHNIQUES

In all volatile collection methods, the used chamber for headspace collection should be free of material that retains volatiles or causes bleeding of compounds that may contaminate the system. Good choices for materials include glass, Teflon, and metal, which are easy to clean and do not show bleeding, whereas materials such as rubber, plastic, glues, adhesives, and wood should be avoided. Details on the materials for the construction of such chambers are discussed by Millar and Sims.[6]

1.2.1 STATIC HEADSPACE SAMPLING TECHNIQUES

Sampling of volatiles from a static headspace has the advantage of an enrichment of volatiles in a closed tube or chamber. The background impurities that can result from a continuous airstream are reduced, which is an advantage when collecting from low-emitting flowers such as *A. thaliana*. However, for longer sampling times in a static airspace, humidity and a lack of gas exchange may interfere with normal physiological processes and affect the emission of volatiles. If volatiles are sampled in the presence of additional light, a temperature increase in the chamber may affect the emission of volatiles. For applications that require a time course with an expected change in volatile emissions, static headspace sampling does not work well because not all of the emitted volatiles are removed at one sampling time and changes in emission are difficult to determine.

1.2.1.1 Solid Phase Microextraction

Solid phase microextraction (SPME) is a very fast, effective, and simple method for collecting volatiles. The method is based on an adsorption-desorption technique using an inert fiber coated with different types of adsorbents that can vary in polarity and thickness and can be selected according to different types of applications. The adsorbent-coated fiber is mounted to a modified syringe and can be extended out of a needle by pushing the plunger and exposing the fiber to volatiles. The SPME device is available from Supelco (Bellefonte, PA). To collect from the static headspace of a flower sample that is enclosed in a glass container sealed with a septum, the needle of the SPME holder is inserted through the septum and the fiber is extended into the headspace (Figure 1.1). The flower volatiles are then adsorbed by the exposed fiber for several minutes to an hour, until equilibrium is reached. After volatile collection, the fiber is retracted into the needle, which is then transferred to a GC injector where the fiber is exposed to thermal desorption of the compounds. Direct desorption of volatiles from the fiber into a GC injector eliminates the need for solvent-mediated desorption, thereby reducing solvent contaminants in the analysis that may obscure some volatile compound peaks. Limitations of SPME sampling are that samples can be injected only once (no repeated injections), and the amount of material obtained from sampling by SPME is generally sufficient for GC analysis, but not for structure elucidation of unknown compounds.

The amounts adsorbed by the SPME fiber depend on the thickness of the fiber and the distribution constant of the analyte, which generally increases with molecular weight and boiling point. For most volatiles, a thick coating is recommended to

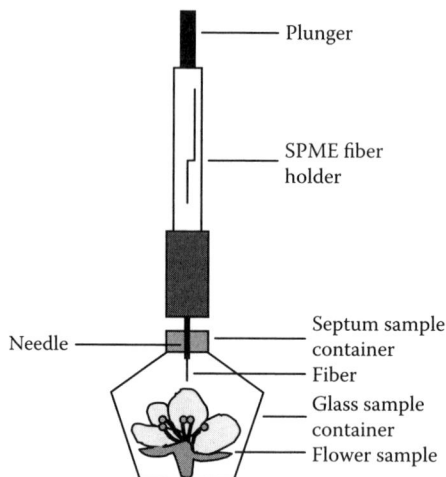

FIGURE 1.1 SPME device to extract volatiles from the headspace of a flower enclosed in a glass sample container. The adsorbent-coated fiber is mounted to an SPME fiber holder, similar to a modified syringe that is injected through the septum of the sample container. By pushing the plunger of the SPME fiber holder, the fiber can be extended out of the needle, exposing the fiber to volatiles. After collection, the fiber is retracted into the needle and the SPME is removed from the container for GC analysis of the absorbed volatiles.

better retain volatile compounds until thermal desorption, whereas semivolatile compounds may be better detected with a thin fiber coating. Thicker coatings desorb analytes more slowly, which increases the risk of carryovers. In general, fibers should be cleaned carefully by heating before reusing them. Thin coatings ensure fast diffusion and release of higher boiling point compounds. The adsorption of volatiles also depends on the polarity and porosity of the surface area. Nonpolar volatile compounds and nonpolar semivolatile compounds are effectively extracted with nonpolar fiber coatings such as polydimethylsiloxane (PDMS). Polar volatiles can be extracted with PDMS/divinylbenzene fibers, and trace level volatiles can be extracted with a PDMS/Carboxen™ fiber.

The effectiveness of SPME extraction is influenced by the volatile concentration relative to the sample volume. At low concentrations, changes in headspace sample volume do not affect responses, because equilibrium is concentration dependent. However, at higher concentrations, the sample volume has a strong effect. In a large sample volume of greater than 5 ml, containing a high concentration of volatile analyte, the amount of analyte removed from the sample is not sufficient to change the concentration. Hence, the response throughout a calibration curve is mostly exponential and is linear only for low concentrations (50 ppb). For the collection of volatiles from flowers, a small volume of 5 ml or less is often not practical. An external or internal calibration for some compounds may be possible, but is often difficult when dealing with a wide range of compounds of different concentrations in one sample. Detailed information on theory, optimization, and different types of fiber adsorbents is available from Supelco. The company also offers a portable field sampler with a Carboxen™/PDMS fiber that has a sealing mechanism to allow storing of samples for later analysis in the laboratory.

Recently the detection of volatiles by SPME has been applied to an increasing number of studies, including a variety of different flowers such as *Ceratonia siliqua*, *Osyris alba*, and different rhododendron species, with a broad range of compounds being extracted from the headspace.[16-18] For a variety of volatile compounds including terpenes and others, we have good experience using a PDMS fiber with a 100 μm film thickness. Although we have observed high selectivity of the fiber for the monoterpene (*E*)-β-ocimene, we found the method very useful for rapid screening of headspace compounds like, for example, from flowers of the butterfly bush (*Buddleja davidii*) (Figure 1.2). However, consistent sampling time, temperature, and sample volume are crucial to obtain comparable results.

1.2.2 DYNAMIC HEADSPACE SAMPLING TECHNIQUES

Sampling of volatiles from a dynamic headspace eliminates some of the problems that are connected to sampling from static headspace. In general, larger amounts of volatiles can be collected over longer time periods by adsorption (see Section 1.3) in a continuous airstream, allowing not only subsequent detection, but also structure elucidation of compounds. In addition, systems with a continuous incoming airstream provide sufficient temperature and gas exchange and avoid accumulation of compounds in the headspace that may affect the volatile release (pull and push-pull systems). Relative humidity can be adjusted in a push-pull system to a desired percentage by adding a humidified airstream to the incoming air and mixing it with dry air. However, care needs to be taken to avoid background impurities by cleaning the incoming air carefully with filters containing, for example, activated charcoal. The problem of background contaminants resulting from continuous incoming air is reduced in closed-loop stripping systems, where a limited air volume is sampled repeatedly.

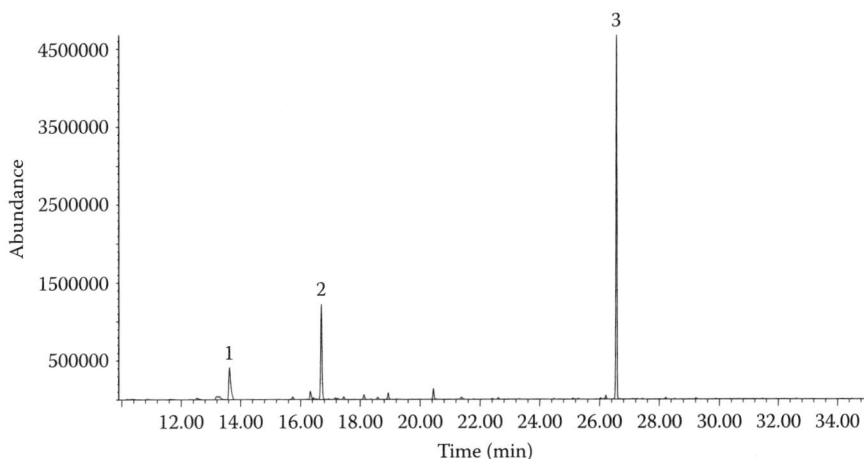

FIGURE 1.2 Volatiles collected by SPME from the headspace of detached flowers of the butterfly bush, *Buddleja davidii* (Scrophulariaceae), enclosed in a glass desiccator for 15 min. Tentative identification of main compounds by GC-MS on a DB-5MS column: (1) (*E*)-β-ocimene; (2) ketoisophorone; (3) (*E,E*)-α-farnesene.

1.2.2.1   Closed-Loop Stripping

In closed-loop stripping systems, volatiles are collected inside closed devices by circulating headspace adsorption. A closed-loop stripping apparatus has been described by Omata et al.[19] for floral scent analysis from oriental orchids. Boland et al.[20] developed a similar system, in which plants or detached plant parts are placed in small glass containers such as 1 l or 3 l desiccators. The top of the container is connected to the odor-collecting device consisting of a circulation pump, stainless steel tubes, and a stainless steel housing containing the volatile trap (Figure 1.3). Headspace air is circulated through the container and the connected trap at flow rates of approximately 3 l/min, allowing continuous quantitative collection of emitted volatiles. The system has been applied not only to the analysis of herbivore-induced volatiles,[21,22] but also to trapping floral volatiles, including collections from flowers with very low emission rates.

Analysis of volatiles from extremely low-scented flowers can be complicated for different reasons. In open headspace systems (see Section 1.2.2.3), often only a proportion of the air passing through the chamber is pulled through the adsorbent filter, thereby reducing the total amount of trapped volatiles. Higher flow rates or extended collection times can compensate for the reduced capture of emitted vola-tiles; however, these can enhance the trapping of contaminants present in the air or the collection system. The increased noise level in the baseline signal of subsequent GC analyses will complicate the detection of less abundant sample volatiles.[7] One example of a plant with low floral volatile emission rates is the model plant *A. thaliana*. *Arabidopsis* plants emit approximately 200-fold less floral volatiles per hour than strongly scented plants such as *Clarkia breweri*.[8] Closed-loop stripping of mono-terpene and sesquiterpene volatiles emitted from *Arabidopsis* flowers was performed by placing 70 to 150 excised inflorescences in small, water-containing, glass beakers inside a sealed 1 l desiccator. Volatiles were collected on traps containing 1.5 mg charcoal or 25 mg Super Q® (Alltech Associates, Deerfield, IL) (see Section 1.3)

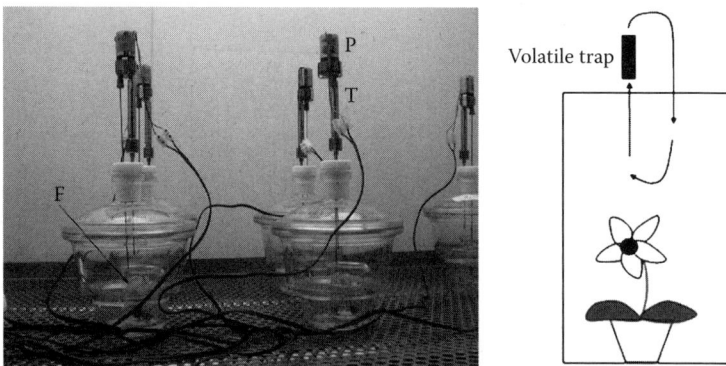

FIGURE 1.3 Volatile collection by the closed-loop stripping procedure. The photograph shows the collection of volatiles emitted from detached snapdragon flowers during feeding with isotope-labeled precursors.[24] P, circulation pump; T, steel housing containing the volatile trap; F, snapdragon flowers.

for 6 to 12 h.[8,23] Floral volatile profiles and compositions analyzed with this method were comparable to those obtained by semiopen headspace volatile trapping from similar numbers of undetached inflorescences, but showed a clearly improved signal:noise ratio, thus allowing a detailed analysis of minor components of the complex terpene volatile mixture.

In addition to qualitative and quantitative analyses of natural floral volatiles, closed-loop stripping procedures have also been used to elucidate the biosynthesis of floral scent components by trapping volatiles from detached flowers treated with biosynthetic inhibitors or supplied with isotope-labeled precursors. For example, cut snapdragon flowers were fed with deuterium-labeled terpene biosynthesis precursors of the mevalonate and methylerythritol phosphate (MEP) pathways, respectively, or inhibitors of these pathways, revealing that mono- and sesquiterpenes emitted from snapdragon flowers are exclusively synthesized via the MEP pathway.[24]

Closed-loop stripping systems can easily be set up in controlled climate chambers. They allow volatile collections from several plants at the same time, which makes them suitable for the analysis of several replicates and screening purposes. A disadvantage of the closed-loop stripping procedure is the strongly reduced air exchange between the inside and outside of the chamber that affects the gas exchange of the plant during longer sampling times. Volatiles that are not adsorbed on the trapping material (e.g., ethylene) or pass through the trap because of a saturated adsorbent surface will accumulate in the chamber and might influence the metabolism of the plant. In addition, transpiration can enhance the relative humidity during collection, requiring intermediate venting of the system between trapping cycles. It is therefore recommended to combine closed-loop stripping analyses with other open headspace sampling techniques for additional result verification and exclusion of potential artifacts.

1.2.2.2　Pull Systems

A simple form of a pull system is an adsorbent trap connected to a vacuum pump that is directly positioned next to a flower.[25,26] This may work well for some flowers that emit large amounts of volatiles; however, the risk is high of trapping ambient air that contains impurities unrelated to the flower and which will obscure minor sample compounds during GC analysis. Enclosure of a flower or flowering branch in a small glass container or a polyacetate cooking bag that releases very little volatiles[3] may reduce the amount of impurities from ambient air. Air enters the container through a purifying filter and is drawn from the chamber by pulling a defined volume of air through an adsorbent trap that can be extracted for further analysis.[7] Although this method is very easy to set up and is portable to the field, it has some drawbacks. Temperature can increase when the chamber is exposed to direct sunlight and the relative humidity inside can increase to nearly 100% within a short time and may lead to condensation on the chamber walls. If a bag is used, it may collapse and damage flower tissue, which may alter volatile emissions. Furthermore, openings in the chamber may provide additional sources of unpurified air entering the system.

1.2.2.3 Push-Pull Systems

In push-pull systems, purified air, which can be humidified to a desired percentage, is pushed into a chamber containing the flowering plant at a controlled flow rate regulated by a flow meter. A defined portion of this air is pulled through a collector trap by a vacuum pump regulated by a second flow meter (Figure 1.4a). Thus a known percentage of the volatiles emitted are collected. To avoid a vacuum or overpressuring of the system, a vent is included in the chamber. Positive-pressure venting prevents ambient air from contaminating the volatiles collected.[27] This positive-pressure venting is also employed in large glass collection chambers on top of a multiport guillotine base. The guillotine base contains concentric gas-sampling ports and two Teflon®-coated removable blades that close the bottom of the chamber around the stem of the plant, leaving an opening for the stem where the blades fit together (Figure 1.4b). This collection system allows for sampling of volatiles from flower parts of intact growing plants, while completely isolating the lower section of the plant including the soil and pot.[9,28,29] Volatiles are subsequently eluted from the trap with solvents such as methylene chloride and analyzed by GC. The technique has been described in detail[30] and was further improved and computer automated to switch traps for drawing samples at predefined time intervals over several days to follow changes in emission of volatiles.[28,31] Such sophisticated automated volatile collection systems can be custom designed by Analytical Research Systems (Gainesville, FL).

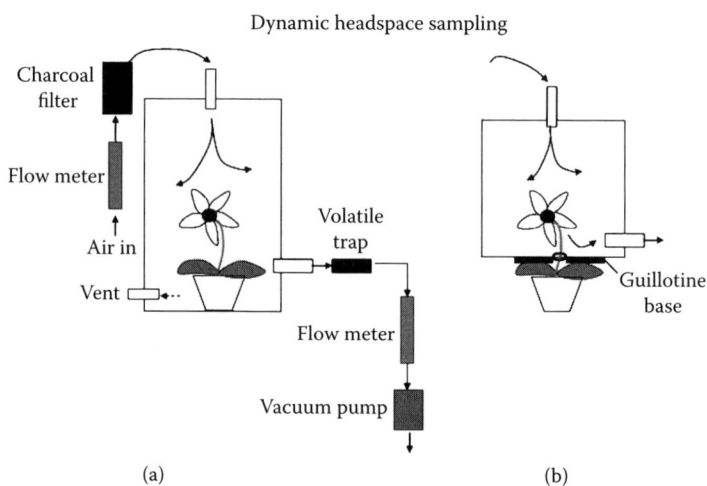

FIGURE 1.4 Example of a push-pull headspace collection system: (a) Air is pushed through a flow meter at a specified rate and cleaned by passing through a charcoal filter before entering the top of the glass chamber containing the flower. After passing over the flower sample, volatiles are pulled through a volatile adsorbent trap on the lower side of the chamber by a vacuum pump at a defined rate, regulated by a second flow meter. Access air for positive pressure venting can escape through the vent on the lower side of the chamber. (b) Example of a push-pull headspace collection chamber following the same principle as (a) for sampling volatiles from parts of a plant. As a modification, Teflon-coated guillotine-like blades close the base of the chamber around the stem of the flower, leaving a small opening for the stem.

Volatiles from flowering *M. jalapa* were identified[28] and their emission over a time course of 48 h was determined with this system.[9] The entire collection chamber can be installed in a greenhouse or in a climate chamber that allows for control of environmental parameters.

1.2.2.4   Online Volatile Collection Systems

Trapping of volatiles on adsorbents and their subsequent elution with organic solvents have been successfully used in many applications (see Section 1.3). However, the sensitivity of the method suffers from the dilution of volatiles by the solvent in the desorption process from the volatile trap. Because only small amounts of the eluted compounds can be analyzed by GC, the sampling times have to be sufficiently long to provide an eluate with enough material for analysis. Therefore precise information on the course of volatile release over short time periods is difficult to obtain. An alternative method is thermal desorption, where volatiles are collected on a trap that is directly inserted into a small oven placed on top of a gas chromatograph. By heating the trap, volatiles are transferred to the analytical column. However, because of interactions between the analytes and the adsorbent surface, higher temperatures may be required for desorption, depending on the compounds. An insufficiently low desorption temperature will result in incomplete release of compounds from the trap, whereas an excessively high desorption temperature can generate artifacts through the formation of degradation products. Tenax® has been successfully used as an adsorbent material in thermal desorption to isolate volatile chemicals[32] (see also Section 1.3), but may lead to interference of degradation products from decomposition of the polymeric skeleton and incomplete desorption of high molecular weight compounds.[33] In comparison to Tenax, the PDMS trapping material, discussed earlier for SPME analysis (Section 1.2.1.1) has been reported to have better properties in preconcentrating the analytes by dissolving them on the bulk of the liquid phase instead of adsorbing them to a porous surface.[34,35] Therefore less energy is necessary to release volatiles from PDMS traps compared to adsorption to Tenax and lower temperatures are sufficient to transfer the sample to a gas chromatography column.

For automated sampling systems, an online thermal desorption method has been described[36,37] that is now commercially available (Gerstel Online-TDS G, Gerstel, Germany). The Gerstel Online-TDS consists of two temperature vaporization chambers placed in series, which are a thermodesorption unit or connected to a temperature-programmable split-splitless injector via a 6-2-way valve mounted on a heated valve chamber. By regulating the mass flow, the system can automatically draw volatile samples with the Online-TDS G. After a cryofocusing step, compounds are flash heated and directly injected on the column. The time resolution of 5 to 60 min depends on the time necessary to collect sufficient amounts of volatiles from the emitting flower for analysis, the compounds themselves, and the time for chromatographic separation on the GC-MS. The TDS G system can be connected to two collecting containers containing the flower samples, as described in detail by Vercammen et al.[35] This method allows for a high time resolution, depending on the compounds, and high sensitivity, but currently requires an extensive laboratory setup.

1.2.3 OTHER HEADSPACE SAMPLING TECHNIQUES

Besides dynamic headspace sampling, vacuum headspace trapping techniques have been frequently applied in the past for floral fragrance analysis.[38] This method is a form of vacuum steam distillation. Flowers are subjected to a vacuum and volatile compounds are distilled off with the water contained in the plant. Subsequent condensation of the compounds at low temperatures results in a fragrance concentrate which, compared to dynamic headspace sampling, often contains higher proportions of higher boiling point compounds. Because of the improved organoleptic quality of this type of odor concentrate compared to concentrates produced by a dynamic method, the vacuum headspace technique is primarily used in perfume applications. However, it is less applicable for studying the role of floral scent in plant pollinator interactions, since the sampling technique destroys the cellular compartmentation of the plant material and can lead to an additional emission of wound-induced volatile compounds.

1.3 ADSORPTION AND DESORPTION OF VOLATILES

Excellent evaluations regarding the choice and application of volatile adsorbents have been published by several authors in recent years[6,7,39]; therefore, this chapter will primarily summarize the most important practical aspects in the use of different adsorbent materials for volatiles collection.

Matrices used for SPME in static headspace collections have already been described in detail in Section 1.2.1.1. For dynamic headspace volatile collections, adsorbents are usually packed in beds of approximately 2 to 50 mm inside glass or metal tubes between glass wool plugs or metal grids. The amount of adsorbent material used depends on the chemical properties of the compounds to be trapped, the adsorbing capacity of the matrix, the sampling volume, and the flow rate of the collection system. In the case of volatile collections from low-scented flowers, the amount of trapping matrix should be kept low to reduce the volume of solvent required for subsequent compound elution. Smaller amounts of adsorbent also minimize potential artifacts and resistance to air flow. However, if insufficient amounts of trapping media are used, volatile compounds might break through the trap because of adsorbent surface saturation. Breakthrough sampling volumes (per gram of adsorbent) are specified by the supplier or need to be determined by using a series of two traps for collection.

A variety of different adsorbents have been applied for floral volatile trapping. The most common media are the polymer-based Porapak Q® (80 to 100 mesh size; Alltech Associates; Supelco, Taufkirchen, Germany) and its refined version Super Q®, Tenax GC® and its cleaner version Tenax TA® (60 to 80 mesh; Alltech Associates), and activated charcoal. Other carbon-based adsorbents are carbon molecular sieves (Carboxen™, Carbosieve™; Supelco) and graphitized carbon blacks (Carbotrap®; Supelco).[39]

Both porous polymers, Tenax (2,6-diphenyl-p-phenylene oxide) and Porapak (ethylvinylbenzene-divinylbenzene), share similar properties. These include a high affinity for lipophilic to medium polarity organic compounds of intermediate molecular weight and low affinity for polar and low molecular weight compounds such as ethanol and water.

Trapped volatiles are usually eluted from the adsorbents into glass vials with pure solvents or mixtures of low boiling point organic solvents such as pentane, hexane, ether, acetone, or dichloromethane, the latter being preferable with Porapak. A defined amount of one or two standard compounds (e.g., 1-bromodecane, n-octane, nonyl acetate) is generally added to the sample for semi-quantitative analysis. Volatile extracts can be further concentrated by solvent evaporation at ambient temperature or under a nitrogen stream before they are stored at freezing temperatures. Under field conditions, extracts are ideally stored in flame-sealed glass ampoules to exclude sample evaporation.[2] After compound elution, adsorbents are reconditioned by rinsing with clean solvent and dried at room temperature or by flushing with clean nitrogen.

While Tenax appears to have a lower capacity for small molecules compared to Porapak,[40] its thermal stability (350°C) is higher than that of the Porapak polymer (250°C). Therefore Tenax is particularly suited for thermal desorption of volatile compounds in GC analysis. Thermal desorption in combination with cryofocusing (see also Sections 1.2.2.4 and 1.4) allows analysis of the total sample and therefore can enhance the detection limit compared to the analysis of aliquots from solvent-eluted samples. Limitations of thermal desorption include the impossibility of repeated sample injections, the degradation of thermally instable compounds, and artifacts produced from the trapping media.[33,41] Artifacts not only occur with thermal desorption, but can be the result of reactions of the adsorbent material itself or reactions of the adsorbed compounds on the polymer surface. For example, ozone reacts with terpenes on Tenax if ambient air is used in the collection chamber.[42,43] Aromatic ketones and alcohols were identified as artifacts from Porapak,[44,45] and benzaldehyde and acetophenone were found from Tenax, particularly under irradiation with sunlight.[44]

In comparison to Tenax and Porapak, activated charcoal is a cheap adsorbent with high adsorbing capacity. It is used in very small traps (1.5 mg) that are commercially available from different companies (e.g., CSLA-Filter, Le Ruisseau de Montbrun, Daumazan sur Arize, France). Traps can be eluted with small volumes (30 to 40 µl) of an organic solvent such as dichloromethane and reconditioned by extensive rinsing with solvents of different polarity. Artifacts observed with charcoal have been described for the adsorption of terpenes such as ocimene that can be oxidized on the active surface of the adsorbent.[38] Charcoal has been reported to be less efficient than Tenax in trapping aromatic aldehydes.[7] Thus combinations of Tenax and activated charcoal have been applied for trapping the full range of floral volatiles emitted from different orchid species.[46]

1.4 GAS CHROMATOGRAPHIC SEPARATION OF VOLATILES

Gas chromatography is the most efficient chromatographic technique for the separation, identification, and quantification of volatile organic compounds, including plant volatiles. Numerous research and review articles have been published describing continuous developments and advances in GC analysis technology.[47-53]

Floral volatiles are usually trapped and preconcentrated on adsorbent matrices prior to GC analysis (see Section 1.3). Samples eluted from adsorbents with organic solvents are injected into the column in a split or splitless mode. Split injection provides the advantage of rapidly transferring a small portion of the analytes to the column, resulting in narrow inlet sample bands and chromatographic peaks. In comparison, in the splitless mode, the entire sample is introduced into the column at lower flow rates. This mode is preferred for high-sensitivity analysis of samples with low concentrations.[54] The temperature of the injection liner is typically adjusted to 230 to 250°C to ensure complete vaporization of all sample components. However, adjustments to lower temperatures have to be considered in case of compound decomposition. For example, a conversion of the sesquiterpene germacrene A to β-elemene can be eliminated at an injection temperature of about 150°C.[55]

When samples are thermally desorbed from adsorbents such as Tenax, the solid material is placed directly in a thermal desorption tube that is heated to 250°C to 300°C. In a two-stage thermal desorber, the thermally released volatiles are then transported with the carrier gas to a cold or cryotrap for preconcentration prior to their injection into the GC column. Thermal desorption units are available from different suppliers (Markes, Perkin-Elmer, Gerstel). Despite their still relatively high price, they can save time and money since no manual sample preparation is needed. This advantage has led to the development of online systems combining volatile trapping with automated thermal desorption (see Section 1.2.2.4).

The separation of volatiles in GC analysis is most frequently achieved by the use of fused-silica capillary columns. The most common stationary phases, bound to the inner surface of the column, are the nonpolar dimethyl polysiloxanes, including DB-1, DB-5, CPSil 5, SE-30, and OV-1, and the more polar polyethylene glycol polymers, including Carbowax™ 20M, DB-Wax, and HP-20M. Columns are usually 30 m long and have a stationary phase film thickness of 0.2 to 0.3 μm and an internal diameter of 0.25 mm correlated with a column efficiency of approximately 5000 plates/m.[56] For different stationary phases, retention index data such as the kovats index system have been developed to facilitate compound characterization and identification. Such retention indices have been determined and summarized for several hundred volatile compounds.[57,59]

1.5 VOLATILE DETECTION AND IDENTIFICATION

For the detection of volatile compounds separated by GC, two different types of detectors can be used. The first type, for example, a flame ionization detector (FID), provides only information on retention times, while detectors of the second type, such as MS and Fourier transform infrared (FT-IR) spectroscopy, allow additional structure evaluation. The FID is the most widely used detector in GC analysis. Organic compounds are ionized in a hydrogen/air flame, producing a signal proportional to the mass flow of carbon. FIDs are primarily employed in quantitative analysis because of their wide linear dynamic range (10^6 to 10^8), a very stable response, and high sensitivity, with detection limits on the order of 0.05 to 0.5 ng per compound.

Since isomers with the same molecular formula and carbon content (e.g., sesqui-terpene hydrocarbons) principally generate the same FID signal response, relative response factors can be calculated for compounds that are not available in pure form for calibration.

Another detector used in quantitative analysis is the thermal conductivity detector (TCD), which operates by differential thermal conductivity of gaseous mixtures. Compared to FIDs, TCDs do not cause sample destruction, but have only moderate sensitivity (5 to 50 ng per compound). Other detectors applied to the analysis of volatiles are the nitrogen phosphorus detectors (NPDs), which show very high sensitivity for nitrogen- and phosphorus-containing compounds,[60] and photoioniza-tion detectors, which have been employed in monitoring plant isoprene and mono-terpene emissions.[61,62]

Besides flame ionization detection, MS is one of the most widely used detection techniques in GC analysis. The most common configuration of bench-top GC-MS systems is a gas chromatograph with a single capillary column directly coupled to a quadrupole mass spectrometer with electron ionization (EI). The operating prin-ciple of MS relies on the generation of positively charged molecules and molecule fragments from compounds exiting the GC column. The produced ion fragments enter the quadrupole mass spectrometer filter where they are selected according to their mass:charge (m:z) ratio by rapid changes in an electromagnetic field. Following detection of the ions with an electron multiplier, a total ion chromatogram is obtained providing information on the retention time of each compound and its mass spectrum that consists of a characteristic ion fragmentation pattern. Ionization is achieved by either electron impact or chemical ionization that causes less massive fragmentation. MS is a highly sensitive detection method with a minimum detectable quantity in the range of 0.1 to 1 ng per compound. The sensitivity can be further increased in the selected ion monitoring (SIM) mode, in which only selected ions representing particular compounds are scanned. The SIM mode can also be applied for quanti-fication by measuring the most abundant base ion unique to each compound.

Because of the popularity of EI-MS for routine analysis of volatile compounds, several comprehensive mass spectral libraries (Wiley, NIST MS Database, 1998) have been established that are used in EI-MS searches to support compound iden-tification. Other databases for mass spectral comparison of volatile compounds have been developed by Adams[59] and König et al.[63] The MassFinder software, by König et al., allows the comparison of GC-MS data with those of the provided mass spectral library as well as retention indices obtained under identical instrumental and exper-imental conditions. The MassFinder library contains approximately 2000 spectra of monoterpene, sesquiterpene, diterpene, aliphatic, and aromatic plant volatiles. In addition, MS data for newly identified compounds can be incorporated into the library. Despite the convenient use of mass spectral library data, an umambiguous identification of a compound can only be achieved by the comparison of its mass spectrum with that of an authentic standard analyzed on the same column and the determination of kovats indices on at least two columns with different polarities.

Tandem MS systems have been established to allow separate analyses of single compounds of complex GC peaks.[47] For example, GC-tandem MS was applied to determine the floral scent composition of *Cucurbita pepo* flowers.[64] Moreover,

GC-MS analysis can be complemented by capillary GC-FT-IR, for example, for the differentiation of closely related isomers with very similar EI mass spectra.[47,65] Since FT-IR provides information on the intact molecular structure, unique spectra even for similar isomers can be obtained. The spectroscopic method has been applied for the identification of different floral volatiles.[66,67] Drawbacks of GC-FT-IR are difficulties in quantification and time-consuming data interpretation, although a growing collection of data is provided by the Sadtler database (Sadtler Division of Bio-Rad, Philadelphia, PA, USA).

1.5.1 ENANTIOSELECTIVE GC, MULTIDIMENSIONAL GC

The chirality of floral scent compounds can be crucial for the olfactory response of pollinators and herbivores. Hence determining the enantiomeric composition of floral volatiles is critical in understanding plant-animal interactions. Since the 1990s, enantioselective capillary columns with chiral phases, such as different hydrophobic cyclodextrin derivatives, have been developed for enantiomer resolution of a variety of chiral volatile compounds, primarily from essential oils.[68,69] As a general rule, polar compounds are better resolved on acylated cyclodextrin derivatives, while non-polar analytes are better separated on prealkylated cyclodextrin derivatives.[69] König et al. have assembled an enormous amount of data for the identification and enantiomeric recognition of hundreds of sesquiterpene hydrocarbons.[70–73] Other examples for the application of cyclodextrin derivatives in flavor and fragrance analysis were documented by Schreier et al.[74]

In situations where complex volatile mixtures cannot be sufficiently separated on a single chiral column, often two-dimensional capillary GC is employed. In this approach compounds are first separated on a conventional column. Then fractions containing compounds eluting from the first column (heartcuts) are directed to the chiral column as the second dimension. The redirected flow might need to be refocused at the start of the second column by cryotrapping. Borg-Karlson et al.[75] applied a multidimensional GC system to determine the enantiomeric purity of linalool oxides in the floral fragrance of the early flowering shrub *Daphne mezereum*. The system consisted of two gas chromatographs with the first chromatograph housing a DB-WAX column and the second containing two enantioselective β-cyclodextrin columns. The two chiral columns can be used in parallel to ensure optimal enantiomeric resolution. Besides their application in chiral analysis, several other multidimensional GC systems or comprehensive GCxGC systems with directly coupled columns have been developed, particularly in the field of essential oil analysis, allowing increased resolution and improved quantitation or identification of volatile components.[49]

1.6 STRUCTURE ELUCIDATION OF VOLATILE COMPOUNDS

For structure elucidation of unknown volatile compounds usually multiple analytical steps need to be considered. According to König and Hochmuth,[69] in only a few cases might simple mass spectra allow a direct derivation of the corresponding structure.

Most often, sufficient amounts of single compounds need to be isolated for one- and two-dimensional nuclear magnetic resonance (NMR) spectroscopic techniques. Preparative isolation can be accomplished by multiple chromatographic steps including preparative-packed GC columns and thick-film capillary columns with highly selective cyclodextrin matrices. Recently, a simple, efficient NMR sample preparation technique for volatile chemicals has been described using a micropreparative GC system.[76] The absolute configuration of a new compound can be determined by comparison to a synthetic reference compound can be chemical correlation using enantioselective GC. Further details regarding structure elucidation are given elsewhere[69] and go beyond the scope of this chapter.

1.7 VOLATILE ANALYSIS TECHNIQUES WITH HIGH TIME RESOLUTION

A detailed understanding of the regulatory mechanisms governing floral volatile biosynthesis and emissions, such as circadian or diurnal control regimes, require analysis techniques that monitor volatile emission changes with appropriate time resolution. Computer-assisted and online dynamic headspace trapping systems are capable of collecting volatiles in hourly or shorter time intervals (see Section 1.2.2.4). However, trapped volatiles are usually desorbed and subsequently analyzed by GC, which presents a time-limiting factor. Recently new automated analytical systems have been developed that allow highly sensitive, ultrafast volatile analyses and hence represent promising tools for continuous monitoring of plant volatile emissions.

1.7.1 FAST AND TRANSPORTABLE GC (zNOSE)

In recent years, efforts have been made to establish faster GC systems as well as miniaturized GC instruments.[77,78] A recently developed portable GC system, the zNose™ (Electronic Sensor Technology, Newbury Park, CA), has been applied to the analysis of plant volatiles including floral scent[79] (Figure 1.5). The zNose separates compounds by fast GC and operates with a highly sensitive surface acoustic wave (SAW) quartz microbalance detector. The detection principle is based on condensation of the analyte on the surface of an oscillating crystal, leading to an increase in oscillator mass and a reduction in the vibrational frequency proportional to the amount of condensate. The temperature of the SAW detector influences the residence time of the compound on the detector surface and is critical for sensitivity and linear detector response.[79] The high sensitivity of the SAW detector (in the ppbv range) drastically reduces the volatile sampling and preconcentration time of the system. Volatiles are sampled in a small air volume for 20 to 40 sec on a Tenax trap. After rapid thermal desorption, compounds are separated on a capillary 1 or 5 m DB-5 GC column by a defined temperature program before they are monitored and quantified by the SAW detector. The short operation time allows the collection of air samples in time intervals as short as 3 min. Sampling, analysis, and storage of data are fully automated, thus volatile analyses over longer time periods are possible without supervision.

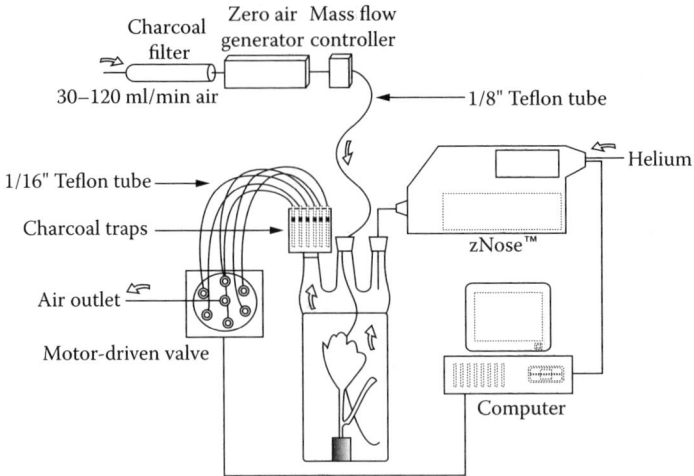

FIGURE 1.5 Volatile monitoring system according to Kunert et al.,[79] combining zNose™ analysis with conventional headspace sampling on activated charcoal traps.

Kunert et al.[79] measured the diurnal emission of volatiles from flowers of the cactus *Rebutia fabrisii* by placing plants in a 2 l glass vessel with a continuous flow of purified air at 30 ml/min. Air samples were taken by the zNose within 40 sec in 30 min intervals. For comparison, volatiles were collected simultaneously in 4 h intervals on charcoal traps for GC-MS analysis. Monitoring the rhythmic emission of floral volatiles with the zNose was comparable to conventional GC analysis, but showed a clearly improved time resolution. An additional advantage of the zNose is its portability, enabling applications not only in the laboratory but also in field experiments. As a drawback of the system, the SAW detector does not allow structure evaluation; therefore, volatiles need to be analyzed by GC-MS prior to calibration of the system with authentic standards. Moreover, the short GC column reduces the compound resolution. Thus monitoring changes of volatiles with similar elution profiles is limited.

In summary, the zNose can be regarded as a tool for quick quantitative estimation of known volatile profiles, making it applicable for high-throughput screenings of natural variants or mutant populations. Given the large time resolution, the system is suitable for monitoring kinetics of volatile emissions from floral and vegetative tissues dependent on diurnal and circadian rhythms and in response to feeding damage or abiotic factors such as light and temperature changes.

1.7.2 PROTON TRANSFER REACTION MASS SPECTROMETRY

Proton transfer reaction mass spectrometry (PTR-MS) analysis technology was developed more than 5 years ago at the University of Innsbruck by Lindinger et al.[80] PTR-MS systems operate independently of GC separation and allow online measurements of volatile organic compounds with concentrations in the pptv range. Originally developed for monitoring changes of volatile organic compounds in the atmosphere, in food control, and in medical analyses, PTR-MS is increasingly applied for real-time analysis of volatile emissions from plants, although no applications have been reported so far for floral scent analysis. PTR-MS instruments are still relatively expensive and their operation requires extended training by experienced researchers.

For detection by PTR-MS, volatiles undergo a chemical ionization by proton transfer reactions with H_3O^+ ions. Differences in proton affinities allow a proton transfer from H_3O^+ ions to a large number of organic volatiles (e.g., alkenes, aldehydes, ketones, alcohols, aromatics, nitriles, sulfides), but prevent a reaction of H_3O^+ ions with the main constituents of the air. The proton transfer reaction takes place under defined conditions in a homogeneous electric field applied to a drift tube (Figure 1.6). Ions exiting the tube are then mass analyzed by a quadrupole mass spectrometer. The soft ionization of compounds by protonation causes only low fragmentation, hence mainly one product ion species occurs for each reactant. The extremely fast time response of the instrument results from the time the volatiles spend in the drift tube, which is less than 1 sec.

The PTR-MS technique has been applied in recent years to measure fluctuations of volatile emissions from various plants. Usually, whole plants or plant parts are enclosed in glass containers, inert bags, or dynamic cuvette systems with a continuous airstream and controlled temperature, humidity, and light conditions, and aliquots of the exiting air are analyzed by PTR-MS. Emissions of volatiles including isoprene and monoterpenes from trees and other plants have been monitored under laboratory and field conditions in response to changes in abiotic factors, such as light and temperature, or biotic factors such as pathogen attack.[81,82]

Besides its use in online qualitative and quantitative analyses of plant volatile organic compounds, PTR-MS has become a valuable tool for investigating the biosynthesis of volatiles using isotope-labeled precursors. For example, carbon sources other than photosynthetically fixed carbon dioxide (CO_2), involved in the biosynthesis of isoprene in poplar leaves, were identified by online measurements of differentially labeled isoprene isotopes during exposure of the plant to ^{13}C-labeled and unlabeled CO_2 and feeding of ^{13}C-labeled and unlabeled glucose.[83] Similar

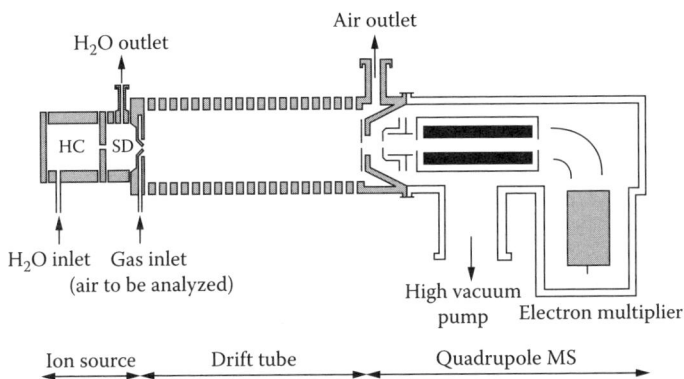

FIGURE 1.6 Principle scheme of the PTR-MS apparatus according to Lindinger et al.[80] (modified). HC, hollow cathode; SD, source drift region.

experimental designs might be applicable to elucidate carbon pools or precursors in biosynthesis pathways of various floral volatiles. In addition, fast metabolic changes in response to enzyme inhibitor applications could be determined.

As a consequence of the ability to perform real-time analyses and measure volatiles with high sensitivity, the PTR-MS system may also qualify as a tool for fast screening of floral emissions from mutants and ecotypes. However, the analysis of volatile mixtures is limited by the ability to determine only the molecular mass of products. Compounds of the same molecular weight cannot be identified separately, therefore additional analysis by GC either in parallel or coupled with the PTR-MS instrument is necessary.[84] Future developments might improve compound identification by combining PTR-MS with an ion trap mass spectrometer, allowing MS/MS performance to distinguish between isomers and other isobars.[85]

1.8 CONCLUSION

The analysis of plant volatiles, including those emitted from floral tissues, has continuously improved over the past 10 to 20 years. The development of materials with high adsorbent capacities for volatile compounds has allowed efficient trapping of volatiles and volatile blends with different chemicophysical properties. Equally important, improvements in gas chromatographic separation, the establishment of highly sensitive spectroscopic detection methods, and expanding mass spectral libraries have laid the foundation for the detection and identification of complex mixtures of volatiles as well as trace amounts of floral volatile components. We have discussed a variety of different volatile collection methods that have been developed, and together they give today's researcher the flexibility of selecting the best method suitable for a particular application. Online analysis techniques are increasingly important for monitoring the kinetics of floral volatile emissions and learning more about the biosynthetic pathways of different scent components. However, greater portability of volatile collection techniques combined with high sensitivity and large time resolution would be desirable to measure floral volatiles in the field in their natural environment, which will further improve our understanding of floral volatile biology.

REFERENCES

1. Bicchi, C. and Joulain, D., Headspace gas chromatographic analysis of medicinal and aromatic plants and flowers, *Flav. Fragr. J.* 14, 185, 1990.
2. Kaiser, R., Trapping, investigation, and reconstitution of flower scents, in *Perfumes: Art, Science, Technology*, Müller, P.M. and Lamparsky, D., Eds., Elsevier Applied Science, London, 1991, p. 213.
3. Dobson, H.E.M., Analysis of flower and pollen volatiles, in *Essential Oils and Waxes: Modern Methods of Plant Analysis*, vol. 12, Linskens, H.F. and Jackson, J.F., Eds., Springer, Berlin, 1991, p. 231.
4. Knudsen, J.T., Tollsten, L., and Bergström, G., Floral scent: a checklist of volatile compounds isolated by headspace techniques, *Phytochemistry* 33, 253, 1993.

5. Jakobsen, H.B., The preisolation phase of *in situ* headspace analysis: Methods and perspectives. In *Essential Oils and Waxes: Modern Methods of Plant Analysis*, vol. 19, Linskens, H.F. and Jackson J.F., Eds., Springer, Berlin, 1997, p. 1.

6. Millar, J.G. and Sims, J.J., Preparation, cleanup, and preliminary fractionation of extracts, in *Methods in Chemical Ecology*, Millar, J.G. and Haynes, K.F., Eds., Kluwer Academic, Boston, 1998, p. 1.

7. Raguso, R.A. and Pellmyr, O., Dynamic headspace analysis of floral volatiles: a comparison of methods, *Oikos* 81, 238, 1998.

8. Chen, F., Tholl, D., D'Auria, J.C., Farooq, A., Pichersky, E., and Gershenzon, J., Biosynthesis and emission of terpenoid volatiles from *Arabidopsis* flowers, *Plant Cell* 15, 481, 2003.

9. Effmert, U., Große, J., Röse, U.S.R., Ehrig, F., Kägi, R., and Piechulla, B., Volatile composition, emission pattern, and localization of floral scent emission in *Mirabilis jalapa* (Nyctaginaceae), *Am. J. Bot.* 92, 2, 2005.

10. Schiestl, F.P., Ayasse, M., Paulus, H.F., Lofstedt, C., Hansson, B.S., Ibarra, F., and Franche, W., Orchid pollination by sexual swindle, *Nature* 399, 421, 1999.

11. Schiestl, F.P. and Ayasse, M., Post-pollination emission of a repellent compound in a sexually deceptive orchid: a new mechanism for maximizing reproductive success?, *Oecologia* 126, 531, 2001.

12. Matile, P. and Altenburger, P., Rhythms of fragrance emission in flowers, *Planta* 174, 242, 1988.

13. Loughrin, J.H., Hamilton-Kemp, T.R., Andersen, R.A., and Hildebrand, D.F., Volatiles from flowers of *Nicotiana sylvestris*, *N. otophora* and *Malus x domestica*: headspace components and day/night changes in their relative concentrations, *Phytochemistry* 29, 2473, 1990.

14. Loughrin, J.H., Hamilton-Kemp, T.R., Andersen, R.A., and Hildebrand, D.F., Circadian rhythm of volatile emission from flowers of *Nicotiana sylvestris* and *N. suaveolens*, *Physiol. Plant.* 83, 492, 1991.

15. Raguso, R.A. and Pichersky, E., Floral volatiles from *Clarkia breweri* and *C. concinna* (Onagraceae): recent evolution of floral scent and moth pollination, *Plant Syst. Evol.* 194, 55, 1995.

16. Tasdemir, D., Demirci, B., Demirci, F., Donmez, A.A., Baser, K.H., and Ruedi, P., Analysis of the volatile components of five Turkish rhododendron species by headspace solid-phase microextraction and GC-MS, *Z. Naturforsch. [C]* 58, 797, 2003.

17. Custodio, L., Nogueira, J.M.F., and Romano, A., Sex and developmental stage of carob flowers affects composition of volatiles, *J. Hort. Sci. Biotechnol.* 79, 689, 2004.

18. Demirci, F. and Baser, K.H.C., The volatiles of fresh-cut *Osyris alba* L. flowers, *Flav. Fragr. J.* 19, 72, 2004.

19. Omata, A. et al. Volatile components of TO-YO-RAN flowers (*Cymbidium faberi* and *Cymbidium virescens*), *Agric. Biol. Chem.* 54, 1029, 1990.

20. Boland, W., Ney, P., Jaenicke, L., and Gassmann, G., A "closed-loop-stripping" technique as a versatile tool for metabolic studies of volatiles, in Schreier, P., Ed., *Analysis of Volatiles*, Walter de Gruyter, New York, 1984, p. 371.

21. Koch, T., Krumm, T., Jung, V., Engelberth, J., and Boland, W., Differential induction of plant volatile biosynthesis in the lima bean by early and late intermediates of the octadecanoid-signaling pathway, *Plant Physiol.* 121, 153, 1999.

22. Engelberth, J., Koch, T., Schüler, G., Bachmann, N., Rechtenbach, J., and Boland, W., Ion channel-forming alamethicin is a potent elicitor of volatile biosynthesis and tendril coiling. Cross talk between jasmonate and salicylate signaling in lima bean, *Plant Physiol.* 125, 369, 2001.

23. Tholl, D., Chen, F., Petri, J., Gershenzon, J., and Pichersky, E., Two sesquiterpene synthases are responsible for the complex mixture of sesquiterpenes emitted from Arabidopsis flowers, *Plant J.* 42, 757, 2005.

24. Dudareva, N., Andersson, S., Orlova, I., Gatto, N., Reichelt, M., Rhodes, D., Boland, W., and Gershenzon, J., The nonmevalonate pathway supports both monoterpene and sesquiterpene formation in snapdragon flowers, *Proc. Natl. Acad. Sci. USA* 102, 933, 2005.

25. Burger, B.V., Munro, Z.M., and Visser, J.H., Determination of plant volatiles 1: Analysis of the insect-attracting allomone of the parasitic plant *Hydnora africana* using Grob-Habich activated charcoal traps, *J. High Resolut. Chromatogr.* 11, 496, 1988.

26. Kaiser, R. and Kraft, P., Neue und ungewöhnliche Naturstoffe faszinierender Blütendüfte, *Chemie in unsere Zeit* 35, 8, 2001.

27. Turlings, T.C.J., Tumlinson, J.H., Heath, R.R., Proveaux, A.T., and Doolittle, R.E., Isolation and identification of allelochemicals that attract the larval parasitoid, *Cotesia marginiventris* (Cresson), to the microhabitat of one of its hosts, *J. Chem. Ecol.* 17, 2235, 1991.

28. Heath, B. and Manukian, A., An automated system for use in collecting volatile chemicals released from plants, *J. Chem. Ecol.* 20, 593, 1994.

29. Röse, U.S.R., Manukian, A., Heath, R.R., and Tumlinson, J.H., Volatile semiochemicals released from undamaged cotton leaves: a systemic response of living plants to caterpillar damage, *Plant Physiol.* 111, 487, 1996.

30. Heath, B. and Manukian, A., Development and evaluation of systems to collect volatile semiochemicals from insects and plants using a charcoal-infused medium for air purification, *J. Chem. Ecol.* 18, 1209, 1992.

31. Manukian, A. and Heath, B., Development of an automated data collection and environmental monitoring system, *Sci. Comput. Autom.* 9, 27, 1993.

32. Agelopoulos, N.G., Hooper, A., Maniar, S., Pickett, J., and Wadhams, L., A novel approach for isolation of volatile chemicals released by individual leaves of a plant in situ, *J. Chem. Ecol.* 25, 1411, 1999.

33. Vercammen, J., Sandra, P., Baltussen, E., Sandra, T., and David, F., Considerations on static and dynamic sorptive and adsorptive sampling to monitor volatiles emitted by living plants, *J. High Resolut. Chromatogr.* 23, 547, 2000.

34. Baltussen, E. et al., Sorption tubes packed with polydimethylsiloxane: a new and promising technique for the preconcentration of volatiles and semivolatiles from air and gaseous samples, *J. High Resolut. Chromatogr.* 21, 333, 1998.

35. Vercammen, J., Pham-Tuan, H., and Sandra, P., Automated dynamic sampling for the on-line monitoring of biogenic emissions from living organisms, *J. Chromatogr. A* 930, 39, 2001.

36. Heiden, A.C. and Wildt, J., Automatisierte Messungen zur Emission biogener Kohlenwasserstoffe in pptV-Mischungsverhältnissen mit Gerstel Online-TDS G, *Gerstel Aktuell* 18, 2–3, 1997.

37. Heiden, A.C., Kobel, K., and Wildt, J., Einfluß verschiedener Streßfaktoren auf die Emission pflanzlicher flüchtiger organischer Verbindungen, Ph.D. dissertation, University Duisburg, 1998.

38. Surburg, H., Guentert, M., and Harder, H., Volatile compounds from flowers. Analytical and olfactory aspects, in Teranishi, R., Buttery, R.G., and Sugisawa, H., Eds., *Bioactive Volatile Compounds from Plants*, American Chemical Society, Washington, DC, 1993, p. 168.

39. Dettmer, K. and Engewald, W., Adsorbent materials commonly used in air analysis for adsorptive enrichment and thermal desorption of volatile organic compounds, *Anal. Bioanal. Chem.* 373, 490, 2002.

40. Williams, A.A., May, H.V., and Tucknott, O.G., Observations on the use of porous polymers for collecting volatiles from synthetic mixtures reminiscent of fermented ciders, *J. Sci. Food Agric.* 29, 1041, 1978.

41. MacLeod, G. and Ames, J.M., Comparative assessment of the artifact background on thermal desorption of Tenax GC and Tenax TA. *J. Chromatogr.* 355, 393, 1986.

42. Stromvall, A.M. and Petersson, G., Protection of terpenes against oxidative and acid decomposition on adsorbent cartridges, *J. Chromatogr.* 589, 385, 1992.

43. Peters, R.J.B., Duivenbode, J.A.D., Duyzer, J.H., and Verhagen, H.L.M., The determination of terpenes in forest air, *Atmos. Environ.* 28, 2413, 1994.

44. Lewis, M.J., and Williams, A.A., Potential artifacts from porous polymers for collecting aroma components, *J. Sci. Food Agric.* 31, 1017, 1980.

45. Sturaro, A., Parvoli, G., and Doretti, L., Artifacts produced by Porapak Q sorbent tubes on solvent desorption, *Chromatographia* 33, 53, 1992.

46. Williams, N.H. and Whitten, W.M., Orchid floral fragrances and male euglossine bees: methods and advances in the last sesquidecade, *Biol. Bull.* 164, 355, 1983.

47. Ragunathan, N., Krock, K.A., Klawun, C., Sasaki, T.A., and Wilkins, C.L., Gas chromatography with spectroscopic detectors, *J. Chromatogr. A* 856, 349, 1999.

48. Lockwood, G.B., Techniques for gas chromatography of volatile terpenoids from a range of matrices, *J. Chromatogr. A* 936, 23, 2001.

49. Marriott, P.J., Shellie, R., and Cornwell, C., Gas chromatographic technologies for the analysis of essential oils, *J. Chromatogr. A* 936, 1, 2001.

50. Dewulf, J. and van Langenhove, H., Analysis of volatile organic compounds using gas chromatography, *Trends Anal. Chem.* 21, 637, 2002.

51. Merfort, I., Review of the analytical techniques for sesquiterpenes and sesquiterpene lactones, *J. Chromatogr. A* 967, 115, 2002.

52. Eiceman, G.A., Gardea-Torresday, J., Overton, E., Carney, K., and Dorman, F., Fundamental reviews: gas chromatography, *Anal. Chem.* 74, 2771, 2002.

53. Eiceman, G.A., Gardea-Torresday, J., Overton, E., Carney, K., and Dorman, F., Fundamental reviews: gas chromatography, *Anal. Chem.* 76, 3387, 2004.

54. Taylor, T., Sample injection systems, in Handley, A.J., and Adlard, E.R., Eds., *Gas Chromatographic Techniques and Applications*, CRC Press, Boca Raton, FL, 2001, p. 52.

55. de Kraker J-W., Franssen, M.C., de Groot, A., Shibata, T., and Bouwmeester, H.J., Germacrenes from fresh costus roots, *Phytochemistry* 58, 481, 2001.

56. Vickers, A.J. and Rood, D., Advances in column technology, in Handley, A.J. and Adlard, E.R., Eds., *Gas Chromatographic Techniques and Applications*, CRC Press, Boca Raton, 2001, p. 91.

57. Jennings, W. and Shibamoto, T., *Qualitative Analysis of Flavor and Fragrance Volatiles by Glass Capillary Gas Chromatography*, Academic Press, New York, 1980.

58. Davies, N.W., Gas-chromatographic retention indexes of monoterpenes and sesquiterpenes on methyl silicone and Carbowax 20 M phases, *J. Chromatogr.* 503, 1, 1990.

59. Adams, R.P., *Identification of Essential Oil Components by Gas Chromatography/Quadrupole Mass Spectrometry*, 3rd ed., Allured Publishing, Carol Stream, IL, 2001.

60. Knudsen, J.T., Tollsten, L., Groth, I., Bergström, G., and Raguso, R.A., Trends in floral scent chemistry in pollination syndromes: floral scent composition in hummingbird-pollinated taxa, *Bot. J. Linn Soc.* 146, 191, 2004.

61. Sharkey, T.D., Loreto, F., and Delwiche, C.F., High carbon dioxide and sun/shade effects on isoprene emission from oak and aspen tree leaves, *Plant Cell Environ.* 14, 333, 1991.

62. Loreto, F., Nascetti, P., Graverini, A., and Mannozzi, M., Emission and content of monoterpenes in intact and wounded needles of the Mediterranean pine, *Pinus pinea*, *Funct. Ecol.* 14, 589, 2000.

63. König, W.A., Joulain, D., and Hochmuth, D.H., Terpenoids and related constituents of essential oils, version 3, 2004, http://www.massfinder.com.

64. Granero, A.M., Gonzalez, F.J., Frenich, A.G., Sanz, J.M., Vidal, J.L., Single step determination of fragrances in *Cucurbita* flowers by coupling headspace solid-phase microextraction low-pressure gas chromatography-tandem mass spectrometry, *J. Chromatogr. A* 1045, 173, 2004.

65. Visser, T., FT-IR detection in gas chromatography, *Trends Anal. Chem.* 21, 627, 2002.

66. Joulain, D., Cryogenic vacuum trapping of scents from temperate and tropical flowers, in *Bioactive Volatile Compounds from Plants*, Teranishi, R., Buttery, R.G., and Sugisawa, H., Eds., American Chemical Society, Washington, DC, 1993, p. 187.

67. Joulain, D. and Tabacchi, R., Two volatile β-chromenes from *Wisteria sinensis* flowers, *Phytochemistry* 37, 1769, 1994.

68. Bicchi, C., D'Amato, A., and Rubiolo, P., Cyclodextrin derivatives as chiral selectors for direct gas chromatographic separation of enantiomers in the essential oil, aroma and flavour fields. *J. Chromatogr. A* 843, 99, 1999.

69. König, W.A. and Hochmuth, D.H., Enantioselective gas chromatography in flavor and fragrance analysis: strategies for the identification of known and unknown plant volatiles, *J. Chromatogr. Sci.* 42, 423, 2004.

70. Joulain, D. and König, W.A., *The Atlas of Spectral Data of Sesquiterpene Hydrocarbons*, E.B.-Verlag, Hamburg, 1998.

71. König, W.A., Collection of enantiomer separation factors obtained by capillary gas chromatography on chiral stationary phases, *J. High Resolut. Chromatogr.* 16, 312, 1993.

72. König, W.A., Collection of enantiomer separation factors obtained by capillary gas chromatography on chiral stationary phases, *J. High Resolut. Chromatogr.* 16, 338, 1993.

73. König, W.A., Collection of enantiomer separation factors obtained by capillary gas chromatography on chiral stationary phases, *J. High Resolut. Chromatogr.* 16, 569, 1993.

74. Schreier, P., Bernreuther, A., and Huffner, A., *Analysis of Chiral Organic Molecules. Methodology and Applications*, de Gruyter, Berlin, 1995.

75. Borg-Karlson, A.-K., Unelius, C.R., Valterova, I., and Nilsson, L.A., Floral fragrance chemistry in the early flowering shrub *Daphne mezereum*, *Phytochemistry* 41, 1477, 1996.

76. Nojima, S., Kiemle, D.J., Webster, F.X., and Roelofs, W.L., Submicro scale NMR sample preparation for volatile chemicals, *J. Chem. Ecol.* 30, 2153, 2004.

77. Yashin, Y.I. and Yashin, Y.A., Miniaturization of gas-chromatographic instruments, *J. Anal. Chem.* 56, 794, 2001.

78. Matisova, E. and Domotorova, M., Fast gas chromatography and its use in trace analysis, *J. Chromatogr. A* 1000, 199, 2003.

79. Kunert, M., Biedermann, A., Koch, T., and Boland, W., Ultrafast sampling and analysis of plant volatiles by a hand-held miniaturised GC with pre-concentration unit: kinetic and quantitative aspects of plant volatile production, *J. Sep. Sci.* 25, 677, 2002.

80. Lindinger, W., Hansel, A., and Jordan, A., On-line monitoring of volatile organic compounds at pptv levels by means of proton-transfer-reaction mass spectrometry (PTR-MS): medical applications, food control and environmental research, *Int. J. Mass Spectrom. Ion Processes* 173, 191, 1998.

81. Hayward, S., Hewitt, C.N., Sartin, J.H., and Owen, S.M., Performance characteristics and applications of a proton transfer reaction-mass spectrometer for measuring volatile organic compounds in ambient air, *Environ. Sci. Technol.* 36, 1554, 2002.

82. Steeghs, M., Bais, H.P., de Gouw, J., Goldan, P., Kuster, W., Northway, M., Fall, R., and Vivanco, J.M., Proton-transfer-reaction mass spectrometry as a new tool for real time analysis of root-secreted volatile organic compounds in Arabidopsis, *Plant Physiol.* 135, 47, 2004.

83. Schnitzler, J.-P., Graus, M., Kreuzwieser, J., Heizmann, U., Rennenberg, H., Wisthaler, A., and Hansel, A., Contribution of different carbon sources to isoprene biosynthesis in poplar leaves, *Plant Physiol.* 135, 152, 2004.

84. Warneke, C., De Gouw, J.A., Kuster, W.C., Goldan, P.D., and Fall, R., Validation of atmospheric VOC measurements by proton-transfer-reaction mass spectrometry using a gas-chromatographic preseparation method, *Environ. Sci. Technol.* 37, 2494, 2003.

85. Prazeller, P., Palmer, P.T., Boscaini, E., Jobson, T., and Alexander, M., Proton transfer reaction ion trap mass spectrometer, *Rapid Commun. Mass Spectrom.* 17, 1593, 2003.

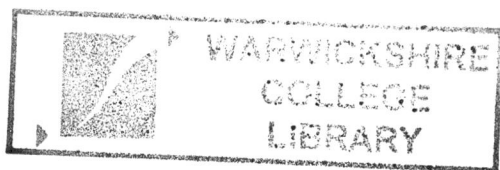

2 The Chemical Diversity of Floral Scent

Jette T. Knudsen and Jonathan Gershenzon

CONTENTS

2.1 INTRODUCTION

Fragrant flowers emit aroma bouquets containing anywhere from a few to more than one hundred different compounds present in varying amounts. The number and diversity of these constituents make floral scent one of the most variable aspects of plant phenotype. Yet, floral scent compounds have a common function in promoting cross pollination, a vital process in the life cycle of most plants. Accompanied by visual and tactile cues, these substances mediate plant–pollinator interactions, which usually benefit both partners. In return for being pollinated, pollinators are rewarded with pollen, nectar, and oils. However, deceptive pollination systems that involve floral scent as an attractant are also well known. Because of its commercial importance in the perfume and flavor industries, the chemistry of floral scents has been studied for a long time. Among the most striking recent examples is a series of papers published during the last 15 years by Roman Kaiser and collaborators at Givaudan Roure. These authors have contributed significantly to our knowledge of floral scents in the orchid family,[1–2] and Kaiser and his collaborators have presented many astonishing new scent molecules from various other plant families.[3–14] Apart from many studies of medicinal plants,[15–17] e.g., most investigators have focused on species from one or a few plant families, either from the perspective of pollination biology and/or chemotaxonomy.

The practical aspects of how to collect, analyze, and identify floral volatiles are treated by Tholl and Röse in Chapter 1. This chapter reviews the structures and abundances of the chemical compound groups reported in Knudsen et al.[18] and dis–cusses various aspects of variation in floral scent chemistry.

2.2   HOW DIVERSE ARE FLORAL SCENTS?

The volatile compounds emitted from flowers belong to several different classes, but are united by their low molecular weight (30 to 300 amu) and vapor pressure sufficient to be released and dispersed into the air under normal temperature regimes. The individual compound classes are widely distributed among the flowers of different species, probably reflecting the fact that the major biosynthetic pathways leading to them are present in all plants.

So far, more than 1700 compounds have been identified in the floral headspace of 990 taxa (most at the species level) belonging to 90 families and 38 orders.[18] The majority of these taxa (78%) belong to the following 19 plant families, in each of which floral scent composition has been characterized from at least 10 or more taxa listed in descending order of taxa (taxa number given in parentheses): Orchidaceae (417), Araceae (55), Arecaceae (40), Magnoliaceae (26), Rosaceae (24), Cactaceae (21), Rutaceae (21), Solanaceae (21), Caryophyllaceae (20), Nyctaginaceae (20), Fabaceae (18), Amaryllidaceae (17), Moraceae (15), Ranunculaceae (14), Asteraceae (13), Lecythidaceae (13), Oleaceae (13), Apiaceae (11), and Rubiaceae (10). For details on species identity and references, see Knudsen et al.[18]

2.2.1   CHEMICAL COMPOUND CLASSES IN FLORAL SCENTS

Most of the 1700 compounds reported in headspace samples of floral scent are lipophilic.[18] The two largest groups (see Table 2.1) are terpenoids, synthesized by the mevalonate or methylerythritol phosphate pathway,[19,20] and aliphatics, synthesized predominantly from fatty acids.[21]

A total of 556 terpenoids have been identified and these include monoterpenes, sesquiterpenes, diterpenes, and irregular terpenes. The monoterpenes are divided about equally between compounds with acyclic or cyclic skeletons; the latter can be mono-, bi-, or tricyclic skeletons (Table 2.1, Figure 2.1, and Figure 2.2). Sesquiterpenes are also characterized by both acyclic and cyclic skeletons, but cyclic skeletons

TABLE 2.1
Distribution of Floral Scent Compounds According to Their Supposed Biosynthetic Origin (Compounds in each class are divided according to either carbon chain length, saturatedness, or skeletal structure.)

Compound Class	No. of Compounds
Aliphatics	
C1	5
C2	59
C3	33
C4	41
C5	59
C6	64
C7	33
C8	39
C9	19
C10	49
C11-C15	76
C16-C20	44
C21-C25	7
Sum	528
Benzenoids and Phenylpropanoids	
C6-C0	23
C6-C1	133
C6-C2	78
C6-C3	83
C6-C4, -C5, -C7	12
Sum	329
C5 Branched-Chain Compounds	
Saturated	40
Unsaturated	53
Sum	93
Terpenoids	
Monoterpenes	
Acyclic	
Regular	136
Irregular	11
Cyclic	
Menthanes	91
Bicyclo[2.2.1]	14

(*Continued*)

TABLE 2.1 (*Continued*)
Distribution of Floral Scent Compounds According to Their Supposed Biosynthetic Origin (Compounds in each class are divided according to either carbon chain length, saturatedness, or skeletal structure.)

Compound Class	No. of Compounds
Bicyclo[3.1.0]	12
Bicyclo[3.1.1]	27
Bicyclo[4.1.0]	3
Tricyclic	1
Sum	295
Sesquiterpenes	
Acyclic	44
Cyclic	114
Sum	158
Diterpenes	
Acyclic	4
Cyclic	2
Sum	6
Irregular terpenes	
Apocarotenoid	52
C8	7
C9	2
C10	8
C11	10
C12	3
C13	5
C14, C16, C18	10
Sum	97
Nitrogen Compounds	
Ammonia	1
Acyclic	41
Cyclic	19
Sum	61
Sulfur Compounds	
Acyclic	37
Cyclic	4
Sum	41
Miscellaneous Cyclic Compounds	
Carbocyclic	60
Heterocyclic	51
Sum	111

are much more common (Table 2.1, Figure 2.3). Because the enzymes producing terpene skeletons, the terpene synthases, often form multiple cyclic and acyclic products from a single substrate, either geranyl diphosphate or farnesyl diphosphate,[19,22] dividing compounds into acyclic and cyclic categories does not necessarily follow a strict biosynthetic criteria. A number of cyclic sesquiterpene skeletons are shown in Figure 2.3C–J. The irregular terpenes include compounds varying in the number of carbon atoms from 8 to 18. Among these are apocarotenoids, which are biodegradation products of carotenoid compounds (C40) like β-carotene,[14,23,24] (Table 2.1, Figure 2.4A–C). Others like the C11 and C16 irregular terpenes 4,8-dimethyl-1,3,7-nonatriene and 4,8,12-trimethyl-1,3,7,11-tridecatetraene are acyclic homoterpenes derived from nerolidol and geranyllinalool, respectively[25,26] (Figure 2.4E, H).

The aliphatics include 528 compounds with chains having between 1 and 25 carbon atoms, the majority having between 2 and 17 carbon atoms (Table 2.1, Figure 2.5). The C6 aliphatic compounds include the well-known green-leaf volatiles, like (Z)-3-hexenyl acetate (Figure 2.5D), found in vegetative as well as floral scents of numerous plants and probably play a role in plant defense.[27] Another large group consists of benzenoids and phenylpropanoids, which are synthesized either from the phenylpropanoid pathway starting with deamination of phenylalanine or from an intermediate of the shikimate pathway prior to phenylalanine.[28,29] A total of 329 benzenoids and phenylpropanoid compounds have been identified in floral scents (Table 2.1, Figure 2.6). Another group is the C5 branched-chain compounds (Table 2.1, Figure 2.7). These are probably derived directly from branched-chain amino acids, but there is little direct evidence to support this assumption.[30] Nitrogen compounds, like indole, and sulfur-containing compounds are also most likely derived from amino acid metabolism.[31] The nitrogen-containing compounds (Table 2.1, Figure 2.8A–F) include cyclic and acyclic compounds, whereas the sulfur-containing compounds are mainly acyclic (Table 2.1, Figure 2.8G–K).

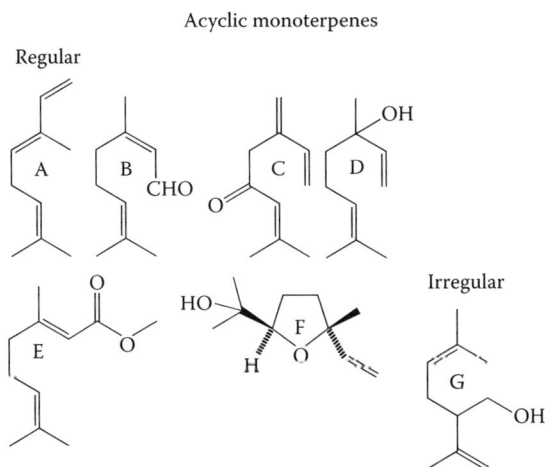

FIGURE 2.1 Acyclic monoterpenes in floral scents: *regular skeleton*, (A) (*E*)-β-ocimene; (B) neral; (C) ipsdienone; (D) linalool; (E) methyl geranate; (F) *trans*-linalool oxide (furanoid); *irregular skeleton*, (G) lavandulol.

Cyclic monoterpenes

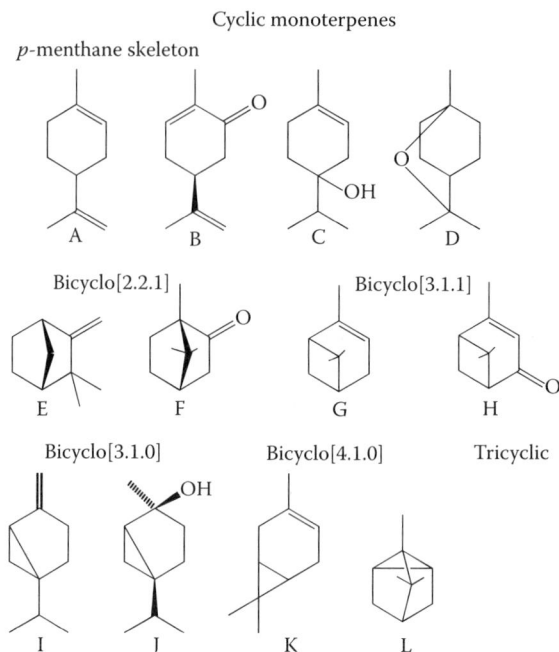

FIGURE 2.2 Cyclic monoterpenes in floral scents: *p-menthane skeleton*, (A) limonene; (B) carvone; (C) terpinen-4-ol; (D), 1,8-cineole; *bicyclo[2.2.1] skeleton*, (E) camphene; (F) camphor; *bicyclo[3.1.1] skeleton*, (G), α-pinene; (H) verbenone; *bicyclo[3.1.0] skeleton*, (I) sabinene; (J) *cis*-sabinene hydrate; *bicyclo[4.1.0] skeleton*, (K) 3-carene; *tricyclic skeleton*, (L) tricyclene.

The 111 miscellaneous compounds are either carbocyclic or heterocyclic, the latter including both carbon and other atoms in the cyclic structures (Table 2.1, Figure 2.9). The group contains compounds of uncertain biosynthetic origin, but many of these are derived from fatty or amino acids.

2.2.2 FUNCTIONAL GROUPS OF FLORAL SCENT COMPOUNDS

Floral scent compounds range from nonpolar (e.g., alkanes, alkenes) to polar. The majority contains an oxygen function and are slightly to moderately polar. Esters, especially esters of aliphatic compounds, benzenoid, and phenylpropanoids, are the most common functional group found, followed by alcohols, ketones, ethers, aldehydes, and acids (Table 2.2, Figures 2.1–2.7 and Figure 2.9). Among nitrogen-containing compounds, amines and oximes (Figure 2.8B, C) are most common, while sulfides are most common among sulfur-containing compounds (Table 2.2, Figure 2.8G). In the miscellaneous cyclic compound group, furans and pyrans are common (Table 2.2, Figure 2.9I–L). Within esters, it is the alcohol part of the compound that is most variable (Table 2.2).

Among the terpenes of floral scent, the percentage of oxygenated compounds is much higher for monoterpenes (81%) than for sesquiterpenes (38%). This difference

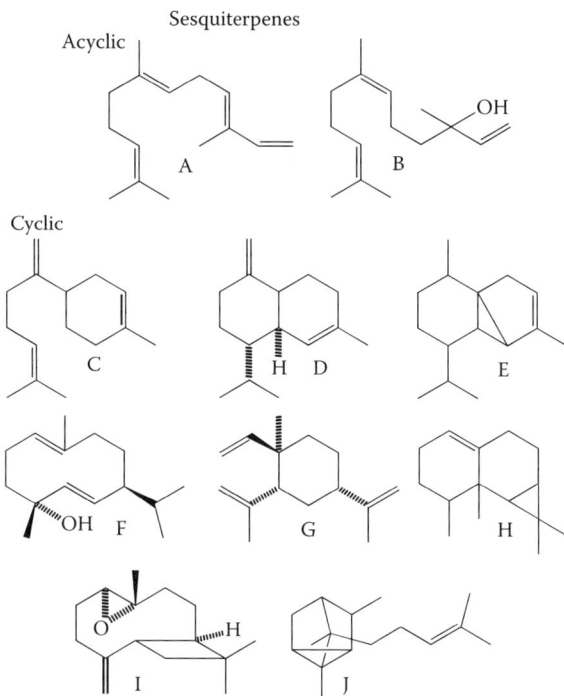

FIGURE 2.3 Sesquiterpenes in floral scents: *acyclic*, (A), (*E,E*)-α-farnesene; (B) (*Z*)-neroli-dol; *cyclic*, (C) β-bisabolene; (D), γ-cadinene; (E) α-cubebene; (F), germacrene D-4-ol; (G), β-elemene; (H) β-gurjunene; (I) caryophyllene oxide; (J) α-santalene.

may reflect the fact that many oxygenated sesquiterpenes are much less volatile than oxygenated monoterpenes and sesquiterpene hydrocarbons. Only monooxygenated sesquiterpenes are likely to have sufficient volatility to be found in floral headspace.[32] The lower frequency of oxygenated sesquiterpenes may also be a consequence of their inefficient collection by standard adsorbent methods and poor desorption with nonpolar solvents like hexane.[33] At the other end of the spectrum, a number of nitrogenous and sulfurous compounds are probably lost during absorption or sample concentration because of their high volatility. Alternatively, they may not be detected by gas chromatography-mass spectrometry (GC-MS) because of their proximity to the solvent peak. In many studies, the mass detector is first switched on after the solvents have eluted.

Most of the common constituents of floral scents are easy to identify by gas chromatographic retention times and mass spectra. However, sesquiterpene hydrocarbons are much more difficult than most others because of the large number of structures with similar mass spectra and retention times (see Adams[34] and Joulain and König[32]). Final confirmation frequently requires a reference standard for comparison, but for many sesquiterpenes, these are not available commercially, and so compounds are often reported as "unknown sesquiterpenes."

Irregular terpenes

Apocarotenoids

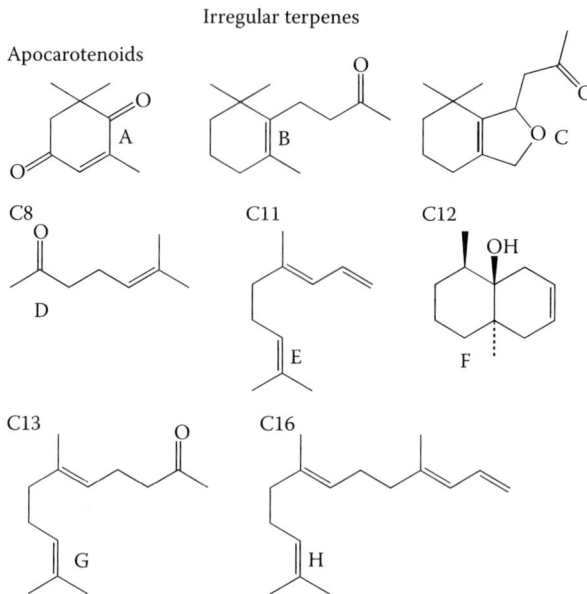

FIGURE 2.4 Irregular terpenes in floral scents: *apocarotenoids*, (A) 2,6,6-trimethyl-2-cyclo-hexene-1,4-dione (4-oxoisophorone); (B) dihydro-β-ionone; (C) 7(11)-oxymegastigma-5(6)-en-9-one; *C8*, (D) 6-methyl-5-hepten-2-one; *C11*, (E) (*E*)-4,8-dimethyl-1,3,7-nonatriene; *C12*, (F) dehydrogeosmin; *C13*, (G) (*E*)-6,10-dimethyl-5,9-undecadien-2-one (geranylacetone); *C16*, (H) (*E,E*)-4,8,12-trimethyl-1,3,7,11-tridecatetraene.

Aliphatics

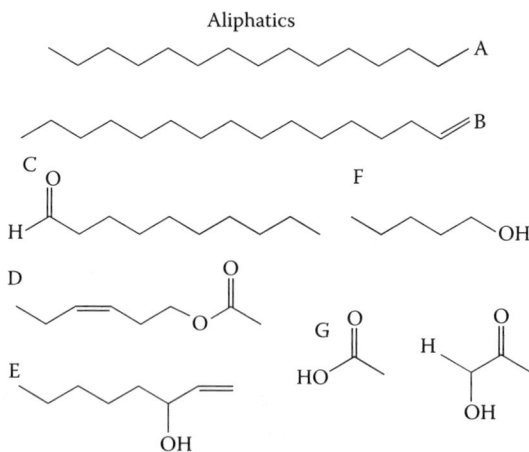

FIGURE 2.5 Aliphatic compounds in floral scents: (A) pentadecane; (B) 1-hexadecene; (C) decanal; (D) (*Z*)-3-hexenyl acetate; (E) 1-octen-3-ol; (F) pentanol; (G) acetic acid; (H) 3-hydroxy-2-butanone.

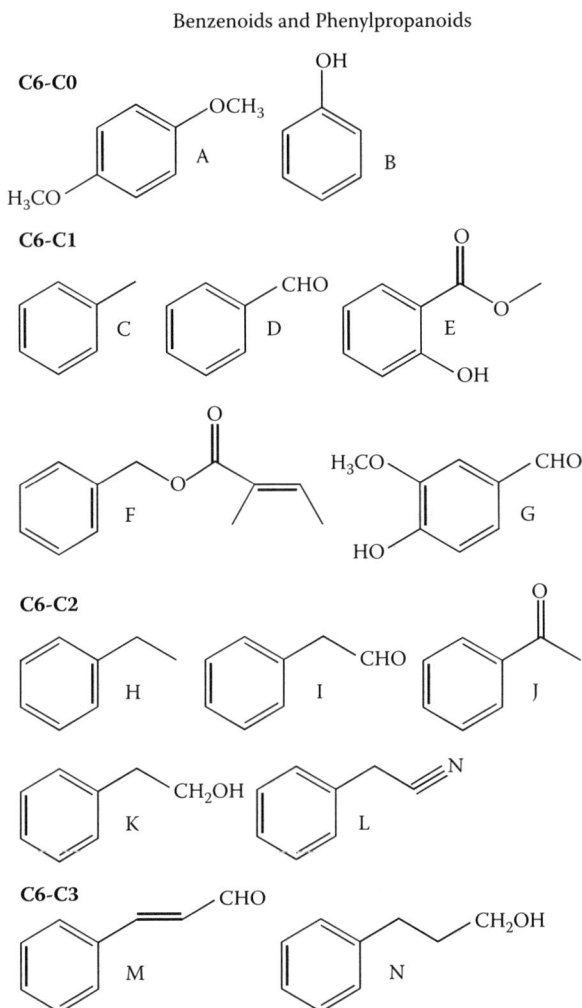

FIGURE 2.6 Benzenoids and phenylpropanoids in floral scents: *C6-C0*, (A) 1,4-dimethoxyben-
zene; (B) phenol; *C6-C1*, (C) methylbenzene; (D) benzaldehyde; (E) methyl 2-hydroxybenzoate
(methyl salicylate); (F) benzyl (*E*)-2-methyl-2-butenoate (benzyl tiglate); (G) 4-hydroxy-3-
methoxybenzaldehyde (vanillin); *C6-C2*, (H) ethylbenzene; (I) phenylacetaldehyde; (J) acetophenone; (K) 2-phenylethanol; (L) 2-phenylacetonitrile; *C6-C3*, (M) (*E*)-cinnamic aldehyde;
(N) phenylpropanol.

2.2.3   STEREOCHEMISTRY OF FLORAL SCENT COMPOUNDS

2.2.3.1   Geometric Isomers

Geometric or (*Z*)- and (*E*)- isomers are common in unsaturated aliphatics,
phenylpropanoids, and terpenoid compounds, and often co-occur in the scent
bouquet of a single species, for example, geranial and neral (Figure 2.1B), which
are the (*E*)- and (*Z*)- of isomer 3,7-dimethylocta-2,6-dienal (citral), respectively.

C5 Branched-chain compounds

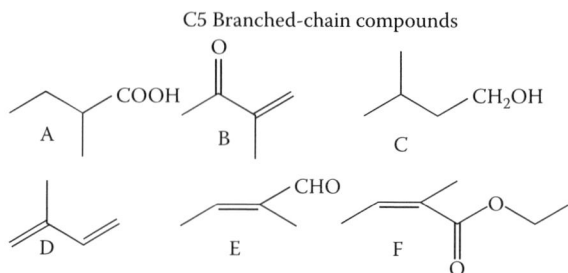

FIGURE 2.7 C5 branched-chain compounds in floral scents: (A) 2-methylbutanoic acid; (B) 3-methyl-3-buten-2-one; (C) 3-methylbutanol; (D) 2-methyl-1,3-butadiene; (E) (*E*)-2-methyl-2-butenal; (F) ethyl (*Z*)-2-methyl-2-butenoate (ethyl angelate).

Nitrogen and sulphur compounds

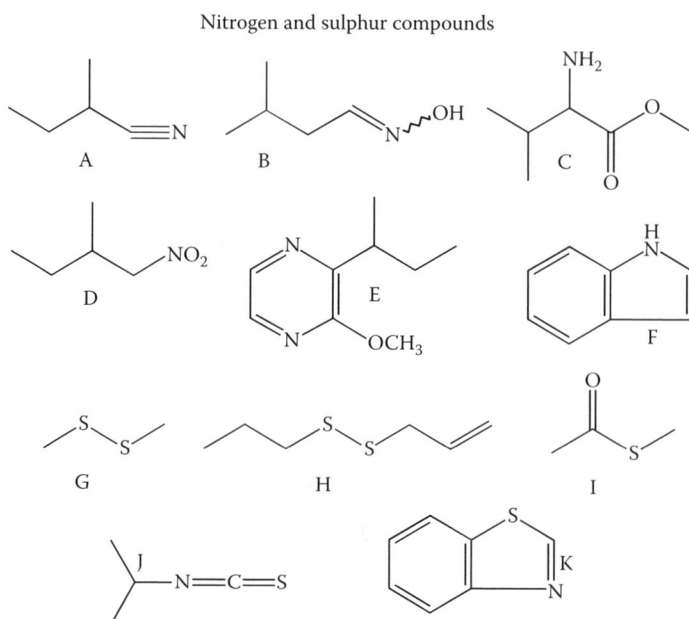

FIGURE 2.8 Nitrogen and sulfur compounds in floral scents: (A) 2-methylbutylnitrile; (B) 3-methylbutylaldoxime; (C) methyl 2-amino-3-methylbutanoate (valine methyl ester); (D) 1-nitro-2-methylbutane; (E) 2-methoxy-3-*sec*-butylpyrazine; (F) indole; (G) dimethyl-disulfide; (H) propyl prop-2-enyl disulfide; (I) methyl thioacetate; (J) isopropyl isothio-cyanate; (K) benzothiazole.

Because (*E*)- and (*Z*)- isomers of a compound usually separate well under standard GC conditions, double-bond geometry is reported often for floral scent compounds. The mass spectra of (*E*)- and (*Z*)- isomers usually differ slightly and the configuration of a compound can be determined with some confidence, especially when both isomers are present. When geometrically pure reference compounds are available, the configuration can be determined unambiguously.

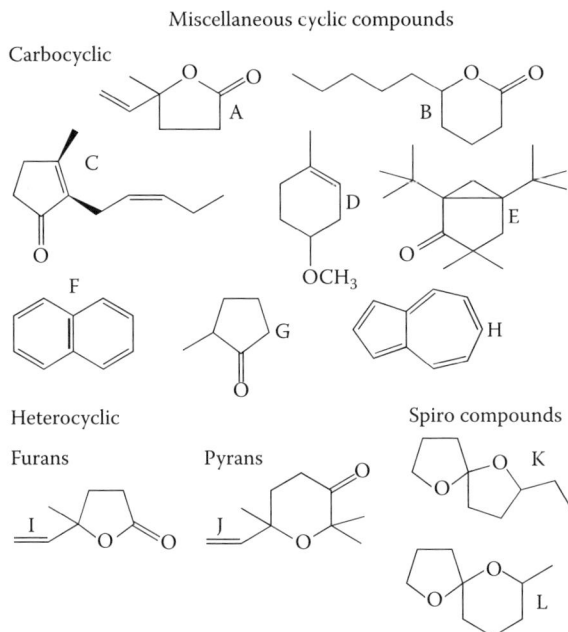

FIGURE 2.9 Miscellaneous cyclic compounds in floral scents: *carbocyclic*, (A) 4-methyl-5-hexen-4-olide (γ-vinyl-γ-valerolactone); (B) δ-decalactone; (C) *cis*-jasmone; (D) 1-methyl-4-methoxy-1-cyclohexene; (E) 1,5-*ditert*-butyl-3,3-dimethylbicyclo[3.1.0]hexan-2-one; (F) naphthalene; (G) 2-methylcyclopentanone; (H) azulene; *heterocyclic, furans*, (I) 5-methyl-5-ethenyldihydro-2(3H)-furanone; *pyrans*, (J) 2,2,6-trimethyl-6-ethenyltetrahydro-3-pyranone; *spiro compounds*, (K) chalcogran; (L) 7-methyl-1,6-dioxaspiro[4.5]decane.

The *cis*- and *trans*- prefixes are preferably not used for double-bond geometry,[32] but rather to indicate through-space relationships of two substituents or two bulky parts of a molecule, for example, *trans*-linalool oxide (furanoid) (Figure 2.1F).

2.2.3.2 Optical Isomers

Many floral scent compounds are chiral molecules with one or more asymmetric carbons. Enantiomers have identical values for all physical constants except that they interact with plane-polarized light in opposite directions. Special protocols using chiral GC columns, multidimensional GC, and enantioselective chemical synthesis are therefore necessary to separate and identify enantiomers.[35–39] Diastereoisomers, unlike enantiomers, have different physical properties and usually separate on standard GC columns, but like enantiomers, they have identical mass spectra.[40]

Despite equal physical properties, enantiomers react differently with other chiral compounds and their biological characteristics may differ. For example, to the human nose (R)-carvone smells like peppermint and (S)-carvone smells like caraway, while (4R)-*cis*-rose oxide has a threshold value 10 times lower than its (4S)-enantiomer.[41]

While most components of floral scent are optically pure, some terpenes occur as mixtures of two enantiomers.[18] As a general rule, enantiomers are likely to be

TABLE 2.2
Distribution Among the Major Chemical Compound Classes of Functional Groups Found in Floral Scent Compounds (In compounds containing more than one functional group priority follows from highest to lowest: acids, aldehydes, ketones, alcohols, esters, ethers.)

	Compound Classes										
Functional Groups	Aliphatics	Benzenoids and Phenyl Propanoids	C5 Branched Compounds	Monoterpenes	Sesquiterpenes	Diterpenes	Irregular Terpenes	Nitrogen Compounds	Sulfur Compounds	Miscellaneous	Total
No. of compounds	528	329	93	295	158	6	97	61	41	111	1719
Hydrocarbons		34	2	56	98	3	10				203
Alkanes	40									6	46
Alkenes	54									4	58
Acids	11	3	2								36
Aldehydes	52	21	6	16	7		3				105
Ketones	60	24	2	36	2		52			12	188
Alcohols	86	47	7	87	31	3	20			5	286
Esters	195	119	74	40	9		7	2		29	475
Ethers	9	48		60	11		4			8	140
Chloro	1	1									2
N-containing		28									28
Benzofurans		1									1
Benzopyrans		3									3
Ammonia								1			1
Amides								5			5
Amines								15			15

Compound class		
Nitriles	5	5
Nitro compounds	3	3
Oximes	11	11
Imidazole	2	2
Indole	3	3
Pyrazine	7	7
Pyrazole	1	1
Pyridine	5	5
Triazine	1	1
Isothiocyanates	3	3
Thiocyanates	1	1
Sulfides	25	25
Thioesters	3	3
Thiols	4	4
Sulfoxides	1	1
Thiazole	2	2
Thiafurans	2	2
Furans	27	27
Pyrans	8	8
Naphthalenes	3	3
Azulenes	1	1
Spiro compounds	8	8
Unknown	1	1

produced by different enzymes. However, some terpene synthases have been described that produce both negative (−) and positive (+) enantiomers of single substances, such as an enzyme from the flowers of *Arabidopsis thaliana* that produces both enantiomers of sabinene and limonene in readily detectable amounts.[22,42]

2.3   VARIATION IN FLORAL SCENT COMPOSITION

The qualitative and quantitative composition of a floral scent varies within and between species and is probably best defined as a product of phylogenetic constraints and balancing selection due to pollinator and florivore attraction, as suggested by Raguso.[43] At the intraspecific level, floral scent composition may vary spatially and temporally within a flower, between flowers on the same plant (see below), between plants, and between populations.[44–53] At the interspecific level, a few groups of plants have shown little variation between closely related taxa,[54–57] but in most groups, each taxon produces its own specific floral scent blend.[58–72]

Floral scent production and composition is genetically determined (see Chapter 3),[53,73–75] but environmental factors may also exert a great impact on the scent emitted. For example, water stress and high or low temperatures may decrease floral scent production or inhibit it completely.[76–80]

2.3.1   Composition and Amount of Emitted Floral Scents

Floral scent bouquets may contain from one to more than 100 compounds,[49,66,81,82] but most species emit between 20 and 60 different compounds. The amounts of floral scent produced vary from the low picogram range to more than 30 μg/h.[2,83,84] The flowers or inflorescences of various beetle- and moth-pollinated species produce the largest amounts of fragrances, while more than half of the hummingbird-pollinated plants analyzed did not produce any detectable scent.[83]

The total amounts reported must be regarded as only approximate because rigorous quantitation of floral scent emission rates is seldom attempted and often not possible. Such an approach requires extended collections under physiologic conditions, use of internal standard to control for losses during collection and sample processing, and dose-response curves to calibrate the detector response of individual compounds (which requires adequate amounts of pure standards).

2.3.1.1   Temporal Variation

Floral scent emission may vary in time because emission follows a circadian rhythm or external stimuli such as light or temperature.[76,85–93] Variation in emission may also be caused by pollination or as part of the developmental aging of the flower itself.[82,94–102] In many hermaphroditic as well as monoecious plant species, there is, on individual plants, a temporal separation of the pistillate and staminate flowering phase with or without overlap in male and female function. In *Peltandra virginica* the temporal separation is accompanied by scent differences and the differences in scent are thought to signal oviposition and food sites, respectively.[97]

2.3.1.2 Spatial Variation

Floral scent emission varies in space because scent may be emitted from specialized regions of floral tissues (see Chapter 6).[103] However, different floral parts, like stamens and pistils, as well as pollen and nectar, may also participate in floral scent emission.[63,86,99,104–110] Even the position within inflorescences may influence the amount and composition of scent emitted by individual flowers,[111,112] although such variation has thus far only been reported in deceptive flowers.

2.3.2 Co-Occurrence of Individual Compounds in Floral Scent

Some compounds in the floral scent of a species are normally found together with one or several other related compounds. This may be a consequence of a shared biosynthetic pathway or enzyme like a terpene synthase.[22,113,114] However, co-occurrence may also be a result of phylogenetic influences and the co-occurring compounds may not necessarily be biosynthetically related.

2.3.2.1 Biosynthetic Relatedness

It is common to find a series of compounds in a single floral scent mixture that are biosynthetically closely related. Among aliphatic compounds, a series of hydrocarbons or methyl ketones with increasing numbers of carbon atoms are found, as are compounds of a single chain length present as aldehydes, ketones, and alcohols[21] (Knudsen JT, unpublished results). Studies of *Clarkia breweri* flowers showed that methyleugenol and methylisoeugenol are produced by the action of a single enzyme.[86] This could be true in other species as well. Methylisoeugenol has been reported in the headspace samples of floral scent in 16 species from nine different plant families (four monocot and five eudicot families), but 12 of these (representing all nine families) also contained methyleugenol, a volatile reported from 45 species (Knudsen JT, unpublished data). In terpenoid compounds, it has been demonstrated that several terpene synthases catalyze the production of more than one product.[19,22,114] Multiproduct terpene synthases may also explain the co-occurrence of some of the most commonly reported monoterpene hydrocarbons in floral scents; for example, limonene (found in 71% of investigated plant families), (*E*)-β-ocimene (71%), myrcene (70%), α-pinene (67%), and β-pinene (59%). Other compounds that occur in more than 50% of the plant families are the terpenoids, linalool (70%), caryophyllene (52%), and 6-methyl-5-hepten-2-one (52%), and the aromatics benzaldehyde (64%), methyl salicylate (57%), benzyl alcohol (56%), and 2-phenylethanol (54%).[18] Geometric or optical isomers of a compound often co-occur, with one of the isomers being much more common than the other.

Some compounds identified in floral scent blends may be produced as artifacts of sample collection. For example, some of the sulfur-containing compounds identified in species of Cactaceae and others probably are rearrangement products of dimethyl disulfide on the adsorbent trap.[7] Another example are the β-ocimenols, which usually occur together with the corresponding β-ocimenes. Brunke et al.[17] showed that these are artifacts produced on the active charcoal surface of one type of charcoal trap.

2.3.2.2 Phylogenetic Constraints

Because major biosynthetic pathways are present in all plant species, there may be few constraints on floral scent distribution at the level of compound classes. However, the musty/earthy odors of geosmin and dehydrogeosmin may be examples of compounds that are phylogenetically constrained to one group of higher plants. Both compounds have been reported in many species of Cactaceae, and geosmin has also been reported in *Dorstenia turneraefolia* (Moraceae).[5,7,115] However, the geosmin in *Dorstenia* may originate from epiphyllic algae[115] and may not be produced by the plant itself.

Reconstructions of plant phylogenies based on floral scent compounds are still few,[58,67,116–119] and in most cases only the outermost branching pattern is consistent with phylogenetic trees based on either morphological or DNA sequence data. Mapping of the major compound classes found in headspace samples of floral scents onto the most parsimonious phylogenetic tree based on DNA sequence data revealed no groupings or discernible patterns.[18] This indicates that, at higher taxonomic levels, floral scent chemicals are too evolutionarily labile to be useful for phylogenetic inference.[117,119] However, this does not exclude the possibility that some floral scent chemicals are patterned phylogenetically at lower taxonomic levels.[117,118,120]

2.4 ADDITIONAL FUNCTIONS OF FLORAL SCENTS

The primary function of floral scent in flowering plants is to attract and guide pollinators.[43,121–124] However, additional functions are ascribed to the presence of volatile chemicals in flowers (reviewed in Pichersky and Gershenzon[125]), including defense and protection against abiotic stresses. These additional functions may help explain some of the abundance and variety of different constituents detected. Because flowers produce pollen and ovules for the next generation, they have a high fitness value to the plant, and the finding of chemicals that defend flowers against herbivores and pathogens is not surprising. Apart from attracting pollinators, floral scent chemicals may also increase the risk of herbivore attack on floral structures, as shown in *Nicotiana attenuata*, and floral tissues may therefore require relatively more protection from enemies.[87] In *N. attenuata*, this was achieved through a herbivore-induced increase in corolla pools of nicotine, a biocidal defense metabolite.

2.4.1 Volatiles in Leaves and Flowers

Most of the plant volatiles identified in defense against insect herbivores[27] have also been identified in floral scents.[18] Whether these compounds originate from floral or green parts of an inflorescence is difficult to determine,[69,118,126] but if they originate from floral parts, their function may be different from that in vegetative parts. Some of the compounds shared by vegetative and floral samples are very common and may have a general function to signal the presence of a plant/flower, while the specific flower chemicals signal the presence of a certain flower species (see the signal-noise discussion in Raguso[127]). Thus chemicals common to the vegetative and floral tissues may have a dual function: defense against herbivores and attraction of pollinators. In the mutualism between *Derelomus chamaeropsis* (Curculionidae) and the dwarf palm *Chamaerops humilis* (Arecaceae), the leaves have overtaken the attractive function of the flowers by producing volatiles that attract the pollinating weevils.[128] The attractive leaf volatiles identified in this mutualism are among those volatiles that have been found both in volatile samples of vegetative and floral tissues of other species. Another example of a dual function is ascribed to chemicals emitted by *Acacia* flowers. In the ant-*Acacia* mutualism, there is a conflict during flowering between the use of guarding ants to deter herbivores and the need to attract pollinators. However, young first-day flowers contain some volatile chemicals that deter the guarding ants, thus allowing pollinating bees to visit the flowers.[129]

2.4.2 FLOWER VOLATILES AS MEDIATORS OF SEX AND BREEDING BEHAVIOR IN INSECTS

Representatives of all principal classes of floral scent chemicals have also been shown to be used by various insects, although not all these compounds are synthesized *de novo* by the insects (see the references in El-Sayed[130]). In insects, the compounds either mediate intraspecific interactions (i.e., they function as pheromones) or interspecific interactions (i.e., they function as allelochemicals). Some plants deceive insects to visit and pollinate their flowers by producing compounds that imitate pheromones or allelochemicals. The foul sulfur- or nitrogen-containing odors and thermogenesis in many species of Araceae imitate allelochemicals that elicit feeding or oviposition behavior in carrion and dung flies.[36,48,131–135] Flowers of many deceptive orchid species produce compounds that mimic the sex pheromones of female insects (usually Hymenoptera) and elicit sexual behavior in males, which effects pollination through attempted copulation with the flowers (see Chapter 10).[136–139] The coproduction of identical volatiles by most other insects and plants is probably just a coincidence, reflecting a limited biosynthetic machinery or limited sensory modalities in perceiving insects. Most of the chemicals probably constitute honest signals with different functions in plants and insects.

In male euglossine bees, floral scent compounds are a primary reward, which the male bees collect during a large part of their lives. The fragrance is somehow exposed during brief premating encounters between the sexes and is suggested to serve as an indicator of male quality. Fragrance collecting behavior could thus have evolved through sexual selection.[140,141] In plants pollinated by scarab beetles (Scarabaeidae: Dynastinae), the production of strong odors and heat coincide with the crepuscular/nocturnal behavior of the beetles.[142–146] The floral volatiles are suggested to function as a sex pheromone or an aggregation cue, attracting both male and female scarab beetles, which produces a high degree of mutual dependence between the plant and its scarab pollinators. Floral scents have only been analyzed in a few scarab-pollinated plant species, but all contained either methoxylated or hydroxylated compounds that otherwise are rare in floral scents, and some have been reported only in scarab-pollinated plants.[18,55,58,65,142] In addition, floral scent is important in brood-site pollination mutualisms, systems in which plants breed their own pollinators,[147] perhaps in parity with the scarab-plant pollination mechanism. In some species of Annonaceae, Araceae, Arecaceae, Cyclanthaceae, Zamiaceae, and other families, both sexes of pollinators are attracted by the floral volatiles produced. Pollinators mate within the inflorescence/flower and females lay their eggs, usually in staminate flowers or floral parts other than the gynoecium, thus avoiding the destruction of developing ovules.[148–153] In the fig-fig wasp mutualism, each *Ficus* species produces its own specific odor, which attracts its specific Agaonid pollinator.[62,154] The occurrence of brood-site mutualisms in both gymnosperms and angiosperms points to an early origin of this interaction. However, further studies are needed to pinpoint the exact role of floral volatiles in the evolution of these mutualisms.

2.5 CONCLUSION

The number of species in which floral scents have been analyzed and the number of compounds that have been identified more than doubled between 1992 and 2002, and more than 1700 floral scent compounds are now known.[18,155] However, only about .25% of the angiosperms have been studied and many families are still totally unknown concerning floral scents. Thus, we expect to encounter more new compounds and perhaps even new compound classes as research continues. To date, the reported compounds can be classified into seven major groups, each of which has a wide distribution in the plant kingdom.

The composition of floral scent varies both quantitatively and qualitatively at many different levels of organization. Hence, in planning an investigation, it is important to standardize environmental factors as well as the collection and analytical methods. Especially in interspecific comparisons, it is important that samples are collected using the same methods and that sufficient replicate samples are collected. All scent collection methods have advantages and disadvantages, and the choice of analytical procedures may introduce bias. In most studies of floral scent, only a limited number of authentic reference compounds have been available to the researchers, so most compounds are tentatively identified by comparison of retention times and mass spectra with those in the literature and in databases. However, such identifications can be quite erroneous[156] because the appearance of a mass spectrum is influenced by its concentration and the kind of instrument it is recorded on. In addition, mass spectra in spectral libraries are sometimes stored in abbreviated form and include only a subset of all mass fragments. Agreement between the Kovats retention index of a known compound (see, e.g., Adams[34]) and that of a tentatively identified compound can help support the identification.[157]

Knowledge of the chemistry of floral scents is critical in understanding their evolution and biological function. Research progress in this area must be a collaboration between biologists, biochemists, and analytical and synthetic chemists. The series of studies on sexually deceptive orchids[100,111,136–139,158,159] serves as an excellent example.

ACKNOWLEDGMENTS

We thank Natalia Dudareva and Eran Pichersky for the invitation to contribute to this book. J. T. Knudsen is grateful to her husband, Bertil Ståhl, who made it possible for her to write this chapter, and J. Gershenzon acknowledges the support of the Max Planck Society.

REFERENCES

1. Kaiser, R., On the scents of orchids, in *Bioactive Volatile Compounds from Plants*, Teranishi, R., Buttery, R.G. and Sugisawa, H., Eds., American Chemical Society, Washington, DC, 1993, p. 240.
2. Kaiser, R., *The Scents of Orchids*, Elsevier, Amsterdam, 1993.
3. Kaiser, R. and Lamparsky, D., Constituants azotés en trace de quelques absolues de fleurs et leurs head spaces correspondants, In *Proceedings of the 8th International Congress of Essential Oils (Cannes, 1980)*, Fedarom, Grasse, France, 1982, p. 287.
4. Kaiser, R., New volatile constituents of *Jasminum sambac* (L.) Aiton, in *Flavors and Fragrances: A World Perspective, Proceedings of the 10th International Congress of Essential Oils (Washington, DC, 1986)*, Lawrence, B.M., Mookherjee, B.D., and Willis, B.J., Eds., Elsevier Science, Amsterdam, 1988, p. 669.
5. Kaiser, R. and Nussbaumer, C., Dehydrogeosmin, a novel compound occurring in the flower scent of various species of Cactaceae, *Helv. Chim. Acta* 73, 133, 1990.
6. Kaiser, R., Trapping, investigation and reconstitution of flower scents, in *Perfumes: Art, Science and Technology*, Müller, P.M. and Lamparsky, D., Eds., Elsevier Applied Science, London, 1991, p. 213.
7. Kaiser, R. and Tollsten, L., An introduction to the scent of cacti, *Flav. Fragr. J.* 10, 153, 1995.
8. Kaiser, R., New and uncommon volatile compounds in floral scents, in *Proceedings of the 13th International Congress of Flavours, Fragrances and Essential Oils (Istanbul, Turkey, 1995)*, Baser, K.H.C., Ed., AREP, Istanbul, 1995, p. 135.
9. Kaiser, R., Environmental scents at the Ligurian coast, *Perfumer Flav.* 22, 7, 1997.
10. Kaiser, R., New or uncommon volatile compounds in the most diverse natural scents, *Rev Ital Eppos* 18, 18, 1997.
11. Kaiser, R., Scents from rain forests, *Chimia* 54, 346, 2000.
12. Schultz, K., Kaiser, R., and Knudsen, J.T., Cyclanthone and derivatives, new natural products in the flower scent of *Cyclanthus bipartitus* Poit, *Flav. Fragr. J.* 14, 185, 1999.
13. Kaiser, R. and Kraft, P., Neue und ungewöhnliche Naturstoffe faszinierender Blütendüfte, *Chem. Zeit* 35, 8, 2001.
14. Kaiser, R., Carotenoid-derived aroma compounds in flower scents, in Winterhalter, P. and Rouseff, R., Eds., *Carotenoid-Derived Aroma Compounds*, American Chemical Society, Washington, DC, 2002, p. 160.
15. Bicchi, C. and Joulain, D., Headspace-gas chromatographic analysis of medicinal and aromatic plants and flowers, *Flav. Fragr. J.* 5, 131, 1990.
16. Bozan, B., Ozek, T., Kurkcuoglu, M., Kirimer, N., and Baser, K.H.C., The analysis of essential oil and headspace volatiles of the flowers of *Pelargonium endlicherianum* used as an anthelmintic in folk medicine, *Plant. Med.* 65, 781, 1999.
17. Brunke, E.J., Hammerschmidt, F.J., and Schmaus, G., Flower scent of some traditional medical plants, *Bioactive Volatile Compounds from Plants*, in Teranishi, R., Buttery, R.G., and Sugisawa, H., Eds., American Chemical Society, Washington, DC, 1993, p. 282.
18. Knudsen, J.T., Eriksson, R., Gershenzon, J., and Ståhl, B., Diversity and distribution of floral scent, *Bot. Rev.*, 72, 1, 2006.
19. Gershenzon, J. and Kreis, W., Biochemistry of terpenoids: monoterpenes, sesquiterpenes, diterpenes, sterols, cardiac glycosides and steroid saponins, in *Biochemistry of Plant Secondary Metabolism*, Wink, M., Ed., Sheffield Academic Press, Sheffield, 1999, p. 222.

20. Rodriguez-Concepcion, M. and Boronat, A., Elucidation of the methylerythritol phosphate pathway for isoprenoid biosynthesis in bacteria and plastids. A metabolic milestone achieved through genomics, *Plant Physiol.* 130, 1079, 2002.

21. Croteau, R. and Karp, F., Origin of natural odorants, in *Perfumes: Art, Science and Technology*, Müller, P.M. and Lamparsky, D., Eds., Elsevier Applied Science, London, 1991, p. 101.

22. Chen, F., Tholl, D., D'Auria, J.C., Farooq, A., Pichersky, E., and Gershenzon, J., Biosynthesis and emission of volatiles from *Arabidopsis* flowers, *Plant Cell* 15, 1, 2003.

23. Eugster, C.H., Hürlimann, H., and Leuenberger, H.J., Crocetindialdehyd und crocetinhalbaldehyd als Blütenfarbstoffe von *Jacquinia angustifolia*, *Helv. Chem. Acta* 52, 89, 1969.

24. Eugster, C.H. and Märki-Fischer, E., The chemistry of rose pigments, *Angew. Chem.* 30, 654, 1991.

25. Donath, J. and Boland, W., Biosynthesis of acyclic homoterpenes in higher plants parallels steroid hormone metabolism, *J. Plant Physiol.* 143, 473, 1994.

26. Donath, J. and Boland, W., Biosynthesis of acyclic homoterpenes: enzyme selectivity and absolute configuration of the nerolidol precursor, *Phytochemistry* 39, 785, 1995.

27. Paré, P.W. and Tumlinson, J.H., Plant volatiles as a defense against insect herbivores, *Plant Physiol.* 121, 325, 1999.

28. Jarvis, A.P., Schaaf, O., and Oldham, N.J., 3-Hydroxy-3-phenylpropanoic acid is an intermediate in the biosynthesis of benzoic acid and salicylic acid but benzaldehyde is not, *Planta* 212, 119, 2000.

29. Wildermuth, M.C., Dewdney, J., Wu, G., and Ausubel, F.M., Isochorismate synthase is required to synthesize salicylic acid for plant defence, *Nature* 414, 562, 2001.

30. Rowan, D.D., Lane, H.P., Allen, J.M., Fielder, S., and Hunt, M., Biosynthesis of 2-methylbutyl, 2-methyl-2-butenyl, and 2-methylbutanoate esters in Red Delicious and Granny Smith apples using deuterium-labeled substrates, *J. Agric. Food Chem.* 44, 3276, 1996.

31. Frey, M., Stettner, C., Pare, P.W., Schmelz, E.A., Tumlinson, J.H., and Gierl, A., An herbivore elicitor activates the gene for indole emission in maize, *Proc. Natl. Acad. Sci. USA* 97, 14801, 2000.

32. Joulain, D. and König, W.A., *The Atlas of Spectral Data of Sesquiterpene Hydrocarbons*, E.B.-Verlag, Hamburg, 1998.

33. Raguso, R. and Pellmyr, O., Dynamic headspace analysis of floral volatiles: a comparison of methods, *Oikos* 81, 238, 1998.

34. Adams, R.P., *Identification of Essential Oils by Ion Trap Mass Spectroscopy*, Academic Press, San Diego, 1989.

35. Bergström, G., Groth, I., Pellmyr, O., Endress, P. K., Thien, L. B., Hübener, A., and Wittko, F., Chemical basis for a highly specific mutualism: chiral esters attract pollinating beetles in Eupomatiaceae, *Phytochemistry* 30, 3221, 1991.

36. Borg-Karlson, A., Englund, F.O., and Unelius, C.R., Dimethyl oligosulphides, major volatiles released from *Sauromatum guttatum* and *Phallus impudicus*, *Phytochemistry* 35, 321, 1994.

37. Borg-Karlson, A.-K., Valterová, I., and Nilsson, A., Volatile compounds from flowers of six species in the family Apiaceae: Boquets for different pollinators, *Phytochemistry* 35, 111, 1994.

38. Borg-Karlson, A.-K., Unelius, C.R., Valterova, I., and Nilsson, L.A., Floral fragrance chemistry in the early flowering shrub *Daphne mezereum*, *Phytochemistry* 41, 1477, 1996.

39. Mori, K., Separation of enantiomers and determination of absolute configuration, in *Methods in Chemical Ecology*, vol. 1, *Chemical Methods*, Millar, J.G. and Haynes, K.F., Eds., Kluwer Academic, Norwell, MA, 1998, p. 295.
40. Francke, W., Hindorf, G., and Reith, W., Mass-spectrometric fragmentation of alkyl-1,6-dioxaspiro[4.5]decanes, *Naturwissenschaften* 66, 619, 1979.
41. Fráter, G., Bajgrowicz, J.A., and Kraft, P., Fragrance chemistry, *Tetrahedron* 54, 7633, 1998.
42. Wise, M.L., Savage, T.J., Katahira, E., and Croteau, R., Monoterpene synthases from common sage (*Salvia officinalis*), *J. Biol. Chem.* 273, 1491, 1998.
43. Raguso, R.A., Floral scent, olfaction, and scent driven foraging behavior, in *Cognitive Ecology of Pollination*, Chittka, L. and Thomson, J.D., Eds., Cambridge University Press, Cambridge, 2001, p. 83.
44. Azuma, H., Toyota, M., and Asakawa, Y., Intraspecific variation of floral scent chemistry in *Magnolia kobus* DC. (Magnoliaceae), *J. Plant Res.* 114, 411, 2001.
45. Dufaÿ, M., Hossaert-McKey, M., and Anstett, M.-C., Temporal and sexual variation in leaf-produced pollinator-attracting odours in the dwarf palm, *Oecologia* 139, 392, 2004.
46. Dötterl, S., Wolfe, L.M., and Jürgens, A., Qualitative and quantitative analyses of flower scent in *Silene latifolia*, *Phytochemistry* 66, 195, 2005.
47. Grison-Pigé, L., Bessière, J.M., Turlings, C.J., Kjellberg, F., Roy, J., and Hossaert-McKey, M., Limited intersex mimicry of floral odour in *Ficus carica*, *Funct. Ecol.* 15, 551, 2001.
48. Kite, G.C., The floral odour of *Arum maculatum*, *Biochem. Syst. Ecol.* 23, 343, 1995.
49. Knudsen, J.T., Variation in floral scent composition within and between populations of *Geonoma macrostachys* (Arecaceae) in the western Amazon, *Am. J. Bot.* 89, 1772, 2002.
50. Olesen, J.M. and Knudsen, J.T., Scent profiles of flower colour morphs of *Corydalis cava* (Fumariaceae) in relation to foraging behaviour of bumblebee queens (*Bombus terrestris*), *Biochem. Syst. Ecol.* 22, 231, 1994.
51. Omata, A., Yomogida, K., Nakamura, S., Ohta, T., Izawa, Y., and Watanabe, S., The odour of *Lotus* (Nelumbonaceae) flower, in *11th International Congress Of Essential Oils*, Bhattacharygas, L., Sen, N. and Sethi, K.L., Eds., Oxford: New Delhi, 1989, p. 43.
52. Pettersson, S. and Knudsen, J.T., Floral scent and nectar production in *Parkia biglobosa* Jacq. (Leguminosae: Mimosoideae), *Bot. J. Linn. Soc.* 135, 97, 2001.
53. Raguso, R. and Pichersky, E., Floral volatiles from *Clarkia breweri* and *C. concinna* (Onagraceae): recent evolution of floral scent and moth pollination, *Plant Syst. Evol.* 194, 55, 1995.
54. Dahl, Å.E., Wassgren, A.-B., and Bergström, G., Floral scents in *Hypecoum* Sect. *Hypecoum* (Papaveraceae): chemical composition and relevance to taxonomy and mating system, *Biochem. Syst. Ecol.* 18, 157, 1990.
55. Ervik, F., Tollsten, L., and Knudsen, J.T., Floral scent chemistry and pollination ecology in phytelephantoid palms (Arecaceae), *Plant Syst. Evol.* 217, 279, 1999.
56. Tollsten, L. and Bergström, L.G., Fragrance chemotypes in *Platanthera* (Orchidaceae)—the result of adaptation to pollinating moths?, *Nord. J. Bot.* 13, 607, 1993.
57. Tollsten, L. and Knudsen, J.T., Floral scent in dioecious *Salix* (Salicaceae)—a cue determining the pollination system?, *Plant Syst. Evol.* 182, 229, 1992.
58. Azuma, H., Toyota, M., Asakawa, Y., Yamaoka, R., García-Franco, J.G., Dieringer, G., Thien, L.B., and Kawano, S., Chemical divergence in floral scents of *Magnolia* and allied genera (Magnoliaceae), *Plant Species Biol.* 12, 69, 1997.
59. Barkman, T.J., Beaman, D.H., and Gage, J.A., Floral fragrance variation in *Cypripedium*: implications for evolutionary and ecological studies, *Phytochemistry* 44, 875, 1997.

60. Gerlach, G. and Schill, R., Fragrance analyses, an aid to taxonomic relationship of the genus *Coryanthes* (Orchidaceae), *Plant Syst. Evol.* 168, 159, 1989.

61. Gregg, K.B., Variation in floral fragrances and morphology: incipient speciation in *Cycnoches?*, *Bot. Gaz.* 144, 566, 1983.

62. Grison-Pigé, L., Hossaert-McKey, M., Greeff, J.M., and Bessière, J.M., Fig volatile compounds—a first comparative study, *Phytochemistry* 61, 61, 2002.

63. Jürgens, A. and Dötterl, S., Chemical composition of anther volatiles in Ranunculaceae: genera-specific profiles in *Anemone, Aquilegia, Caltha, Pulsatilla, Ranunculus*, and *Trollius* species, *Am. J. Bot.* 91, 1669, 2004.

64. Knudsen, J.T., Floral scent chemistry in geonomoid palms (Palmae: Geonomeae) and its importance in maintaining reproductive isolation, *Mem. N. Y. Bot. Gard.* 83, 141, 1999.

65. Knudsen, J.T., Tollsten, L., and Ervik, F., Flower scent and pollination in selected neotropical palms, *Plant Biol.* 3, 642, 2001.

66. Levin, R.A., Raguso, R.A., and McDade, L.A., Fragrance chemistry and pollinator affinities in Nyctaginaceae, *Phytochemistry* 58, 429, 2001.

67. Lindberg, A. B., Knudsen, J. T., and Olesen, J. M., Independence of floral morphology and scent chemistry as trait groups in a set of Passiflora species, in *The Scandinavian Association for Pollination Biology honours Knut Fgri*. Totland, Ö., Ed., *Det Norske Videnskaps-Akademi. I. Mat.-Naturv. Klasse. Ny Ser.*, Oslo.

68. Jürgens, A., Witt, T., and Gottsberger, G., Flower scent composition in *Dianthus* and *Saponaria* species (Caryophyllaceae) and its relevance for pollination biology and taxonomy, *Biochem. Syst. Ecol.* 31, 345, 2003.

69. Raguso, R.A., Levin, R.A., Foose, S.E., Holmberg, M.W., and McDade, L.A., Fragrance chemistry, nocturnal rhythms and pollination "syndromes" in *Nicotiana*, *Phytochemistry* 63, 265, 2003.

70. Thien, L.B., Bernhardt, P., Gibbs, G.W., Pellmyr, O.M., Bergstrom, G., Groth, I., and McPherson, G., The pollination of *Zygogynum* (Winteraceae) by a moth, *Sabatinca* (Micropterigidae): an ancient association?, *Science* 227, 540, 1985.

71. Toyoda, T., Nohara, I., and Sato, T., Headspace analysis of volatile compounds from various citrus blossoms, *Bioactive Volatile Compounds from Plants*, in Teranishi, R., Buttery, R.G., and Sugisawa, H., Eds., American Chemical Society, Washington, DC, 1993, pp. 205.

72. Whitten, W.M. and Williams, N.H., Floral fragrances of *Stanhopea* (Orchidaceae), *Lindleyana* 7, 130, 1992.

73. Pichersky, E. and Gang, D.R., Genetics and biochemistry of secondary metabolites in plants: an evolutionary perspective, *Trends Plant Sci.* 5, 439, 2000.

74. Ishizaka, H., Yamada, H., and Sasaki, K., Volatile compounds in the flowers of *Cyclamen persicum, C. purpurascens* and their hybrids, *Sci. Hort.* 94, 125, 2002.

75. Loper, G.M., Differences in alfalfa flower volatiles among parent and F1 plants, *Crop Sci.* 16, 107, 1976.

76. Knudsen, J.T., Andersson, S., and Bergman, P., Floral scent attraction in *Geonoma macrostachys*, an understorey palm of the Amazonian rain forest, *Oikos* 85, 409, 1999.

77. Hansted, L., Jacobsen, H.B., and Olsen, C.E., Influence of temperature on the rhythmic emission of volatiles from *Ribes nigrum* in situ, *Plant Cell Environ.* 17, 1069, 1994.

78. Jakobsen, H.B. and Olsen, C.E., Influence of climatic factors on emission of volatiles in situ, *Planta* 192, 365, 1994.

79. Loper, G.M., and Berdel, R.L., Seasonal emanation of ocimene from alfalfa flowers with three irrigation treatments, *Crop Sci.* 18, 447, 1978.

80. Nielsen, J.K., Jakobsen, H.B., Hansen, P.F.K., Moller, J., and Olsen, C.E., Asynchronous rhythms in the emission of volatiles from *Hesperis matronalis* flowers, *Phytochemistry* 38, 847, 1995.

81. Gerlach, G. and Schill, R., Composition of orchids scents attracting euglossine bees, *Bot. Acta* 104, 379, 1991.

82. Omata, A., Yomogida, K., Nakamura, S., Ohta, T., Izawa, Y. and Watanabe, S., The scent of *Lotus* flowers, *J. Essent. Oil Res.* 3, 221, 1991.

83. Knudsen, J.T., Tollsten, L., Groth, I., Bergström, G., and Raguso, R.A., Trends in floral scent chemistry in pollination syndromes: Floral scent composition in hummingbird-pollinated taxa, *Bot. J. Linn. Soc.* 146, 191, 2004.

84. Knudsen, J.T. and Tollsten, L., Trends in floral scent chemistry in pollination syndromes: floral scent composition in moth-pollinated taxa, *Bot. J. Linn. Soc.* 113, 263, 1993.

85. Altenburger, R. and Matile, P., Circadian rhythmicity of fragrance emission in flowers of *Hoya carnosa* R. Br., *Planta* 174, 248, 1988.

86. Dudareva, N., Piechulla, B., and Pichersky, E., Biogenesis of floral scents, *Hort. Rev.* 24, 31, 2000.

87. Euler, M. and Baldwin, I.T., The chemistry of defense and apparency in the corollas of *Nicotiana attenuata*, *Oecologia* 107, 102, 1996.

88. Helsper, J.P.F., Davies, J.A., Bouwmeester, H.J., Krol, A.F., and van Kampen, M.H., Circadian rhythmicity in emission of volatile compounds by the flowers of *Rosa hybrida* L. cv. Honesty, *Planta* 207, 88, 1998.

89. Hills, H.G. and Williams, N.H., Fragrance cycle of *Clowesia rosea*, *Orquídea (Méx.)* 12, 19, 1990.

90. Kuanprasert, N., Kuehnle, A.R., and Tang, C.S., Floral fragrance compounds of *Anthurium* (Araceae) species and hybrids, *Phytochemistry* 49, 521, 1998.

91. Loper, G.M. and Lapioli, A.M., Photoperiodic effects on the emanation of volatiles from alfalfa (*Medicago sativa* L.) florets, *Plant Physiol.* 49, 729, 1971.

92. MacTavish, H.S., Davies, N.W., and Menary, R.C., Emission of volatiles from brown *Boronia* flowers: some comparative observations. *Ann. Bot.* 86, 347, 2000.

93. Pott, M.B., Pichersky, E., and Piechulla, B., Evening specific oscillations of scent emission, SAMT enzyme activity, and SAMT mRNA in flowers of *Stephanotis floribunda*, *J. Plant Physiol.* 159, 925, 2002.

94. Awano, K., Honda, T., Ogawa, T., Suzuki, S., and Matsunaga, Y., Volatile components of *Phalaenopsis schilleriana* Rehb., *Flav. Fragr. J.* 12, 341, 1997.

95. Lewis, J.A., Moore, C.J., Fletcher, M.T., Drew, R.A., and Kitching, W., Volatile compounds from flowers of *Spathiphyllum cannaefolium*, *Phytochemistry* 27, 2755, 1988.

96. Matsumoto, F., Idetsuki, H., Harada, K., Nohara, I., and Toyoda, T., Volatile components of *Hedychium coronarium* Koenig flowers, *J. Essent. Oil Res.* 5, 123, 1993.

97. Patt, J.M., French, J.C., Schal, C., Lech, J., and Hartman, T.G., The pollination biology of tuckahoe, *Peltandra virginica* (Araceae), *Am. J. Bot.* 82, 1230, 1995.

98. Pham-Delegue, M.H., Etievant, P., Guichard, E., and Masson, C., Sunflower volatiles involved in honeybee discrimination among genotypes and flowering stages, *J. Chem. Ecol.* 15, 329, 1989.

99. Sazima, M., Vogel, S., Cocucci, A., and Hauser, G., The perfume flowers of *Cyphomandra* (Solanaceae): pollination by euglossine bees, bellows mechanism, osmophores, and volatiles, *Plant Syst. Evol.* 187, 51, 1993.

100. Schiestl, F.P., Ayasse, M., Paulus, H.F., Erdmann, D., and Francke, W., Variation of floral scent emission and postpollination changes in individual flowers of *Ophrys sphegodes* subsp. *sphegodes*, *J. Chem. Ecol.* 23, 2881, 1997.

101. Stránsky, K. and Valterová, I., Release of volatiles during the flowering period of *Hydrosme riviera*, *Phytochemistry* 52, 1387, 1999.
102. Tollsten, L., A multivariate approach to post-pollination changes in the floral scent of *Platanthera bifolia* (Orchidaceae), *Nord. J. Bot.* 13, 495, 1993.
103. Vogel, S., *The Role of Scent Glands in Pollination*, Amerind Publishing, New Delhi, 1990.
104. Bergström, G., Dobson, H.E.M., and Groth, I., Spatial fragrance patterns within the flowers of *Ranunculus acris* (Ranunculaceae), *Plant Syst. Evol.* 195, 221, 1995.
105. Dobson, H.E.M., Groth, I., and Bergström, G., Pollen advertisement: chemical contrasts between whole-flower and pollen odors, *Am. J. Bot.* 83, 877, 1996.
106. Dobson, H.E.M. and Bergström, G., The ecology and evolution of pollen odors, *Plant Syst. Evol.* 222, 63, 2000.
107. Ecroyd, C.E., Franich, R.A., Kroese, H.W., and Steward, D., Volatile constituents of *Dactylanthus taylorii* flower nectar in relation to flower pollination and browsing by animals, *Phytochemistry* 40, 1387, 1995.
108. Flamini, G., Cioni, P.L., and Morelli, I., Differences in the fragrances of pollen and different floral parts of male and female flowers of *Laurus nobilis*, *J. Agric. Food Chem.* 50, 4647, 2002.
109. Knudsen, J.T. and Tollsten, L., Floral scent and intrafloral scent differentiation in *Moneses* and *Pyrola* (Pyrolaceae), *Plant Syst. Evol.* 177, 81, 1991.
110. Raguso, R.A., Why are some floral nectars scented?, *Ecology* 85, 1486, 2004.
111. Ayasse, M., Schiestl F.P., Paulus, H.F., Lofstedt, C., Hansson, B., Ibarra, F., and Francke, W., Evolution of reproductive strategies in the sexually deceptive orchid *Ophrys sphegodes*: how does flower-specific variation of odor signals influence reproductive success?, *Evolution* 54, 1995, 2000.
112. Moya, S. and Ackerman, J.D., Variation in the floral fragrance of *Epidendrum ciliare* (Orchidaceae), *Nord. J. Bot.* 13, 41, 1993.
113. Bohlmann, J., Martin, D., Oldham, N.J., and Gershenzon, J., Terpenoid secondary metabolism in *Arabidopsis thaliana*: cDNA cloning, characterization, and functional expression of a myrcene/(*E*)-β-ocimene synthase, *Arch. Biochem. Biophys.* 375, 261, 2000.
114. Tholl, D., Chen, F., Petri, J., Gershenzon, J., and Pichersky, E., Two sesquiterpene synthases are responsible for the complex mixture of sesquiterpenes emitted from *Arabidopsis* flowers, *Plant J.* 42, 757, 2005.
115. Schlumpberger, B.O., Dehydrogeosmin produzierende Kakteen: Untersuchungen zur Verbreitung, Duftstoff-Produktion und Bestäubung, dissertation, Universität Tübingen, Verlag Grauer, Stuttgart, 2002.
116. Azuma, H., Thien, L.B., and Kawano, S., Molecular phylogeny of *Magnolia* (Magnoliaceae) inferred from cpDNA sequences and evolutionary divergence of the floral scents, *J. Plant Res.*, 112, 291, 1999.
117. Barkman, T.J., Character coding of secondary chemical variation for use in phylogenetic analyses, *Biochem. Syst. Ecol.* 29, 1, 2001.
118. Levin, R.A., McDade, L.A., and Raguso, R.A., The systematic utility of floral and vegetative fragrance in two genera of Nyctaginaceae, *Syst. Biol.* 52, 334, 2003.
119. Williams, W.M. and Whitten, N.H., Molecular phylogeny and floral fragrances of male euglossine bee-pollinated orchids: a study of *Stanhopea*, *Plant Species Biol.* 14, 143, 1999.
120. Lozada, T.M., Borchsenius, F., Knudsen, J.T., and Frydenberg, J., Reproductive isolation of sympatric forms of the neotropical understory palm *Geonoma macrostachys* var. *macrostachys* in western Amazonia, *Plant Syst. Evol.* submitted.
121. Dobson, H.E.M., Floral volatiles in insect biology, in *Insect-Plant Interactions*, Bernays, E.A., Ed., CRC Press, Boca Raton, FL, 1994, p. 47.

122. Robacker, D.C., Meeuse, B.J.D., and Erickson, E.H., Floral aroma: how far will plants go to attract pollinators?, *BioScience* 38, 390, 1988.
123. Metcalf, R.L., Plant volatiles as insects attractants, *CRC Crit. Rev. Plant Sci.* 45, 251, 1987.
124. Williams, N.H., Floral fragrances as cues in animal behavior, in *Handbook of Experimental Pollination Biology*, Jones, C.E. and Little, R.J., Eds., Van Nostrand Reinhold, New York, 1983, p. 50.
125. Pichersky, E. and Gershenzon, J., The formation and function of plant volatiles: perfumes for pollinator attraction and defense, *Curr. Opin. Plant Biol.* 5, 237, 2002.
126. Andersson, S., Nilsson, L.A., Groth, I., and Bergström, G., Floral scents in butterfly-pollinated plants: possible convergence in chemical composition, *Bot. J. Linn. Soc.* 140, 129, 2002.
127. Raguso, R.A., Olfactory landscapes and deceptive pollination: signal, noise and convergent evolution in floral scent, in *Insect Pheromone Biochemistry and Molecular Biology*, Blomquist, G.J. and Vogt, R., Eds., Academic Press, New York, 2003, p. 631.
128. Dufaÿ, M., Hossaert-McKey, M., and Anstett, M.C., When leaves act like flowers: how dwarf palms attract their pollinators, *Ecol. Lett.* 6, 28, 2003.
129. Willmer, P.G. and Stone, G.N., How aggressive ant-guards assist seed-set in *Acacia* flowers, *Nature* 388, 165, 1997.
130. El-Sayed, A.M., The Pherobase: database of insect pheromones and semiochemicals, available at http://www.pherobase.com, 2004.
131. Angioy, A.-M., Stensmyr, M.C., Urru, I., Puliafito, M., Collu, I., and Hansson, B.S., Function of the heater: the dead horse arum revisited, *Biol. Lett. Suppl.* 271(pt. 3), S13, 2004.
132. Seymour, R.S. and Schultze-Motel, P., Heat-producing flowers, *Endeavour* 21, 125, 1997.
133. Seymour, R.S., White, C.R., and Gibernau, M., Heat reward for insect pollinators, *Nature* 426, 243, 2003.
134. Seymour, R.S., Gibernau, M., and Ito, K., Thermogenesis and respiration of inflorescences of the dead horse arum *Helicodiceros muscivorus*, a pseudo-thermoregulatory aroid associated with fly pollination, *Funct. Ecol.* 17, 886, 2003.
135. Stensmyr, M.C., Urru, I., Collu, I., Celander, M., Hansson, B.S., and Angioy, A.M., Rotting smell of dead-horse arum florets, *Nature* 420, 625, 2002.
136. Schiestl, F.P. and Ayasse, M., Do changes in floral odor cause speciation in sexually deceptive orchids?, *Plant Syst. Evol.* 234, 111, 2002.
137. Schiestl, F., Ayasse, M., Paulus, H.F., Löfstedt, C., Hansson, B., Ibarra, F., and Francke, W., Orchid pollination by sexual swindle, *Nature* 399, 421, 1999.
138. Ayasse, M., Schiestl, F., Paulus, H.F., Erdmann, D., and Francke, W., Chemical communication in the reproductive biology of *Ophrys sphegodes*, *Mitt. Dtsch. Ges. Allg. Angew. Ent.* 11, 473, 1997.
139. Ayasse, M., Schiestl, F., Paulus, H.F., Ibarra, F., and Francke, W., Pollinator attraction in a sexually deceptive orchid by means of unconventional chemicals, *Proc. R. Soc. Lond. B.* 270, 517, 2003.
140. Eltz, T., Whitten, W.M., Roubik, D.W., and Linsenmair, K.E., Fragrance collection, storage, and accumulation by individual male orchid bees, *J. Chem. Ecol.* 25, 157, 1999.
141. Eltz, T., Roubik, D.W., and Whitten, W.M., Fragrances, male display and mating behaviour of *Euglossa hemichlora*: a flight cage experiment, *Physiol. Entomol.* 28, 251, 2003.
142. Ervik, F. and Knudsen, J.T., Scarabs and water lilies: faithful partners for the past 100 million years?, *Linn. J. Biol.* 80, 539, 2003.

143. Gottsberger, G., Flowers and beetles in the South American tropics, *Bot. Acta* 103, 360, 1990.

144. Gottsberger, G. and Silberbauer-Gottsberger, I., Olfactory and visual attraction of *Erioscelis emarginata* (Cyclocephalini, Dynastinae) to the inflorescences of *Philodendron selloum* (Araceae), *Biotropica* 23, 23, 1991.

145. Prance, G.T. and Arias, J.R., A study of the floral biology of *Victoria amazonica* (Poepp.) Sowerby (Nymphaeaceae), *Acta Amazon.* 5, 109, 1975.

146. Schatz, G.E., Some aspects of pollination biology in Central American forests, in *Reproductive Ecology of Tropical Plants*, Bawa, K.S. and Hadley, M., Eds., Parthenon, New York, 1990, p. 69.

147. Sakai, S., A review of brood-site pollination mutualisms: plants providing breeding sites for their pollinators, *J. Plant Res.* 115, 161, 2002.

148. Anstett, M.C., An experimental study of the interaction between the dwarf palm (*Chamaerops humilis*) and its floral visitor *Derelomus chamaeropsis* throughout the life cycle of the weevil, *Acta Oecol.* 20, 551, 1999.

149. Bernal, R. and Ervik, F., Floral biology and pollination of the dioecious palm *Phytelephas seemannii* in Colombia: an adaptation to Staphylinid beetles, *Biotropica* 28, 682, 1996.

150. Eriksson, R., The remarkable weevil pollination of the neotropical Carludovicoideae (Cyclanthaceae). *Plant Syst. Evol.* 189, 75, 1994.

151. Gottsberger, G., Pollination and evolution in neotropical Annonaceae, *Plant Species Biol.* 14, 143, 1999.

152. Miyake, T. and Yafuso, M., Floral scents affect reproductive success in fly-pollinated *Alocasia odora* (Araceae), *Am. J. Bot.* 90, 370, 2003.

153. Terry, I., Moore, C.J., Walter, G.H., Forster, P.I., Roemer, R.B., Donaldson, J., and Machin, P., Association of cone thermogenesis and volatiles with pollinator specificity in *Macrozamia* cycads. *Plant Syst. Evol.* 243, 233, 2004.

154. Grison-Pigé, L., Bessière, J.-M., and Hossaert-McKey, M., Specific attraction of fig-pollinating wasps: role of volatile compounds released by tropical figs. *J. Chem. Ecol.* 28:283–295, 2002.

155. Knudsen, J.T., Tollsten, L., and Bergström, G., A review: Floral scents — a check list of volatile compounds isolated by head-space techniques. *Phytochemistry* 33, 253, 1993.

156. Webster, F.X., Millar, J.G., and Kiemle, D.J., Mass spectrometry, in *Methods in Chemical Ecology*, vol. 1, *Chemical Methods*, Millar, J.G. and Haynes, K.F., Eds., Kluwer Academic, Norwell, MA, 1998, p. 127.

157. Heath, R.R., and Dueben, B.D., Analytical and preparative gas chromatography, in *Methods in Chemical Ecology*, vol. 1, *Chemical Methods*, Millar, J.G. and Haynes, K.F., Eds., Kluwer Academic, Norwell, MA, 1998, p. 85.

158. Schiestl, F.P., Ayasse, M., Paulus, H.F., Löfstedt, C., Hansson, B.S., Ibarra, F., and Francke, W., Sex pheromone mimicry in the early spider orchid (*Ophrys sphegodes*): patterns of hydrocarbons as the mechanism for pollination by sexual deception, *J. Comp. Physiol. A* 186, 567, 2000.

159. Schiestl, F.P. and Ayasse, M., Post-pollination emission of a repellent compound in a sexually deceptive orchid: a new mechanism for maximizing reproductive success, *Oecologia* 126, 531, 2001.

Section II

Biochemistry and Molecular Biology of Floral Scent

3 Floral Scent Metabolic Pathways: Their Regulation and Evolution

Natalia Dudareva and Eran Pichersky

CONTENTS

3.1 INTRODUCTION

Floral scents are a diverse blend of low molecular weight, mostly lypophilic compounds (Chapter 2). Although there are more than 1000 known volatiles, most of them are produced by only a few major biochemical pathways (isoprenoid, lipoxygenase, and phenylpropanoid/benzenoid pathways). Their diversity is mainly derived from specific enzymatic derivatizations, which increase the volatility of compounds at the final step of their formation.[1] In the past decade or so, significant progress has been made in understanding the biochemical routes to floral volatiles and the molecular mechanisms that regulate their formation. Several model plant species with powerful floral scents, such as *Clarkia breweri*, snapdragon (*Antirrhinum majus*), *Petunia hybrida*, rose (*Rosa* spp.), *Stephanotis floribunda*, and *Nicotiana suaveolens* have been used to isolate and characterize enzymes and genes involved in the biosynthesis of floral volatiles. Moreover, investigations of volatile production in vegetative tissues and fruits, as well as genes expressed in *Arabidopsis* flowers, have also significantly contributed to our understanding of the formation of floral volatiles.

3.2 BIOCHEMICAL PATHWAYS

3.2.1 BIOSYNTHESIS OF VOLATILE TERPENES

Terpenoids, the largest class of floral volatiles (Knudsen and Gershenzon, 2005; see Chapter 2) and plant secondary metabolites in general, include such well-known common constituents of floral scents as monoterpenes, linalool, limonene, myrcene, ocimene, geraniol, sesquiterpenes, farnesene, neralidol, caryophyllene, and germacrene. All terpenoids originate through the condensation of the universal five-carbon building blocks, isopentenyl diphosphate (IPP) and dimethylallyl diphosphate (DMAPP), which are derived from two alternative pathways localized in different cellular compartments. In the cytosol, IPP is synthesized from the classical mevalonic acid (MVA) pathway that starts with the condensation of acetyl-CoA,[2,3] whereas in plastids, IPP is formed from pyruvate and glyceraldehyde 3-phosphate via the methylerythritol phosphate (MEP, or nonmevalonate) pathway.[4-7] However, metabolic "crosstalk" between these two different IPP biosynthetic pathways has recently been documented,[8-12] particularly in the direction from plastids to cytosol.[13,14] Moreover, it has been shown that the plastid-localized MEP pathway provides a major IPP and DMAPP source for sesquiterpene biosynthesis and that the MVA pathway, localized in the cytosol, appears to be inactive in the petal tissue of open snapdragon flowers.[14] Such trafficking of isoprenoid intermediates could be mediated by a specific metabolite transporter, which was recently characterized in spinach.[15]

   In both compartments, IPP and DMAPP are then used by prenyltransferases in condensation reactions to produce prenyl diphosphates. In plastids, a head-to-tail condensation of IPP and DMAPP catalyzed by prenyltransferase geranyl diphosphate (GPP) synthase (GPPS; EC 2.5.1.1) yields GPP, the immediate precursor of all monoterpenes.[16,17] Also, in the same compartment, geranylgeranyl diphosphate synthase (GGPPS; EC 2.5.1.30) adds three molecules of IPP to DMAPP to form GGPP, the C_{20} diphosphate precursor of diterpenes. In the cytosol, condensation of two IPP molecules with one DMAPP by the action of farnesyl diphosphate (FPP) synthase (FPPS; EC 2.5.1.10) generates FPP, the C_{15} diphosphate precursor for sesquiterpene biosynthesis.[18] Genes encoding GPPS, FPPS, and GGPPS have been isolated from a diverse range of plant species.[19-22] The sequences of these proteins are all related to each other, as well as to prenyltransferases from animals, fungi, and bacteria,[23] thus forming a family of short-chain prenyltransferase proteins. While both FPPS and GGPPS are functional homodimers, the situation with GPPS is more complex. The GPPSs of *Arabidopsis*[24] and *Abies grandis*[25] are also homodimers, but those reported from peppermint leaves[21] and the flowers of snapdragon and *C. breweri*[22] are unusual heterodimeric enzymes, in which both subunits are absolutely required for prenyltransferase activity.

After the formation of these prenyl diphosphate precursors, the various monoterpenes, sesquiterpenes, and diterpenes are generated through the action of a large family of enzymes known as terpene synthases.[26,27] One of the most exceptional properties of these enzymes is their tendency to make multiple products from a single substrate. Out of 556 terpenoids identified (Knudsen and Gershenzon, 2005; see Chapter 2) in floral scents, only a few are represented by volatile diterpenes, which have high vapor pressure at ambient temperature and volatilize easily. This list includes geranyl linalool, isophytol, and phytol, which were found in jasmine floral scent,[28] and cembrene and neophytadiene emitted from *Nicotiana tabacum* flowers.[29]

Many of the terpene volatiles found in floral bouquets are direct products of terpene synthases (Figure 3.1A,B), while others are formed through alterations of the primary terpene skeletons formed by TPSs by hydroxylation, dehydrogenation, acylation, and other reaction types.[1] The example of the cytochrome P450-catalyzed 3-hydroxylation of a monoterpene skeleton includes the conversion of limonene to *trans*-isopiperitenol, a volatile flavor compound found in mint.[30] A 6-hydroxylation of limonene by another P450 enzyme is the first step in the biosynthesis of another volatile spice, carvone, in the caraway (*Carum carvi*) fruit.[31] A P450 enzyme is also responsible for the conversion of the sesquiterpene 5-epi-aristolochene to capsidiol, a dihydroxylated volatile compound.[32] Nonspecific dehydrogenases convert some terpene alcohols such as geraniol and carveol to aldehydes.[33,34] Geranial and neral (which are coproduced by the oxidation of geraniol; the mixture is termed citral) have a "lemony" aroma and are found in many plants, while carvone gives caraway (*C. carvi*) its distinct flavor. On the other hand, an acetylation of geraniol by acetyltransferase generates geranyl acetate,[35] a compound with a fruity rose scent reminiscent of pear and slightly of lavender that occurs in many plant species.[36]

The scent bouquet of some flowers often contains irregular volatile C_{13} terpenoid compounds, including β-ionone (a C_{13} cyclohexone) and geranyl acetone (a C_{13} aldehyde).[37,38] Although these compounds are present at very low concentrations in the scent, they contribute significantly to the overall scent bouquet because of their low odor threshold.[39] These apocarotenoids are the products of oxidative cleavage of carotenoids, catalyzed by carotenoid cleavage dioxygenases (CCDs), which were recently isolated and characterized.[38,40]

3.2.2 BIOSYNTHESIS OF VOLATILE FATTY ACID DERIVATIVES

Volatile fatty acid derivatives, including saturated and unsaturated short-chain alcohols, aldehydes, and esters, represent the second largest class of floral volatiles and originate from membrane lipids. The products of the lipoxygenase pathway, they mainly derive from the degradation of C_{18} fatty acids (linolenic and linoleic acids), which, after being transformed to a hydroperoxide by a lipoxygenase, are cleaved

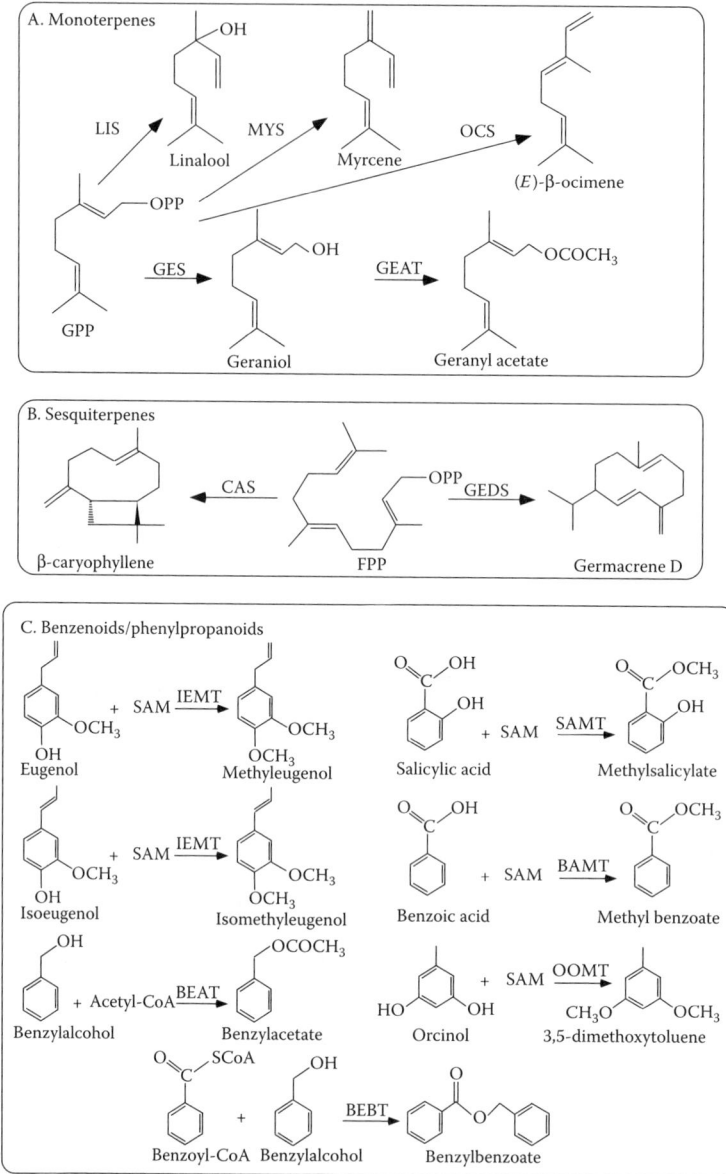

FIGURE 3.1 Examples of biochemical reactions that produce floral volatiles and the enzymes that catalyze these reactions. (A) Biosynthesis of monoterpenes and monoterpene esters. GEAT, geraniol:acetyl CoA acetyltransferase; GES, geraniol synthase; GPP, geranyldiphosphate; LIS, linalool synthase; OCS, ocimene synthase. (B) Biosynthesis of sesquiterpenes. CAS, β-caryophyllene synthase; GEDS, germacrene D synthase. (C) Biosynthesis of terpenoid and benzenoid volatiles. BAMT, benzoic acid methyltransferase; BEAT, acetyl-CoA:benzyl alcohol acetyltransferase; BEBT, benzoyl-CoA:benzyl alcohol benzoyl transferase; IEMT, isoeugenol/eugenol O-methyltransferase; OOMT, orcinol O-methyltransferase; SAMT, salicylic acid methyltransferase.

into C_{12} and C_6 components by hydroperoxide lyase.[41] Depending on the C_{18} substrate, hydroperoxide lyase produces either 3-*cis*-hexenal or hexanal, which are the common constituents of floral and green-leaf volatiles.[42] These short-chain aldehydes can undergo further processing by alcohol dehydrogenase and acyltransferase and be converted to the corresponding alcohols (3-*cis*-hexenol or hexanol) or 3-hexenyl acetate.[43] In the past several years, many genes involved in the lipoxygenase pathway have been isolated and characterized[44]; however, the expression of these genes has not yet been examined in floral tissue.

3.2.3 BIOSYNTHESIS OF VOLATILE PHENYLPROPANOIDS/BENZENOIDS

Phenylpropanoids constitute a large class of secondary metabolites in plants and are derived from phenylalanine via a complex series of branched pathways. While most of these aromatic compounds are usually nonvolatile, those that are reduced at the C9 position (to either aldehyde, alcohol, or alkane/alkene) or contain alkyl additions to the hydroxyl groups of the phenyl ring or to the carboxyl group (i.e., ethers and esters) are volatile. In addition, many benzenoid compounds, which lack the three-carbon chain and originate from *trans*-cinnamic acid as a side branch of the general phenylpropanoid pathway, are also volatile. These volatile phenylpropanoids/benzenoids are common constituents of floral scent.[42]

The first committed step in the biosynthesis of some phenylpropanoid and benzenoid compounds is catalyzed by the well-known and widely distributed enzyme L-phenylalanine ammonia-lyase (PAL; EC 4.3.1.5). PAL catalyzes the deamination of L-phenylalanine (Phe) to produce *trans*-cinnamic acid. The volatile phenylpropenes, such as eugenol, methyleugenol, chavicol, and methylchavicol share the first few biosynthetic steps with the lignin pathway; however, the entire biochemical pathway leading to these compounds has not yet been elucidated.[45]

The formation of benzenoids from cinnamic acid requires the shortening of the side chain by a C_2 unit, for which several routes have been proposed. The side chain shortening could occur via a CoA-dependent β-oxidative pathway, CoA-independent non-β-oxidative pathway, or via a combination of these two routes. The CoA-dependent-oxidative pathway is analogous to that underlying β-oxidation of fatty acids and proceeds through the formation of four CoA-ester intermediates. The CoA-independent non-β-oxidative pathway involves hydration of the free *trans*-cinnamic acid to 3-hydroxy-3-phenylpropionic acid and side-chain degradation via a reverse aldol reaction with formation of benzaldehyde, which is then oxidized to benzoic acid by an NADP[+]-dependent aldehyde dehydrogenase. Recent *in vivo* stable isotope labeling and computer-assisted metabolic flux analysis has revealed that both the CoA-β-dependent-oxidative and CoA-independent non-β-oxidative pathways are involved in the formation of benzenoid compounds in *P. hybrida*.[46]

Formation of phenylpropanoid-related compounds such as phenylacetaldehyde and phenylethanol from L-phenylalanine does not occur via cinnamic acid.[46] Moreover, a quantitative explanation of the labeling kinetics of phenylacetaldehyde and phenylethanol from deuterium-labeled Phe suggests that phenylacetaldehyde is not the only precursor of phenylethanol, and the major flux to the latter goes through a different route, possibly through phenylpyruvate and phenyllactic acid, as has been recently reported in rose flowers.[47,46] Although we still know very little about the enzymes and genes responsible for the metabolic steps leading to phenylpropanoids/benzenoids, significant progress has been made in the discovery of common modifications, such as hydroxylation, acetylation, and methylation of downstream products.[1]

3.2.4 MODIFICATION REACTIONS

A large portion of floral volatiles contain a methylated hydroxyl group (i.e., a methoxyl group). For example, methyleugenol and methyl chavicol are the results of the 4-hydroxyl methylation of eugenol and chavicol, respectively, catalyzed by two separate, but very similar enzymes, eugenol and chavicol O-methyltransferases (OMTs), which use S-adenosyl-L-methionnine (SAM) as the methyl donor (Figure 3.1C).[48–50] Two successive methylations of orcinol (3,5-dioxytoluene), also catalyzed by two very similar MTs, orcinol OMTs (OOMT1 and OOMT2), lead to the formation of dimethoxyorcinol, a major scent compound in many hybrid roses.[51,52] Interestingly, both enzymes can carry out both reactions; however, OOMT1 is more catalytically efficient with orcinol, while OOMT2 is more catalytically efficient with 3-methoxy,5-hydroxytoluene.[51] Chinese rose (*Rosa chinensis*) flowers make a similar compound with three methoxyl groups, 1,3,5-trimethoxybenzene, which is synthesized from 1,3,5-trihydroxybenzene. OOMT1 and OOMT2 can catalyze the methylation of the second and third intermediates (1-methoxy,3,5-dihydroxybenzene and 1,3-dimethoxy,5-hydroxybenzene), but not the methylation of 1,3,5-trihydroxybenzene, also known as phloroglucinol.[51,52] The enzyme that methylates this compound, phloroglucinol OMT (POMT), and the gene encoding this protein were recently isolated and characterized from rose petals.[53]

Another widespread group of methylated fragrant compounds includes methyl esters such as methyl benzoate, methyl cinnamate, methyl jasmonate, and methyl salicylate, which are formed via the transfer of the methyl group of SAM to a free carboxyl group of corresponding acids (Figure 3.1C). Enzymes capable of methylating benzoic, jasmonic, and salicylic acids have been identified from several plant species. While benzoic acid carboxyl methyltransferase (BAMT), the enzyme responsible for the snapdragon floral volatile methyl benzoate, methylates only benzoic acid,[54,55] the SAM:benzoic acid/salicylic acid carboxyl methyltransferases (BSMTs) and SAM:salicylic acid carboxyl methyltransferases (SAMTs) are able to methylate both salicylic and benzoic acids.[56]

Acylation of alcohols most often with an acetyl moiety, but also with larger acyls such as butanoyl or benzoyl acyls, leads to the formation of volatile esters, which are also common in floral scents. Enzymes directly involved in the volatile synthesis include benzyl alcohol acetyl-CoA transferase (BEAT) from *C. breweri* flowers, which produces benzyl acetate[57]; benzyl alcohol benzoyl-CoA transferase (BEBT), which produces benzyl benzoate in flowers of *Clarkia*[43]; and benzyl alcohol/phenyl ethanol benzoyl-CoA transferase (BPBT), which produces benzyl benzoate and phenylethyl benzoate in petunia flowers (*P. hybrida*).[46] These acyltransferases often show wide substrate specificity for both the acyl moiety and the alcohol moiety. For example, the petunia BPBT enzyme can also transfer an acetyl moiety to the alcohol phenylethanol, producing phenylethylacetate.[46] Similarly, an acyltransferase from ripening strawberry (*Fragaria* spp.) fruit can use a series of acyl moieties such as acetyl, butyryl, and hexanoyl, and transfer them to various alcohols such as hexenol, octanol, and geraniol.[58,59]

Volatile alcohols and aldehydes are common constituents of floral scents. Their interconversion in plant tissue is catalyzed by NADP/NAD-dependent oxidoreductases, which often have broad substrate specificity. It has been shown that benzyl alcohol, a major floral scent component in many Nicotianeae species[60] and elsewhere, is derived from benzaldehyde in a reversible reaction catalyzed by a member of the NADP/NAD-dependent oxidoreductases family.[46]

3.3 GENES RESPONSIBLE FOR SCENT PRODUCTION

In the past decade, investigations into floral scent in many laboratories have resulted in the characterization of a large number of genes encoding enzymes responsible for the synthesis of scent compounds. The initial breakthrough began in 1996 when the (*S*)-linalool synthase (LIS) gene encoding an enzyme responsible for the formation of the acyclic monoterpene linalool was isolated from *C. breweri* flowers using a classical biochemical approach through enzyme purification from petal tissues with the highest activity.[61,62] Thereafter, four additional genes responsible for the biosynthesis of floral volatiles were isolated using the same protein-based cloning strategy. These include *S*-adenosyl-L-methionine (SAM):(iso)eugenol *O*-methyltransferase (IEMT),[48] acetyl-coenzyme A:benzyl alcohol acetyltransferase (BEAT),[57] and *S*-adenosyl-L-methionine:salicylic acid carboxyl methyltransferase (SAMT),[63] all from *C. breweri*, and *S*-adenosyl-L-methionine:benzoic acid carboxyl methyltransferase (BAMT) from *A. majus*,[54,55] which encodes the enzymes responsible for the formation of methyl isoeugenol, benzyl acetate, methyl salicylate, and methyl benzoate, respectively.

Development of functional genomic technology in recent years allowed the isolation of more genes responsible for scent production. This list includes myrcene synthase and ocimene synthase from snapdragon (*A. majus*),[64] germacrene D synthase from roses (*Rosa hybrida*),[65] (*S*)-LIS and caryophyllene synthase from *Arabidopsis thaliana* flowers,[66] geraniol/citronellol acetyl transferase from *R. hybrida*,[35] SAMT from Madagascar jasmine (*S. floribunda*),[67] BSMT from *P. hybrida*, *A. thaliana*, and tobacco (*N. suaveolens*),[68–70] BEAT and BEBT from *C. breweri*,[43,57] BPBT from *P. hybrida*,[46] POMT from *R. chinensis*,[53] OOMT from *R. hybrida*,[51] and 9,10(9′,10′) carotenoid cleavage dioxygenase (PhCCD1) from *P. hybrida*.[40]

Many of these genes have turned out to belong to gene families, groups of genes that encode evolutionarily and structurally related enzymes (Table 3.1). For example, LIS is a member of the terpene synthase family, which includes monoterpene, sesquiterpene, and diterpene synthases. Some members of this gene family encode for proteins involved in the biosynthesis of vegetative volatiles.[1] Other scent genes belong to families that make some nonvolatile compounds. For example, BEAT belongs to the BAHD acyltranferases, which includes members involved in the biosynthesis of anthocyanin pigments and some phytoalexins.[71] SAMT and BAMT belong to the SABATH family of methyltransferases, which includes several membranes involved

TABLE 3.1
Examples of Gene Families with Members Encoding Floral Scent Biosynthetic Enzymes

Family	Example of Enzyme	Product	Plant
Terpene synthases	Linalool synthase (LIS)	Linalool	*Arabidopsis thaliana*
Methyltransferases (type I)	Orcinol *O*-methyltransferase (OOMT)	Dimethoxytoluene	Rose (*Rosa hybrida*)
Methyltransferases (type II)	Benzoic acid methyltransferase (BAMT)	Methyl benzoate	Snapdragon (*Antirrhinum majus*)
Acyltransferases (BAHD)	Acetyl-CoA:benzyl alcohol acetyltransferase (BEAT)	Benzylacetate	*Clarkia breweri*
Carotenoid cleavage dioxygenases (CCD)	Carotenoid cleavage dioxygenases	β-Ionone	*Petunia hybrida*

in the biosynthesis of the defense compound caffeine and the pigment compound bixin.[72] Even with all the progress made recently, most genes and enzymes involved in the biosynthesis of scent compounds are still unknown.

3.4 REGULATION OF SCENT BIOSYNTHETIC PATHWAYS

Substantial progress in the past decade in the isolation and characterization of genes responsible for the formation of floral volatiles has facilitated investigations into regulation of their biosynthesis. It has been found that floral volatiles are synthesized *de novo* in the tissues from which they are emitted and their production in plants is under both spatial and temporal control.

3.4.1 SPATIAL REGULATION

Of the plant organs in scented species, flowers produce the most diverse and the greatest amount of volatile compounds, which peak when the flowers are ready for pollination. Within the flowers, the petals are the principal emitters of volatiles, although various other parts of the flower may also participate in fragrance emission.[73,74] While the same floral scent components are often emitted from all parts of the flower (although not necessarily at the same amount or rate), sometimes specific compounds may be emitted from only a subset of the floral organs.[75–79] In addition, as discussed in Chapter 6, in some species (e.g., orchids), floral volatiles are emitted from highly specialized "scent glands" (i.e., osmophores) within the flower.[80,81] It is not yet clear how prevalent scent glands are among other scented flowers. So far, investigations of scent glands have been conducted mostly on the anatomic level. The question of whether such glands represent sites of emission only, or also of synthesis of scent volatiles, has not yet been addressed.

Identification of the enzymes responsible for the formation of some floral volatiles allowed us to determine how the levels of enzymatic activities are distributed in different floral parts. The activity levels of four enzymes (LIS, IEMT, BEAT, and SAMT, which are responsible for the formation of linalool, methyl isoeugenol, benzyl acetate, and methyl salicylate, respectively) were calculated per total weight of each organ of *C. breweri* flowers. The highest levels of activity of all these enzymes were found in the petals,[74] although other flower organs also contained detectable levels of activity. In the case of LIS, the stigma actually contained higher levels of LIS-specific activity than the petals. However, because the mass of the stigma of *C. breweri* is so small compared to the mass of the petals, LIS activity in the petals comprised the majority of activity present in the flower.[77] In snapdragon flowers, the majority of total BAMT activity, which is responsible for the formation of methyl benzoate, was found in the upper and lower lobes of the petals, with much less activity present in the tube and anthers. None of the remaining floral organs (pistils, sepals, and ovaries) or leaves were found to contain BAMT activity. Thus production of volatile compounds in snapdragon flowers was found to be limited mostly to the upper and lower lobes, which appear to make almost equal contribution to the whole-flower fragrance.[54]

After being synthesized, scent volatiles have to move to the exterior of the cell and evaporate (see Chapter 7). Until recently, it was not known whether these compounds were synthesized at the surface or whether they were transported there from adjacent cells. *In situ* hybridization and immunolocalization studies performed with LIS, IEMT, and BAMT have demonstrated that the biosynthesis of volatile products of these enzymes occurs almost exclusively in cells of the epidermal layer of petals and of other flower parts from which they can easily escape into the atmosphere.[62,82,83] Interestingly, in snapdragon flowers there is a higher concentration of the BAMT enzyme in the parts of the petals that are closer to the path that bees take to reach the nectar, including the "hairs" (i.e., unicellular glands) found in the center of the basal petal.[83] To date, little is known about the subcellular localization of the biosynthesis of scent compounds, although it has been shown that methyl benzoate is made in the cytosol.[83]

Before release from the flower surface, floral volatiles must move from their intracellular sites of biosynthesis through the outermost cuticle membrane. A detailed analysis of snapdragon petal cuticular wax amount and composition, cuticle thickness, and the amounts of internal and emitted methyl benzoate during 12 days of flower development revealed no changes in cuticle characteristics, indicating that the cuticle provides little diffusive resistance to volatile emission.[84] However, chemical analysis of cuticular waxes identified the unique wax composition of snapdragon petal cuticles, which are rich in branched alkanes and hydroxy esters. High relative proportions of branched alkanes and hydroxy esters could provide high permeability and thus permit rapid volatile emission by scented flowers.[84]

3.4.2 TEMPORAL REGULATION

During the lifespan of the flower, production and emission of scent volatiles are developmentally regulated. Volatile emission in flowers of different plant species follows similar developmental patterns, increasing during the early stages of flower development, peaking when the flowers are ready for pollination, and decreasing thereafter.[35,46,48,51,54,57,64,65,77,82] Analysis of the activities of enzymes responsible for the formation of scent compounds revealed two different developmental patterns. The activities of the first group of enzymes, represented by *C. breweri* LIS and SAMT,[82] snapdragon BAMT,[54] and rose geraniol acetyltransferase,[35] increase in young flowers and decline in old flowers, but remain relatively high (30% to 50% of the maximum level) even though emissions of corresponding volatile compounds had practically ceased. The activities of the second group of enzymes, represented by *C. breweri* IEMT, BEAT,[82] and BEBT,[43] and petunia BPBT,[46] show little or no decline at the end of the life span of the flower, although again, emissions of corresponding volatile compounds did decline substantially. The only enzyme identified so far that does not follow these two developmental patterns is rose OOMT, which is responsible for the formation of orcinol dimethyl ether. Its activity peaks sharply during flower maturation and decreases to almost undetectable levels in old flowers.[51]

The causes and consequences of high levels of scent biosynthetic enzyme activity in old flowers without the concomitant emission of the corresponding volatile product were unknown until recently. In the case of snapdragon, it has been found that the level of the enzyme responsible for the final step of the biosynthesis of a particular volatile is not the only limiting factor and that the target for the regulation of developmental production of volatile compounds includes the level of supplied substrate in the cell. Thus, the low emission of methyl benzoate in old flowers is due to low levels of benzoic acid in petal tissue.[54]

Analysis of the expression of genes encoding scent biosynthetic enzymes revealed its temporal and spatial regulation. In different plant species, the highest level of expression was found in the petal tissue for the majority of scent genes analyzed,[35,40,46,48,51,54,57,64,65,67,68,70] with the exception of *C. breweri* LIS and BEBT genes, for which the highest level of transcripts was detected in the stigma.[43,62] While the expression of some genes like petunia BSMT[68] and BPBT,[46] *S. floribunda* SAMT,[67] snapdragon BAMT,[54] ocimene synthase, and myrcene synthase[64] was restricted to petals, mRNA transcripts for *C. breweri* LIS,[62] IEMT,[48] BEAT,[57] and BEBT,[48] *N. suaveolens* BSMT,[70] and petunia PhCCD1[40] were found in other floral organs, including stigma, stamens, sepals, and ovaries in different combinations. Interestingly, out of all the genes analyzed, only two, petunia PhCCD1[40] and *C. breweri* BEBT,[43] were expressed in leaves, although at low levels.

During flower development, expression of genes encoding scent biosynthetic enzymes peaks 1 to 2 days ahead of enzyme activity and emission of the corresponding compound. The concurrent temporal changes in activities of enzymes responsible for the final steps of volatile formation, enzyme protein levels, and the expression of corresponding structural genes suggest that the developmental biosynthesis of volatiles is regulated largely at the level of gene expression.[35,43,46,51,54,57,62,64,65,67] It is still unclear to what extent transcriptional, post-transcriptional, translational, post-translational, and other events contribute to this process.

In general, more than one biochemical pathway is responsible for a blend of volatile compounds released from different plant tissues. A comparative analysis of the regulation of benzenoid and monoterpene emission in snapdragon (*A. majus*) flowers revealed that the orchestrated emission of phenylpropanoid and isoprenoid compounds is regulated upstream of individual metabolic pathways and includes the coordinated expression of genes that encode enzymes involved in the final steps of scent biosynthesis.[54,64] However, transcription factors that regulate multiple biosynthetic pathways leading to the formation of odor bouquet have not yet been discovered. Recently the first transcription factor *ODORANT1* (*ODO1*) involved in the regulation of production of benzenoid volatiles was identified in *P. hybrida* cv. Mitchell.[85] It belongs to an R2R3-type *MYB* family and its down-regulation in transgenic petunia plants leads to a strong reduction of volatile benzenoid levels by decreasing the level of precursors from the shikimate pathway.

3.4.3 REGULATION OF RHYTHMIC EMISSION OF SCENT COMPOUNDS

Emission of floral volatiles from some plant species changes rhythmically during a 24 h period, whereas other flowers continuously emit volatiles at a constant level. In addition, some plants emit one set of compounds during the day and another at night.[86,87] Moreover, it has also been found that within the flower, some compounds are emitted in a rhythmic manner during a 24 h period, while others are not, suggesting that different mechanisms regulate the biosynthesis and emission of these volatiles.[88,89] The rhythmic release of scent is often correlated with the corresponding temporal activity of flower pollinators and is controlled by a circadian clock or regulated by light.[90–93] Nocturnally pollinated plants exhibit a circadian, endogenously controlled rhythmicity in their nocturnal emission patterns. This rhythmicity is maintained upon exposure to continuous light or dark.[86,88,94,95] In contrast, diurnal rhythmicity in the emission of volatile compounds by plants was reported to be noncircadian and controlled by irradiation levels.[90,95,96] However, recent investigations in *R. hybrida* cv. Honesty and snapdragon *A. majus* revealed that their scent emissions display "free-running" cycles in the absence of environmental cues (in continuous dark or continuous light), indicating that diurnal rhythmicity in the emission of floral volatiles in these species is controlled by a circadian clock.[91–92]

A detailed time-course analysis during a 48 h period of the activity of BAMT, an enzyme catalyzing the final step in methyl benzoate formation in snapdragon flowers, revealed that there were similar levels of activity at night as well as during the day, suggesting that the activity of this enzyme is not an oscillation-determining factor in snapdragon petals. Oscillations during the daily light and dark cycle, which were also retained in continuous dark, were found in the amount of benzoic acid, the immediate precursor of methyl benzoate, indicating the involvement of a circadian clock in the control of substrate levels.[92] A similar scenario was found for the regulation of nocturnal emission of methyl benzoate in tobacco and petunia flowers. These results show that rhythmic emission of volatiles can be regulated at the cellular level of the substrate for the final step in the formation of the volatile compound.[92] Moreover, regulation of rhythmic emission can also include transcriptional and post-translational control of the expression of a gene responsible for the final step of volatile production, as was shown for nocturnally emitting *S. floribunda* flowers.[93]

3.4.4 REGULATION OF FLORAL SCENT AFTER POLLINATION

The scent of many flowers is markedly reduced soon after pollination. Such quantitative or qualitative postpollination changes in floral bouquets, shown mostly in orchids,[97–100] lower the attractiveness of these flowers, as well as increase the overall reproductive success of the plant by directing pollinators to the unpollinated flowers. This is particularly important for plants with a low visitation rate, where reproductive success is mostly pollinator limited.[101] Investigation of the molecular mechanisms responsible for postpollination changes in floral scent emission in snapdragon and petunia flowers revealed that the decrease in emission begins only after pollen tubes reach the ovary, suggesting that fertilization is a prerequisite for the reduction of floral scent after pollination.[68] Using methyl benzoate as an example, it has been shown that in snapdragon, the decrease in ester emission after pollination is the result of down-regulation of both the methylation index (the ratio of S-adenosyl-L-methionine to S-adenosyl-L-homocysteine) and activity of the enzyme responsible for methyl benzoate formation (BAMT). In petunia, the BAMT gene expression is suppressed by ethylene, which is produced in response to pollination.[68]

3.4.5 ROLE OF SUBSTRATES IN THE REGULATION OF SCENT FORMATION

In some cases, the nature of the product and the efficiency of its formation is determined by the availability of substrates for the final reaction, especially when that final reaction is catalyzed by an enzyme with broad substrate specificity (e.g., some carboxyl methyltransferases and acyltransferases).[46,68,70] Within the past 5 years, six benzenoid carboxyl methyltransferases, which are responsible for the formation of methyl esters (e.g., methyl benzoate and methyl salicylate) of floral scent, were isolated and characterized from several plant species.[55,56,63,67,68,70] With the exception of A. majus BAMT, which synthesizes only methyl benzoate, all other characterized carboxyl methyltransferases can produce both methyl salicylate and methyl benzoate. However, the analysis of the scent profiles in these plant species revealed that the biochemical properties of the enzymes do not always determine the flower scent composition, which also depends on the plant's cellular pools of available substrates. For example, in petunia flowers, which emit only methyl benzoate, the apparent catalytic efficiencies (k_{cat}/K_m ratio) of the isolated BSMTs were 40- to 75-fold higher with salicylic acid than with benzoic acid, indicating that salicylic acid was the preferred substrate.[68] Although a high carboxyl methyltransferase activity toward salicylic acid was also detected in petunia petals, there was a very small internal pool of free salicylic acid (approximately 10 times lower than the apparent K_m values of these enzymes for salicylic acid), indicating that the enzymes could not produce methyl salicylate in planta due to the lack of substrate. On the other hand, the level of benzoic acid (approximately 7 mM) was in the range of K_m values for benzoic acid, suggesting that these enzymes are involved in methyl benzoate emission.[68]

In the case of N. suaveolens, which emits both methyl benzoate and methyl salicylate, the isolated enzyme exhibits a higher catalytic efficiency with benzoic acid than with salicylic acid, partially reflecting the ratio of two methyl esters in the floral bouquet. However, with the benzoic acid level exceeding that of salicylic acid in petal tissue, it is likely that the isolated enzyme from *N. suaveolens* flowers is primarily involved in the synthesis of methyl benzoate.[70]

The role of substrate in regulating the biosynthesis of volatile compounds was also recently confirmed by metabolic engineering. When the LIS gene was introduced under the control of the cauliflower mosaic virus (CaMV) 35S constitutive promoter into *P. hybrida* W115, the differences between organs in the amount of the synthesized linalool or its glycoside depended more on the availability of the substrate GPP in the tissue than on expression of the LIS gene.[102]

When enzymes competed for the same substrate, as in the case of three monoterpene synthases (γ-terpinene cyclase, (+)-limonene cyclase, and (−)-β-pinene cyclase) introduced into tobacco (*N. tabacum*) plants, the magnitude of monoterpene emission in leaves was close to that predicted based on the K_m values of the enzymes for GPP, while the emission levels in flowers were comparable, suggesting that the GPP pool did not limit monoterpene production.[103] The regulation of GPP formation can occur at the level of GPP synthase, as was shown in snapdragon, where the small subunit can play a key role in GPP biosynthesis.[22] Feedback regulation of GPP synthase byproduct and substrate inhibition may also contribute to the regulatory control of the flux to GPP and subsequently to monoterpene production.[22]

The results discussed above show that although significant progress has been made in our understanding of the regulation of the final steps of volatile biosynthesis, a detailed understanding of the regulation of the flux through the entire biochemical pathway is essential for complete understanding of the production and emission of secondary volatile compounds.

3.5 EVOLUTION OF GENES AND PATHWAYS

The striking diversity of floral scents is something we are so accustomed to that we may even take it for granted. This diversity is achieved operationally by the flowers in two major ways. One is the sheer number of compounds that can be synthesized by the various species. The second is the specific combination of volatile compounds found in each flower as well as the relative amounts of each component. The effect is almost an endless number of distinct combinations of scents that can be recognized by both animals (including humans) and insects.

Both quantitative and qualitative changes in the amount of a given compound can happen by regulation of the pathway on multiple levels—regulation of existing genes encoding the biosynthetic enzymes (as discussed in earlier sections of this chapter), changes in the coding information (i.e., mutations), post translation modifications, or in the steps that effect the availability of substrate, stability of the product, and the process of its emission. The loss of a specific scent compound could also occur by mutations in genes specifying its biosynthesis. On the other hand, the creation of new scent compounds can also occur by mutation, when new enzymes are created with new substrate specificities or new products, and when mutations block existing pathways, leading to an increase in flux in related pathways and an increase in biosynthesis of a volatile compound. In this section we describe the various examples that have come to light regarding the molecular mechanisms by which various plant species have evolved the ability to add a new compound to their scent, substitute one scent compound for another, or lose the ability to make a certain scent compound.

3.5.1   EVOLUTION OF A NEW SCENT COMPOUND

3.5.1.1   Changes at the Level of Gene Regulation

Two clear examples of regulation at the gene level come from *C. breweri*. Linalool is a major scent component in the flowers of this species. The flowers of *C. breweri*'s closest relative, *C. concinna*, are mostly scentless (as are flowers of all the other species in the genus). As described previously, the LIS gene of *C. breweri* has been shown to be highly expressed in the stigmas, with slightly lower levels in the petals.[62] This LIS gene has also been isolated from *C. concinna* and has been shown to encode an identical protein.[104] However, in *C. concinna*, the gene is not expressed at all in the petals, but is expressed in the stigma at a much lower level than that of *C. breweri*.

   Although the *C. breweri* LIS gene is expressed more highly in the stigma than in the petals, the stigma does not emit linalool, but instead emits linalool oxides, which are derived from the further oxidation of linalool. While the gene is expressed at lower levels in the petals, the petals are relatively large and, overall, copious amounts of linalool are synthesized and emitted (1.6 μg/flower/h). In *C. concinna*, because LIS is not expressed in the petals, no linalool is emitted from the petals, and the small amount of linalool that is produced in the stigma is converted to linalool oxide, and none is emitted.

   Phylogenetic analysis indicates that scented *C. breweri* is a derived species in the overall nonscented genus, and it therefore appears that *C. breweri* has acquired its ability to produce and emit linalool from the petals after the split from the *C. concinna* lineage. Therefore the most likely explanation is that LIS in the ancestor of *C. breweri* and *C. concinna* was expressed only in the stigma at relatively low levels, perhaps because linalool oxide has some defense function, as do many terpenes.[105] Eventually a genetic change in the *C. breweri* lineage resulted in the up-regulation of the LIS gene, so that it is now expressed at higher levels in the stigma and also at an appreciable level in the petals. It is even possible that the ability to produce floral scent in that lineage was a factor in the speciation event that gave rise to *C. breweri* and *C. concinna*. Certainly today each species is pollinated by different insects—*C. breweri* by hawkmoths and *C. concinna* mostly by bees.[106]

   Another example of a change in regulation that led to the synthesis of a new scent component was also recorded in *C. breweri*. The gene encoding BEBT is expressed in the flowers of this plant, and the resulting enzyme is responsible for biosynthesis of the floral volatile benzyl benzoate. However, the expression of the gene for BEBT in *C. breweri*, as well as in other plant species, is also induced in leaf tissue upon pathogen infection or damage, and the resulting benzyl benzoate synthesized in the leaf is believed to act as a microbicide.[43] It appears that here, too, BEBT was recruited to produce a floral volatile in the *C. breweri* lineage.

   The scent of *C. breweri* also shows within-species variation. For example, some plant populations emit both eugenol and methyleugenol, while others emit only eugenol. It was shown that the plants that emit methyleugenol express the gene encoding the enzyme eugenol methyltransferase (IEMT), which converts eugenol to methyleugenol, while the plants that do not emit methyleugenol do not express this gene.[48] Other quantitative differences have been observed in the *C. breweri* scent. For example, one line emits twice as much benzyl acetate as another line.[106] However, the molecular basis for this difference has not yet been elucidated.

3.5.1.2 Creation of a New Gene

Since relatively little research has been done on the molecular biology of floral scent, examples of new scent genes arising during evolution are difficult to document. Such work requires a careful analysis of related DNA sequences in related plant lineages to firmly demonstrate that a gene encoding a new enzyme for the biosynthesis of a floral volatile is "new." Nonetheless, two of the examples described above are likely candidates. The LIS gene of *Clarkia* is not closely related to other LISs from plants in other lineages, and it has been postulated that it arose from other types of terpene synthases by gene duplications followed by domain swapping or simply by divergence.[104] Also, the *Clarkia* IEMT is not closely related to the enzymes that methylate similar compounds in basil, but instead to a methylase enzyme of the lignin biosynthetic pathway, indicating that it arose recently by duplication of a gene for primary, nonvolatile metabolite following sequence divergence.[50]

3.5.2 Substitution of Scent Compounds

It is generally assumed that new genes arise after gene duplication followed by divergence because divergence of a nonduplicated gene to encode a new biochemical function will eliminate the original function, which is likely to be disadvantageous to the organism. While this is a reasonable assumption in primary metabolism, a change in the function of an enzyme responsible for the synthesis of one scent component allowing the synthesis of a different scent component might be advantageous for the plant and carry no negative consequences.[107] In this case, we would expect orthologous genes from different species to encode different enzymes. Since few scent genes have been studied and few plant genomes are known in detail, proving orthology is not easy. However, examination of closely similar enzymes from closely related species would be informative. For example, *S. floribunda* is a moth-pollinated species whose scent is rich in methyl benzoate, but also contains some methyl salicylate. A related species, *Hoya carnosa*, emits only methyl salicylate from its flowers. Enzymes that use SAM as the methyl donor to make the methyl esters of benzoic acid and the closely related salicylic acid (2-hydroxybenzoic acid) have been identified in these two species.[67,70] The *H. carnosa* enzyme is specific for salicylic acid. The *S. floribunda* enzyme prefers salicylic acid, but can also methylate benzoic acid. It appears that the *S. floribunda* enzyme has evolved a wider substrate specificity, which allows it to make a new methyl ester (this is based on the observation that all plants have an enzyme that is specific for salicylic acid and makes methyl salicylate in the leaves in response to injury[69]). Since the substrate benzoic acid is much more abundant in the *S. floribunda* flowers, more methyl benzoate than methyl salicylate is produced.

3.5.3   LOSS OF SCENT COMPONENTS

Arabidopsis thaliana flowers emit monoterpenes and sesquiterpenes (see Chapter 4). In a screening of 37 ecotypes, several have been found to lack a group of four sesquiterpenes (including β-caryophyllene) that are all produced by a single enzyme. The gene for this enzyme is expressed in these ecotypes; however, the coding regions contain a deletion and several mutations that render the enzyme inactive.[105] A similar situation was found in basil, where the gene for LIS in one cultivar is normal, but in another it contains a frame-shift mutation.[108] Both LIS genes are expressed mostly in glands as well as in flowers. The basil cultivar with the mutated LIS gene cannot make linalool.

3.5.4   SCENT CHANGES DURING CLASSICAL AND MOLECULAR BREEDING

Plants cultivated for their flowers are a major human endeavor. People throughout history have noted certain flowers, such as roses, as particularly "sweet smelling" or otherwise distinctly scented. The original perfume industry arose from the observation that floral chemicals can be isolated and concentrated into "essential oils" (from essence, or smell). Products such as attar of roses are still manufactured and sold, although the bulk of perfumes today are produced from synthetic compounds.

Recent commercial plant breeding programs in the "cut flower" industry have resulted in many new cultivars of formerly scented species that substantially lack scent. The reasons for this are unclear. Traits such as color, visual attractiveness, and long shelf life have been targeted, without any attention to whether the selected lines are still scented. It may be the case that scent has been lost simply due to the fact that the few selected individuals per generation in each breeding program happened to have allelic compositions in the scent gene loci that were not optimal for quality and quantity of scent. It is also possible that production of scent is antagonistic for the optimization of the selected traits, such as long shelf life.[109] While the genetic mechanism for the loss of scent in cultivated species is still unclear, two examples can provide some intriguing clues.

First, the scent of *R. chinensis* is rich in 1,3,5-trimethoxybenzene, but most modern roses, which are believed to be hybrids obtained by crossing *R. chinensis* with other rose species, do not emit this compound. The methyltransferase enzymes responsible for the last steps in its synthesis are present in modern roses,[51,53] leading to the hypothesis that hybrid roses lack the ability to synthesize 1,3,5-trihydroxy-toluene, the substrate of these methyltransferases. However, whether and how this pathway was interrupted during the breeding history of modern roses has not yet been determined.

The second example is the carnation variety cv. Eilat. Its orange-red petal color is a result of the synthesis of an anthocyanin pigment that is derived from phenyla-lanine. In an attempt to produce white carnations, a group of investigators trans-formed the plant with a gene construct designed to suppress the expression of a key enzyme in biosynthesis of the pigment.[110] This approach was almost completely successful, resulting in flowers that had only traces of the pigment. Interestingly, the plants were much more fragrant, with a particularly significant increase in emission of methyl benzoate. It is believed that benzoic acid, the precursor of methyl benzoate, is also derived from phenylalanine.[46] Clearly the blockage in the flux of the pathway from phenylalanine to anthocyanin resulted in an increase in the flux from phenylalanine to methyl benzoate. Again, while the exact mechanism of this phenomenon has not yet been elucidated, it is easy to hypothesize that the opposite —selection for more pigment at the expense of less scent—has often occurred in the process of breeding more visually attractive flowers.

3.6   CONCLUSION

Plants depend on external factors, either biotic or abiotic, to effect cross-pollination. Floral scent plays an important role in attracting insect and animal pollinators, and therefore contributes to the successful reproduction of many plants. Scent provides long-distance information to pollinators about the identity of the flower. Plants have developed different strategies to attract and compete for both generalist and specialist pollinators that involve the complexity and intensity of floral scent. Small differences in scent composition often distinguish closely related species and, by attracting different pollinators, ensure their genetic isolation. Molecular dissection of the genes, enzymes, and pathways of scent biosynthesis and regulation has begun to identify the molecular mechanisms that bring about such diverse outcomes and how genetic changes lead to changes in scent profiles and intensities in related plant species.

ACKNOWLEDGMENTS

The authors' laboratory work is supported by the U.S. National Science Foundation (grants MCB-0212802 [to N.D.], MCB-0312466, and IBN-0211697 [to E.P.]), U.S. Department of Agriculture (grant 2003-35318-13619 [to N.D.]), U.S.-Israel Binational Agriculture Research and Development Fund (grant US-3437-03 [to N.D.] and IS-3332-02 [to E.P.]), and the Fred Gloeckner Foundation, Inc. (to N.D.).

REFERENCES

1. Dudareva, N., Pichersky, E., and Gershenzon, J., Biochemistry of plant volatiles, *Plant Physiol.* 135, 1893, 2004.
2. Qureshi, N. and Porter, J.W., Conversion of acetyl-coenzyme A to isopentenyl pyrophosphate, in *Biosynthesis of Isoprenoid Compounds*, Porter, J.W. and Spurgeon, S.L., Eds., John Wiley & Sons, New York, 1981, p. 47.
3. Newman, J.D. and Chappell, J., Isoprenoid biosynthesis in plants: carbon partitioning within the cytoplasmic pathway, *Crit. Rev. Biochem. Mol. Biol.* 34, 95, 1999.
4. Eisenreich, W., Schwarz, M., Cartayrade, A., Arigoni, D., Zenk, M.H., and Bacher, A., The deoxyxylulose phosphate pathway of terpenoid biosynthesis in plants and microorganisms, *Chem. Biol.* 5(9), R221, 1998.
5. Lichtenthaler, H.K., The 1-deoxy-D-xylulose-5-phosphate pathway of isoprenoid biosynthesis in plants, *Annu. Rev. Plant Physiol. Plant Mol. Biol.* 50, 47, 1999.
6. Rohmer, M., The discovery of a mevalonate-independent pathway for isoprenoid biosynthesis in bacteria, algae and higher plants, *Nat. Prod. Rep.* 16, 565, 1999.

7. Rodriguez-Concepcion, M., and Boronat, A., Elucidation of the methylerythritol phosphate pathway for isoprenoid biosynthesis in bacteria and plastids. A metabolic milestone achieved through genomics, *Plant Physiol.* 130, 1079, 2002.

8. Piel, J., Donath, J., Bandemer, K., and Boland, W., Mevalonate-independent biosynthesis of terpenoid volatiles in plants: induced and constitutive emission of volatiles, *Angew. Chem. Int. Ed.* 37, 2478, 1998.

9. Adam, K.P., Thiel, R., and Zapp, J., Incorporation of 1-[1-C-13]deoxy-D-xylulose in chamomile sesquiterpenes, *Arch. Biochem. Biophys.* 369, 127, 1999.

10. Jux, A., Gleixner, G., and Boland, W., Classification of terpenoids according to the methylerythritolphosphate or the mevalonate pathway with natural $^{12}C/^{13}C$ isotope ratios: dynamic allocation of resources in induced plants, *Angew. Chem. Int. Ed.* 40, 2091, 2001.

11. Hemmerlin, A., Hoeffler, J.-F., Meyer, O., Tritsch, D., Kagan, I.A., Grosdemange-Billiard, C., Rohmer, M., and Bach, T.J., Cross-talk between the cytosolic mevalonate and the plastidial methylerythritol phosphate pathways in tobacco bright yellow-2 cells, *J. Biol. Chem.* 278, 26666, 2003.

12. Schuhr, C.A., Radykewicz, T., Sagner, S., Latzel, C., Zenk, M.H., Arigoni, D., Bacher, A., Rohdich, F., and Eisenreich, W., Quantitative assessment of crosstalk between the two isoprenoid biosynthesis pathways in plants by NMR spectroscopy, *Phytochem. Rev.* 2, 3, 2003.

13. Laule, O., Fürholz, A., Chang, H.-S., Zhu, T., Wang, X., Heifetz, P.B., Gruissem, W., and Lange, M., Crosstalk between cytosolic and plastidial pathways of isoprenoid biosynthesis in *Arabidopsis thaliana*, *Proc. Natl. Acad. Sci. USA* 100, 6866, 2003.

14. Dudareva, N., Andersson, S., Orlova, I., Gatto, N., Reichelt, M., Rhodes, D., Boland, W., and Gershenzon, J., The nonmevalonate pathway supports both monoterpene and sesquiterpene formation in snapdragon flowers, *Proc. Natl. Acad. Sci. USA* 102, 933, 2005.

15. Bick, J.A. and Lange, B.M., Metabolic cross talk between cytosolic and plastidial pathways of isoprenoid biosynthesis: unidirectional transport of intermediates across the chloroplast envelope membrane, *Arch. Biochem. Biophys.* 415, 146, 2003.

16. Poulter, C.D. and Rilling, H.C., Prenyl transferases and isomerase, in *Biosynthesis of Isoprenoid Compounds*, Porter, J.W. and Spurgeon, S.L., Eds., John Wiley & Sons, New York, 1981, p. 161.

17. Ogura, K. and Koyama, T., Enzymatic aspects of isoprenoid chain elongation, *Chem. Rev.* 98, 1263, 1998.

18. McGarvey, D.J. and Croteau, R., Terpenoid metabolism, *Plant Cell* 7, 1015, 1995.

19. Wang, K. and Ohnuma, S., Chain-length determination mechanism of isoprenyl diphosphate synthases and implications for molecular evolution, *Trends Biochem. Sci.* 24, 445, 1999.

20. Gershenzon, J. and Kreis, W., Biochemistry of terpenoids: monoterpenes, sesquiterpenes, diterpenes, sterols, cardiac glycosides and steroid saponins, in *Biochemistry of Plant Secondary Metabolism*, Wink, M., Ed., CRC Press, Boca Raton, FL, 1999, p. 222.

21. Burke, C.C., Wildung, M.R., and Croteau, R., Geranyl diphosphate synthase: cloning, expression, and characterization of this prenyltransferase as a heterodimer, *Proc. Natl. Acad. Sci. USA* 96, 13062, 1999.

22. Tholl, D., Kish, C.M., Orlova, I., Sherman, D., Gershenzon, J., Pichersky, E., and Dudareva, N., Formation of monoterpenes in *Antirrhinum majus* and *Clarkia breweri* flowers involves heterodimeric geranyl diphosphate synthases, *Plant Cell* 16, 977, 2004.

23. Chen, A., Kroon, P.A., and Poulter, C.D., Isoprenyl diphosphate synthases: protein-sequence comparisons, a phylogenetic tree, and predictions of secondary structure, *Protein Sci.* 3, 600, 1994.

24. Bouvier, F., Suire, C., D'Harlingue, A., Backhaus, R,A., and Camara, B., Molecular cloning of geranyl diphosphate synthase and compartmentation of monoterpene synthesis in plant cells, *Plant J.* 24, 241, 2000.
25. Burke, C. and Croteau, R., Geranyl diphosphate synthase from *Abies grandis*: cDNA isolation, functional expression, and characterization, *Arch. Biochem. Biophys.* 405, 130, 2002.
26. Cane, D.E., Sesquiterpene biosynthesis: cyclization mechanisms, in *Comprehensive Natural Products Chemistry*, vol. 2, *Isoprenoids Including Carotenoids and Steroids*, Cane, D.E., Ed., Pergamon Press, Oxford, 1999, p. 155.
27. Wise, M.L. and Croteau, R., Monoterpene biosynthesis, in *Comprehensive Natural Products Chemistry*, vol. 2, *Isoprenoids Including Carotenoids and Steroids*, Cane, D.E., Ed., Pergamon Press, Oxford, 1999, p. 97.
28. Kaiser, R., New volatile constituents of *Jasminum sambac* (L.) Aiton, in *Flavors and Fragrances: A World Perspective, Proceedings of the 10th International Congress of Essential Oils* (Washington, DC, 1986), Lawrence, B.M., Mookherjee, B.D., and Willis, B.J., Eds., Elsevier Science, Amsterdam, 1988, p. 669.
29. Loughrin, J.H., Hamilton-Kemp, T.R., Andersen, R.A., and Hildebrand, D.F., Volatiles from flowers of *Nicotiana sylvestris, N. otophora* and *Malus x domestica*: headspace components and day/night changes in their relative concentrations, *Phytochemistry* 29, 2473, 1990.
30. Lupien, S., Karp, F., Wildung, M., and Croteau, R., Regiospecific cytochrome P450 limonene hydroxylases from mint (Mentha) species: cDNA isolation, characterization, and functional expression of (–)-4S-limonene-3-hydroxylase and (–)-4S-limonene-6-hydroxylase, *Arch. Biochem. Biophys.* 368, 181, 1999.
31. Bouwmeester, H.J., Konings, M.C.J.M., Gershenzon, J., Karp, F., and Croteau, R., Cytochrome P-450 dependent (+)-limonene-6-hydroxylation in fruits of caraway (*Carum carvi*), *Phytochemistry* 50, 243, 1999.
32. Ralston, L., Kwon, S.T., Schoenbeck, M., Ralston, J., Schenk, D.J., Coates, R.M., and Chappell, J., Cloning, heterologous expression, and functional characterization of 5-epi-aristolochene-1,3-dihydroxylase from tobacco (*Nicotiana tabacum*), *Arch. Biochem. Biophys.* 393, 222, 2001.
33. Bouwmeester, H.J., Gershenzon, J., Konings, M.C.J.M., and Croteau, R., Biosynthesis of the monoterpenes limonene and carvone in the fruit of caraway. I. Demonstration of enzyme activities and their changes with development, *Plant Physiol.* 117, 901, 1998.
34. Hallahan, D.L., West, J.M., Wallsgrove, R.M., Smiley, D.W.M., Dawson, G.W., Pickett, J.A., and Hamilton, J.G.C., Purification and characterization of an acyclic monoterpene primary alcohol:NADP+ oxidoreductase from catmint (*Nepeta racemosa*), *Arch. Biochem. Biophys.* 318, 105, 1995.
35. Shalit, M., Guterman, I., Volpin, H., Bar, E., Tamari, T., Menda, N., Adam, Z., Zamir, D., Vainstein, A., Weiss, D., Pichersky, E., and Lewinsohn, E., Volatile ester formation in roses. Identification of an acetyl-coenzyme A geraniol/citronellol acetyltransferase in developing rose petals, *Plant Physiol.* 131, 1868, 2003.
36. Baucr, K., Garbe, D., and Surburg, H., *Common Fragrance and Flavor Materials*, Wiley-VCH Velagsgesellschaft mbH, Weinheim, Germany, 2001, p. 44.
37. Cooper, C., Davies, N.W., and Menary, R.C., C-27 apocarotenoids in the flowers of *Boronia megastigma* (Nees), *J. Agric. Food Chem.* 51, 2384, 2003.
38. Simkin, A.J., Schwartz, S.H., Auldridge, M., Taylor, M.G., and Klee, H.J., The tomato carotenoid cleavage dioxygenase 1 genes contribute to the formation of the flavor volatiles beta-ionone, pseudoionone, and geranylacetone, *Plant J.* 40, 882, 2004.

39. Baldwin, E.A., Scott, J.W., Shewmaker, C.K., and Schuch, W., Flavor trivia and tomato aroma: biochemistry and possible mechanisms for control of important aroma components, *HortScience* 35, 1013, 2000.

40. Simkin, A.J., Underwood, B.A., Auldridge, M., Loucas, H.M., Shibuya, K., Schmelz, E.A., Clark, D.G., and Klee, H.J., Circadian regulation of the PhCCD1 carotenoid cleavage dioxygenase controls emission of beta-ionone, a fragrance volatile of petunia flowers, *Plant Physiol.* 136, 3504, 2004.

41. Feussner, I. and Wasternack, C., Lipoxygenase catalyzed oxygenation of lipids, *Fett/Lipid* 100(4–5), 146, 1998.

42. Knudsen, J.T., Tollsten, L., and Bergstrom, G., Floral scents: a checklist of volatile compounds, isolated by head-space techniques, *Phytochemistry* 33, 253, 1993.

43. D'Auria, J.C., Chen, F., and Pichersky, E., Characterization of an acyltransferase capable of synthesizing benzylbenzoate and other volatile esters in flowers and damaged leaves of *Clarkia breweri*, *Plant Physiol.* 130, 466, 2002.

44. Feussner, I. and Wasternack, C., The lipoxygenase pathway, *Annu. Rev. Plant Biol.* 53, 275, 2002.

45. Gang, D.R., Wang, J.H., Dudareva, N., Nam, K.H., Simon, J.E., Lewinsohn, E., and Pichersky, E., An investigation of the storage and biosynthesis of phenylpropenes in sweet basil, *Plant Physiol.* 125, 539, 2001.

46. Boatright, J., Negre, F., Chen, X., Kish, C.M., Wood, B., Peel, G., Orlova, I., Gang, D., Rhodes, D., and Dudareva, N., Understanding in vivo benzenoid metabolism in petunia petal tissue, *Plant Physiol.* 135, 1993, 2004.

47. Watanabe, S., Hayahi, K., Yagi, K., Asai, T., MacTavish, H., Picone, J., Turnbull, C., and Watanabe, N., Biogenesis of 2-phenylethanol in rose flowers: incorporation of [$^2H_8$]L-phenylalanine into 2-phenylethanol and its beta-D-glucopyranoside during the flower opening of *Rosa* "Hoh-Jun" and *Rosa damascena* Mill, *Biosci. Biotechnol. Biochem.* 66, 943, 2002.

48. Wang, J., Dudareva, N., Bhakta, S., Raguso, R.A., and Pichersky, E., Floral scent production in *Clarkia breweri* (Onagraceae). II. Localization and developmental modulation of the enzyme S-adenosyl-L-methionine:(iso)eugenol O-methyltransferase and phenylpropanoid emission, *Plant Physiol.* 114, 213, 1997.

49. Lewinsohn, E., Ziv-Raz, I.I., Dudai, N., Tadmor, Y., Lastochkin, E., Larkov, O., Chaimovitsh, D., Ravid, U., Putievsky, E., Pichersky, E., and Shoham, Y., Biosynthesis of estragole and methyl-eugenol in sweet basil (*Ocimum basilicum* L). Developmental and chemotypic association of allylphenyl O-methyltransferase activities, *Plant Sci.* 160, 27, 2000.

50. Gang, D.R., Lavid, N., Zubieta, C., Chen, F., Beuerle, T., Lewinsohn, E., Noel, J.P., and Pichersky, E., Characterization of phenylpropene O-methyltransferases from sweet basil: facile change of substrate specificity and convergent evolution within a plant OMT family, *Plant Cell* 14, 505, 2002.

51. Lavid, N., Wang, J., Shalit, M., Guterman, I., Bar, E., Beuerle, T., Menda, N., Shafir, S., Zamir, D., Adam, Z., Vainstein, A., Weiss, D., Pichersky, E., and Lewinsohn, E., O-methyltransferases involved in the biosynthesis of volatile phenolic derivatives in rose petals, *Plant Physiol.* 129, 1899, 2002.

52. Scalliet, G., Journot, N., Jullien, F., Baudino, S., Magnard, J.L., Channeliere, S., Vergne, P., Durmas, C., Bendahmane, M., Cock, J.M., and Hugueney, P., Biosynthesis of the major scent components 3,5-dimethoxytoluene and 1,3,5-trimethoxybenzene by novel rose O-methyltransferases, *FEBS Lett* 523, 113, 2002.

53. Wu, S.Q., Watanabe, N., Mita, S., Dohra, H., Ueda, Y., Shibuya, M., Ebizuka, Y., The key role of phloroglucinol O-methyltransferase in the biosynthesis of *Rosa chinensis* volatile 1,3,5-trimethoxybenzene, *Plant Physiol.* 135, 95, 2004.

54. Dudareva, N., Murfitt, L.M., Mann, C.J., Gorenstein, N., Kolosova, N., Kish, C.M., Bonham, C., and Wood, K., Developmental regulation of methyl benzoate biosynthesis and emission in snapdragon flowers, *Plant Cell* 12, 949, 2000.

55. Murfitt, L.M., Kolosova, N., Mann, C.J., and Dudareva, N., Purification and characterization of S-adenosyl-L-methionine: benzoic acid carboxyl methyltransferase, the enzyme responsible for biosynthesis of the volatile ester methyl benzoate in flowers of *Antirrhinum majus*, *Arch. Biochem. Biophys.* 382, 145, 2000.

56. Effmert, U., Saschenbrecker, S., Ross, J., Negre, F., Fraser, C.M., Noel, J.P., Dudareva, N., and Piechulla, B., Floral benzenoid carboxyl methyltransferases: from *in vitro* to *in planta* function, *Phytochemistry* 66, 1211, 2005.

57. Dudareva, N., D'Auria, J.C., Nam, K.H., Raguso, R.A., and Pichersky, E., Acetyl-CoA:benzyl alcohol acetyltransferase—an enzyme involved in floral scent production in *Clarkia breweri*, *Plant J.* 14, 297, 1998.

58. Aharoni, A., Keizer, L.C.P., Bouwmeester, H.J., Sun, Z., Alvarez-Huerta, M., Verhoeven, H.A., Blaas, J., van Houwelingen, A.M.M.L., De Vos, R.C.H., van der Voet, H., Jansen, R.C., Guis, M., Mol, J., Davis, R.W., Schena, M., van Tunen, A.J., and O'Connell, A.P., Identification of the *SAAT* gene involved in strawberry flavor biogenesis by use of DNA microarrays, *Plant Cell* 12, 647, 2000.

59. Beekwilder, J., Alvarez-Huerta, M., Neef, E., Verstappen, F.W., Bouwmeester, H.J., and Aharoni, A., Functional characterization of enzymes forming volatile esters from strawberry and banana, *Plant Physiol.* 135, 1865, 2004.

60. Raguso, R.A., Levin, R.A., Foose, S.E., Holmberg, M.W., and McDade, L.A., Fragrance chemistry, nocturnal rhythms and pollination "syndromes" in Nicotiana, *Phytochemistry* 63, 265, 2003.

61. Pichersky, E., Lewinsohn, E., and Croteau, R., Purification and characterization of S-linalool synthase, an enzyme involved in the production of floral scent in *Clarkia breweri*, *Arch. Biochem. Biophys.* 316, 803, 1995.

62. Dudareva, N., Cseke, L., Blanc, V.M., and Pichersky, E., Evolution of floral scent in *Clarkia*: novel patterns of S-linalool synthase gene expression in the *C. breweri* flower, *Plant Cell* 8, 1137, 1996.

63. Ross, J.R., Nam, K.H., D'Auria, J.C., and Pichersky, E., S-adenosyl-L-methionine: salicylic acid carboxyl methyltransferase, an enzyme involved in floral scent production and plant defense, represents a new class of plant methyltransferases, *Arch. Biochem. Biophys.* 367, 9, 1999.

64. Dudareva, N., Martin, D., Kish, C.M., Kolosova, N., Gorenstein, N., Faldt, J., Miller, B., and Bohlmann, J., (*E*)-β-Ocimene and myrcene synthase genes of floral scent biosynthesis in snapdragon: function and expression of three terpene synthase genes of a new TPS-subfamily, *Plant Cell* 15, 1227, 2003.

65. Guterman, I., Shalita, M., Menda, N., Piestun, D., Dafny-Yelin, M., Shalev, G., Bar, E., Davydov, O., Ovadis, M., Emanuel, M., Wang, J., Adam, Z., Pichersky, E., Lewinsohn, E., Zamir, D., Vainstein, A., and Weiss, D., Rose scent: genomics approach to discovering novel floral fragrance-related genes, *Plant Cell* 14, 2325, 2002.

66. Chen, F., Tholl, D., D'Auria, J.C., Farooq, A., Pichersky, E., and Gershenzon, J., Biosynthesis and emission of terpenoid volatiles from Arabidopsis flowers, *Plant Cell* 15, 481, 2003.

67. Pott, M.B., Pichersky, E., and Piechulla, B., Evening-specific oscillation of scent emission, SAMT enzyme activity, and mRNA in flowers of *Stephanotis floribunda*, *J. Plant Physiol.* 159, 925, 2002.

68. Negre, F., Kish, C.M., Boatright, J., Underwood, B., Shibuya, K., Wagner, C., Clark, D.G., and Dudareva, N., Regulation of methylbenzoate emission after pollination in snapdragon and petunia flowers, *Plant Cell* 15, 2992, 2003.

69. Chen, F., D'Auria, J.C., Tholl, D., Ross, J.R., Gershenzon, J., Noel, J.P., and Pichersky, E., An *Arabidopsis thaliana* gene for methylsalicylate biosynthesis, identified by a biochemical genomics approach, has a role in defense, *Plant J.* 36, 577, 2003.

70. Pott, M., Hippauf, F., Saschenbrecker, S., Chen, F., Ross, J., Kiefer, I., Slusarenko, A., Noel, J.P., Pichersky, E., Effmert, U., and Piechulla B., Biochemical and structural characterization of benzenoid carboxyl methyltransferases involved in floral scent production in *Stephanotis floribunda* and *Nicotiana suaveolens*, *Plant Physiol.* 135, 1946, 2004.

71. Suzuki, H., Sawada, S., Watanabe, K., Nagae, S., Yamaguchi, M., Nakayama, T., and Nishino, T., Identification and characterization of a novel anthocyanin malonyltransferase from scarlet sage (*Salvia splendens*) flowers: an enzyme that is phylogenetically separated from other anthocyanin acyltransferases, *Plant J.* 38, 994, 2004.

72. Bouvier, F., Dogbo, O., and Camara, B., Biosynthesis of the food and cosmetic plant pigment bixin (annatto), *Science* 300, 2089, 2003.

73. Dobson, H.E.M., Floral volatiles in insect biology, in Bernays, E., Ed., *Insect-Plant Interactions*, vol. 5, CRC Press, Boca Raton, FL, 1994, p. 47.

74. Dudareva, N., Piechulla, B., and Pichersky, E., Biogenesis of floral scent, *Hort. Rev.* 24, 31, 1999.

75. Dobson, H.E.M., Bergstrom, G., and Groth, I., Differences in fragrance chemistry between flower parts of *Rosa rugosa* Thunb (Rosaceae), *Isr. J. Bot.* 39, 143, 1990.

76. Dobson, H.E.M., Groth, I., and Bergstrom, G., Pollen advertisement: chemical contrasts between whole-flower and pollen odors, *Am. J. Bot.* 83, 877, 1996.

77. Pichersky, E., Raguso, R.A., Lewinsohn, E., and Croteau, R., Floral scent production in *Clarkia* (Onagraceae). I. Localization and developmental modulation of monoterpene emission and linalool synthase activity, *Plant Physiol.* 106, 1533, 1994.

78. Mactavish, H.S. and Menary, R.C., Volatiles in different floral organs, and effect of floral characteristics on yield of extract from *Boronia megastigma* (Nees), *Ann. Bot.* 80, 305, 1997.

79. Verdonk, J.C., Ric de Vos, C.H., Verhoeven, H.A., Haring, M.A., van Tunen, A.J., and Schuurink, R.C., Regulation of floral scent production in petunia revealed by targeted metabolomics, *Phytochemistry* 62, 997, 2003.

80. Stern, W.L., Curry, K.J., and Pridgeon, A.M., Osmophores of *Stanhopea anfracta* (Orchidaceae), *Am. J. Bot.* 74, 1323, 1987.

81. Curry, K.J., Initiation of terpenoid synthesis in osmophores of *Stanhopea anfracta* (Orchidaceae): a cytochemical study, *Am. J. Bot.* 74, 1332, 1987.

82. Dudareva, N. and Pichersky, E., Biochemical and molecular genetic aspects of floral scent, *Plant Physiol.* 122, 627, 2000.

83. Kolosova, N., Sherman, D., Karlson, D., and Dudareva, N., Cellular and subcellular localization of *S*-adenosyl-L-methionine:benzoic acid carboxyl methyltransferase, the enzyme responsible for biosynthesis of the volatile ester methylbenzoate in snapdragon flowers, *Plant Physiol.* 126, 956, 2001.

84. Goodwin, S.M., Kolosova, N., Kish, C.M., Wood, K.V., Dudareva, N., and Jenks, M.A., Cuticle characteristics and volatile emissions of petals in *Antirrhinum majus*, *Physiol. Plant.* 117, 435, 2003.

85. Verdonk, J.C., Haring, M.A., van Tunen, A.J., and Schuurink, R.C., *ODORANT1* regulates fragrance biosynthesis in petunia flowers, *Plant Cell* 17, 1612, 2005.

86. Matile, P. and Altenburger, R., Rhythms of fragrance emission in flowers, *Planta* 174, 242, 1988.

87. Loughrin, J.H., Hamilton-Kemp, T.R., Burton, H.R., Andersen, R.A., and Hilderbrand, D.F., Glycosidically bound volatile components of *Nicotiana sylvestris* and *N. suaveolens* flowers, *Phytochemistry* 31, 1537, 1992.

88. Loughrin, J.H., Potter, D.A., and Hamilton-Kemp, T.R., Circadian rhythm of volatile emission from flowers of *Nicotiana sylvestris* and *N. suaveolens*, *Plant Physiol.* 83, 492, 1991.

89. Nielsen, J.K., Jakobsen, H.B., Hansen, P.F.K., Moller, J., and Olsen, C.E., Asynchronous rhythms in the emission of volatiles from *Hesperis matronalis* flowers, *Phytochemistry* 38, 847, 1995.

90. Jakobsen, H.B. and Olsen, C.E., Influence of climatic factors on rhythmic emission of volatiles from *Trifolium repens* L. flowers in situ, *Planta* 192, 365, 1994.

91. Helsper, J.P.F.G., Davies, J.A., Bouwmeester, H.J., Krol, A.F., and van Kampen, M.H., Circadian rhythmicity in emission of volatile compounds by flowers of *Rosa hybrida* L. cv. Honesty, *Planta* 207, 88, 1998.

92. Kolosova, N., Gorenstein, N., Kish, C.M., and Dudareva, N., Regulation of circadian methylbenzoate emission in diurnally and nocturnally emitting plants, *Plant Cell* 13, 2333, 2001.

93. Pott, M.B., Effmert, U., and Piechulla, B., Transcriptional and post-translational regulation of S-adenosyl-L-methionine: salicylic acid carboxyl methyltransferase (SAMT) during *Stephanotis floribunda* flower development, *J. Plant Physiol.* 160, 635, 2003.

94. Altenburger, R. and Matile, P., Circadian rhythmicity of fragrance emission in flowers of *Hoya carnosa* R. Br., *Planta* 174, 248, 1988.

95. Altenburger, R. and Matile, P., Further observations on rhythmic emission of fragrance in flowers, *Planta* 180, 194, 1990.

96. Jakobsen, H.B., Friis, P., Nielsen, J.K., and Olsen, C.E., Emission of volatiles from flowers and leaves of *Brassica napus* in situ, *Phytochemistry* 37, 695, 1994.

97. Arditti, J., Aspects of the physiology of orchids, in *Advances in Botanical Research*, vol. 7, Woolhouse, H.W., Ed., Academic Press, London, 1979, p. 422.

98. Tollsten, L. and Bergstrom, G., Headspace volatiles of whole plants and macerated plant parts of *Brassica* and *Sinapis*, *Phytochemistry* 27, 4013, 1989.

99. Tollsten, L., A multivariate approach to post-pollination changes in the floral scent of *Platanthera bifolia* (Orchidaceae), *Nord. J. Bot.* 13, 495, 1993.

100. Schiestl, F.P., Ayasse, M., Paulus, H.F., Erdmann, D., and Francke, W., Variation of floral scent emission and post pollination changes in individual flowers of *Ophrys sphegodes* subsp. *sphegodes*, *J. Chem. Ecol.* 23, 2881, 1997.

101. Neiland, M.R.M. and Wilcock, C.C., Fruit set, nectar reward, and rarity in the Orchidaceae, *Am. J. Bot.* 85, 1657, 1998.

102. Lucker, J., Bouwmeester, H.J., Schwab, W., Blaas, J., van Der Plas, L.H., and Verhoeven, H.A., Expression of *Clarkia* S-linalool synthase in transgenic petunia plants results in the accumulation of S-linalyl-beta-D-glucopyranosid, *Plant J.* 27, 315, 2001.

103. Lucker, J., Schwab, W., van Hautum, B., Blaas, J., van der Plas, L.H., Bouwmeester, H.J., and Verhoeven, H.A., Increased and altered fragrance of tobacco plants after

metabolic engineering using three monoterpene synthases from lemon, *Plant Physiol.* 134, 510, 2004.

104. Cseke, L., Dudareva, N., and Pichersky, E., Structure and evolution of linalool synthase, *Mol. Biol. Evol.* 15, 1491, 1998.

105. Tholl, D., Chen, F., Petri, J., Gershenzon, J., and Pichersky, E., Two sesquiterpene synthases are responsible for the complex mixture of sesquiterpenes emitted from Arabidopsis flowers, *Plant J.* 42, 757, 2005.

106. Raguso, R.A. and Pichersky, E., Floral volatiles from *Clarkia breweri* and *C. concinna* (Onagraceae): recent evolution of floral scent and moth pollination, *Plant Syst. Evol.* 194, 55, 1995.

107. Pichersky, E. and Gang, D.R., Genetics and biochemistry of secondary metabolites in plants: an evolutionary perspective, *Trends Plant Sci.* 5, 439, 2000.

108. Iijima, Y., Davidovich-Rikanati, R., Fridman, E., Gang, D.R., Bar, E., Lewinsohn, E., and Pichersky, E., The biochemical and molecular basis for the divergent pattern in the biosynthesis of terpene and phenylpropenes in the peltate glands of three cultivars of sweet basil, *Plant Physiol.* 136, 3724, 2004.

109. Vainstein, A., Lewinsohn, E., Pichersky, E., and Weiss, D., Floral fragrance—new inroads into an old commodity, *Plant Physiol.* 127, 1383, 2001.

110. Zuker, A., Tzfira, T., Ben-Meir, H., Ovadis, M., Shklarman, E., Itzhaki, H., Forkmann, G., Martens, S., Neta-Sharir, I., Weiss, D., and Vainstein, A., Modification of flower color and fragrance by antisense suppression of the flavanone 3-hydroxylase gene, *Mol Breed.* 9, 33, 2002.

4 Biosynthesis of Volatile Terpenes in the Flowers of the Model Plant *Arabidopsis thaliana*

Dorothea Tholl and Eran Pichersky

CONTENTS

4.1 INTRODUCTION

Flowers whose scents have been studied over the years have tended to be, for obvious reasons, highly scented to the human nose. However, the range of compounds that can be detected by humans and insects does not completely overlap, even for compounds that can be detected by both, because of differences in the sensitivity of the olfactory systems of humans and insects (and other animal pollinators). Nonetheless, the initial selection of flowers for chemical analysis based on human sensory perception has yielded a wealth of information regarding the type of scent compounds emitted by flowers.

When investigators first began to study the molecular and biochemical basis for the synthesis of floral scent compounds, highly scented species such as *Clarkia breweri* and snapdragon were chosen as model systems.[1,2] This approach was appropriate at the time, since the methods available then to identify enzymes and genes involved in scent biosynthesis required the initial purification of biosynthetic enzymes, which are naturally more abundant in flowers that synthesize high levels of scent compounds. Recent developments in the areas of genomics and bioinformatics have opened up new avenues for studying the molecular basis of many biochemical, physiological, and ecological processes in plants. In particular, the DNA sequences of entire genomes of two plant species, *Arabidopsis* and rice, have been completely elucidated, and several other plant genomes are being determined at the present time. The availability of whole genome sequences makes it possible for researchers to try to elucidate the types of processes in an opposite direction than has traditionally been done; instead of from the phenotype to the genotype, it can now be done from the gene to the phenotype (this new approach is sometimes referred to as "reverse genetics"). This new approach provides so many advantages over the older approach that it can also be applied to species whose floral scent is not particularly strong.

In this chapter, we describe the use of *Arabidopsis thaliana* as a model plant for floral scent studies by focusing on investigations involving the large family of genes encoding terpene synthases. *A. thaliana* flowers emit very little scent, consisting mostly of monoterpenes and sesquiterpenes, and this scent mixture is barely detected by humans. However, by employing the genetic and genomic tools available for *Arabidopsis* and the latest technologies in expression and metabolite profiling, we have been able to explore the physiological and ecological significance of the *A. thaliana* floral scent terpenes and the basic principles of the regulation and evolution of scent biosynthesis. Although terpene biosynthesis has been studied in numerous plant species and the biochemistry of the basic pathways is well known, a comprehensive and detailed understanding of the regulation of biosynthesis and the biological roles of this large class of secondary metabolites will most likely come from investigating model plant species that provide extensive genetic and genomic resources.

4.2 THE TERPENE SYNTHASE GENE FAMILY
OF *A. THALIANA*

The basic pathways leading to the biosynthesis of volatile terpenes (mostly monoterpenes and sesquiterpenes) are covered in Chapter 3. The immediate precursor of all monoterpenes, which are produced in the plastids, is geranyl diphosphate (GPP), and the immediate precursor of sesquiterpenes is farnesyl diphosphate (FPP). Previous research on certain terpene-accumulating species such as resin-producing gymnosperm trees or the herbs in the Lamiaceae family has resulted in the identification of a family of structurally related genes encoding mono-, sesqui-, and diterpene synthases.[3,4]

With the completion of the sequencing of the *A. thaliana* genome, it became possible to examine this species for the presence of terpene synthase (*TPS*) genes, even though the presence of mono-, sesqui-, or diterpenes (other than gibberellic acid [GA] derivatives) had not previously been reported. Using standard homology search methods, Aubourg et al.[5] showed that the *Arabidopsis* genome contains more than 30 *TPS* genes (*AtTPSs*), distributed over all five chromosomes. Our own detailed analysis (Figure 4.1), as well as a similar analysis performed by Aubourg et al.,[5] showed the presence of three classes. Seven of the genes form one clade and the six proteins they encode (two genes encode identical proteins) are most similar to monoterpene synthases from other angiosperm species. These seven genes also appear to encode proteins with a transit peptide for plastidial targeting. The two genes previously determined to encode GA biosynthetic enzymes in the plastid[6,7] form a separate clade, together with a third *TPS* gene. Finally, a large clade contains

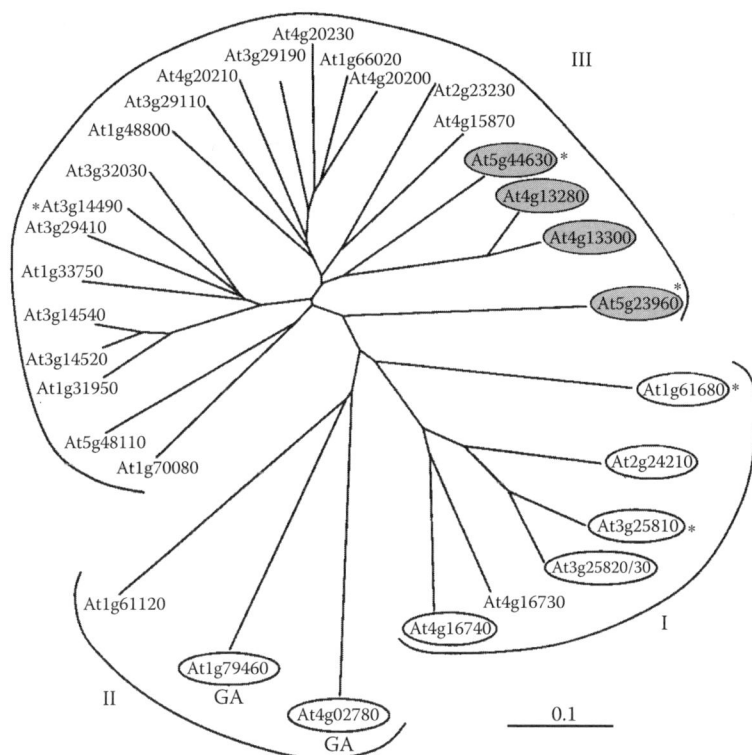

FIGURE 4.1 *Arabidopsis thaliana TPS* genes and their expression in flowers. A neighbor-joining tree is based on the degree of sequence similarity between the members of the *Arabidopsis* terpene synthase (*TPS*) gene family. *AtTPS* genes form three major clades. Members of clades I and II encode proteins with high similarity to monoterpene or diterpene synthases of other angiosperms, respectively. Proteins encoded by genes of the large clade III are all likely to function as sesquiterpene synthases (highlighted in grey) or monoterpene/diterpene synthases based on the absence and presence, respectively, of a plastidial transit peptide. Asterisks indicate genes predominantly or exclusively expressed in flowers. Functionally characterized *AtTPS* genes are marked with circles. GA, diterpene synthases involved in gibberellic acid biosynthesis.

all other *AtTPS* genes, some of which encode proteins with a plastid-targeting sequence (and therefore may be diterpene or perhaps monoterpene synthases) and some genes which encode proteins with no transit peptide (and therefore are probably all sesquiterpene synthases).

4.3 TERPENE BIOSYNTHESIS IN FLOWERS OF *A. THALIANA*

4.3.1 EXPRESSION OF *AtTPS* GENES IN *A. THALIANA* FLOWERS

Although the human nose does not typically detect scent emission from *A. thaliana* flowers, the presence of many *TPS* genes in the *A. thaliana* genome prompted us to examine whether any of these genes could possibly be involved in floral scent production. We therefore conducted a detailed analysis of the expression of all *Arabidopsis TPS* genes in the main organs of the plant (flowers, leaves, stems, roots, and siliques) using a semiquantitative reverse transcription polymerase chain reaction (RT-PCR) approach. Our results indicated that most of the *AtTPS* genes are expressed in one or more organs under normal growth conditions.[8] In particular, several *AtTPS* genes are expressed in flowers and some exclusively so (Figure 4.1).

4.3.2 EMISSION OF MONOTERPENES AND SESQUITERPENES FROM *A. THALIANA* FLOWERS

This observation led us to examine whether *Arabidopsis* flowers emit terpene volatiles. However, standard volatiles collection and analysis techniques did not result in readily detectable levels of terpenes. We therefore adapted a closed-loop stripping method developed initially by Donath and Boland[9] for the detection of *Arabidopsis* volatiles.[8] Details of this method are presented in Chapter 1. Alternatively, a slightly less sensitive semiopen dynamic headspace sampling system was applied in which purified air was pumped into a 4 l glass jar containing the plant, and 90% of the air was actively pulled out through a charcoal filter, while the remaining air was vented through the top of the glass container.

Using these methods in combination with gas chromatography-mass spectrometry (GC-MS) of the collected volatiles, we were able to detect the emission of a number of monoterpenes as well as a large group of sesquiterpenes from whole *Arabidopsis* (Columbia ecotype) plants. To determine which part of the plant was responsible for the emission of each of these terpenes, we removed inflorescences or siliques and conducted headspace collections of the isolated plant parts and the remaining vegetative tissue. Comparative analysis of the emitted volatiles showed that inflorescences were the main source of monoterpenes and most sesquiterpenes, together comprising more than 60% of the total amount of floral volatiles. In total, 3 monoterpenes (β-myrcene, linalool, and limonene) and more than 20 sesquiterpene hydrocarbons (divided into group A and group B sesquiterpenes) were consistently detected, with (*E*)-β-caryophyllene as the predominant terpene volatile (Figure 4.2). The sesquiterpene volatiles showed a high structural diversity, including acyclic, mono-, di-, and tricyclic compounds. All monoterpenes and more than

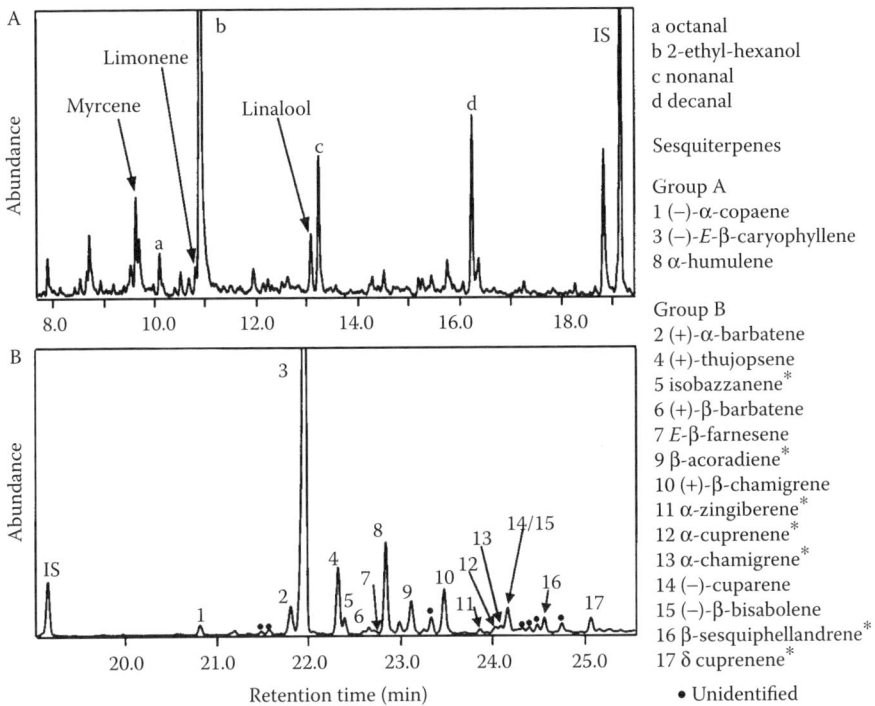

FIGURE 4.2 Monoterpene and sesquiterpene compounds emitted from inflorescences of *A. thaliana* (Columbia ecotype). (A) GC-MS analysis of monoterpenes collected from 70 cut inflorescences during 8 h of closed-loop stripping. (B) A later portion of the same chromatogram as in (A) shows the sesquiterpene hydrocarbon region. Sesquiterpenes are divided into group A sesquiterpenes, produced by TPS At5g23960, and group B products of the TPS enzyme At5g44630. The number of each compound corresponds to the numbered peaks in (B). Chirality was determined for all sesquiterpenes except those labeled with asterisks. Linalool and limonene were found as a mixture of (+) and (–) enantiomers. Coelution of compounds is indicated by the separation of numbers with slashes. IS, internal standard, nonyl acetate.

15 sesquiterpenes were identified with certainty by mass spectra and comparison with authentic standards. Other volatile compounds emitted from *A. thaliana* flowers and vegetative tissues were primarily aliphatic aldehydes and alcohols.

4.3.3 VARIATION OF MONOTERPENE AND SESQUITERPENE EMISSIONS AMONG *A. THALIANA* ECOTYPES

A survey of 40 *A. thaliana* accessions, including ecotypes from various geographical regions, revealed distinct qualitative and quantitative differences in floral terpene emission. Significant quantitative differences were observed in the emission of (*E*)-β-caryophyllene and some of the monoterpenes such as linalool and (*E*)-ocimene. However, the majority of these ecotypes contained virtually the same set of compounds in their floral scent. The quantitative variation of terpene volatile emissions among different ecotypes may reflect the result of selective adaptation to particular pollinator populations or other biotic or abiotic factors present in a specific habitat. Interestingly, flowers of the ecotypes Can-0 and CVI, found in islands near the North African coast, had extremely reduced floral sesquiterpene emissions. This observation may be significant, as theory predicts that plant species introduced into a new ecosystem that lacks the natural pollinator of the species are more likely to establish a population if the original founder is highly self-pollinating.[10] Indeed, small islands could be depauperate in specific taxa of pollinators. However, in general, no obvious correlations were evident between the geographical habitat and the amount or composition of floral volatiles of particular ecotypes.

In a few cases, qualitative differences between cultivars in the sesquiterpene profiles were also observed. Some cultivars did not emit any sesquiterpenes belonging to group A, and in other ecotypes the group B sesquiterpenes were missing. Such intraspecific variations provide an excellent opportunity to study the mechanisms regulating natural variation and evolution of volatile terpene biosynthesis.

4.4 IDENTIFICATION OF THE FUNCTION OF THE FLOWER-SPECIFIC *AtTPS* GENES

To determine which genes are responsible for the synthesis of the floral terpene volatiles that we observed, we used RT-PCR to obtain full-length complementary DNA (cDNA) clones of the *AtTPS* genes shown to be expressed in flowers and predicted to encode mono- and sesquiterpene synthases. We then ligated these cDNAs into a bacterial expression vector carrying the T7 viral promoter and expressed them in *Escherichia coli*. The *E. coli*-produced AtTPS proteins were tested for activity with GPP and FPP, the universal precursors of monoterpenes and sesquiterpenes, respectively. The results indicated that the enzyme encoded by *At3g25810* is responsible for the synthesis of the floral monoterpenes β-myrcene and limonene, and *At1g61680* is responsible for the synthesis of the floral monoterpene linalool. Thus, these two enzymes can account for almost all the monoterpenes usually observed in the scent of *A. thaliana* flowers. Two other *AtTPSs*, *At2g24210* and *At4g16740*, which are also expressed in flowers (but not exclusively), were shown to encode monoterpene synthases that catalyze the formation of myrcene and ocimene, respectively.[11,12] Ocimene is also occasionally observed in *A. thaliana* floral scent.

The enzyme encoded by *At5g23960* was found to catalyze the formation of the main floral sesquiterpenes (*E*)-β-caryophyllene and also α-humulene and α-copaene (group A sesquiterpenes; Figure 4.2), whereas heterologous expression of *At5g44630* showed that the encoded enzyme is responsible for the production of most, if not all, of the other floral sesquiterpene hydrocarbons (group B sesquiterpenes; Figure 4.3).[8,13] Together these two genes could account for all the sesquiterpenes found in the *A. thaliana* floral scent. Because a single gene appeared to be responsible for the production of many sesquiterpene synthases, the assignment of gene function was also tested by analyzing *A. thaliana* mutant lines that had transfer DNA (T-DNA) insertions in genes *At5g23960* and *At5g44630* (the insertions inactivated the genes). Lines containing an insertion in gene *At5g23960* lacked group A sesquiterpene

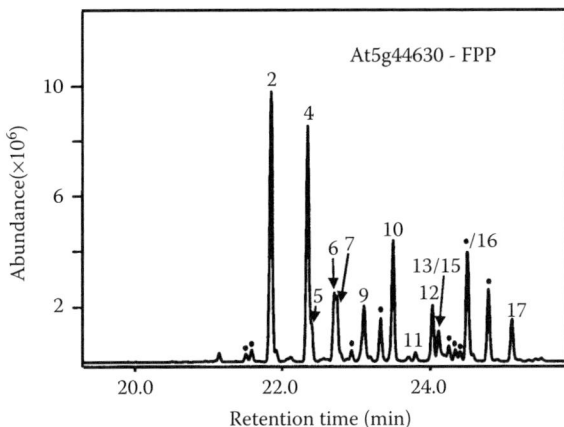

FIGURE 4.3 Identification of the sesquiterpene products of the enzyme encoded by *At5g44630*. GC-MS chromatogram of sesquiterpene products of the *E. coli*-expressed *At5g44630* protein. Only a portion of the chromatogram is shown. The number of individual peaks refers to the number of compounds in Figure 4.2. Dots indicate unidentified sesquiterpene products.

emission, while flowers of a line with an insertion in gene *At5g44630* lacked the emission of group B sesquiterpenes. These results confirmed the function assigned to each of these two genes in the biochemical experiments. The formation of multiple enzymatic products from a single substrate is a characteristic feature of terpene synthases and can be ascribed to multiple reaction paths of the initially formed carbocationic intermediate, including differential internal electrophilic additions, hydride shifts, rearrangements, deprotonations, or addition of water.[3,14,15]

4.4.1 LOCALIZATION OF THE EXPRESSION OF THE *AtTPS* GENES IN FLORAL ORGANS

To study the tissue-specific expression of floral *AtTPS* genes, we used an approach in which promoter regions of these genes were fused to the coding region of the *E. coli* β-glucuronidase (*GUS*) gene, and the entire construct was inserted into the *Arabidopsis* genome by *Agrobacterium*-mediated transformation.[16] The *GUS* reporter gene encodes an enzyme that catalyzes the formation of a blue-colored precipitating product by hydrolysis of the colorless substrate X-Gluc (5-bromo-4-chloro-3-indoyl-β-D-glucuronic acid). *In vivo* staining of transgenic plants allows for the observation of the tissue(s) in which the promoter being tested is active.[17]

Experiments with several *AtTPS* genes showed staining in various parts of the flower, verifying that these promoters are active in floral tissues. GUS expression driven by the promoter of the sesquiterpene synthase gene *At5g44630* was mainly detected at the base or receptacle of young and mature flowers and the abscission zone of siliques. Additional staining was observed in the ovules or developing seeds (Figure 4.4A). In contrast, GUS activity under the control of the promoter of the monoterpene synthase gene *At3g25810* was observed in sepals, stigma, anther

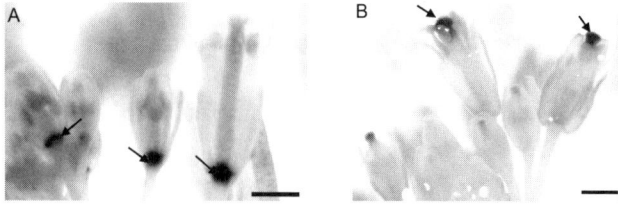

FIGURE 4.4 Identification of the sites of expression of the two sesquiterpene synthase genes expressed in *A. thaliana* flowers. (A) Expression patterns of the *At5g44630* promoter *GUS* gene in *A. thaliana* flowers. GUS activity was mainly observed at the base of young and old flowers and the abscission zone of floral organs. Additional GUS staining was detected in ovaries and developing seeds. (B) Expression patterns of the *At5g23960* promoter *GUS* gene in *A. thaliana* flowers. GUS activity was primarily observed in the stigma of unopened and opened flowers. GUS staining is indicated by arrows. Bars = 1 mm.

filaments, and the receptacle of the mature flower bud, as well as the young and mature open flower.[8] A similar GUS staining pattern with pronounced GUS activity in the stigma was observed under the control of the promoter of the sesquiterpene synthase gene *At5g23960* (Figure 4.4B).

These results suggest several functions for the volatile terpenes in *Arabidopsis* flowers. The obvious function of terpenes released from floral tissues is the attraction of pollinators.[1] Specifically, the observation of activity of the promoters of several *AtTPS* genes in sepals, filaments, and receptacles suggests such a function, since several flower tissues are involved. Interestingly, no expression of the *AtTPS* genes investigated thus far has been observed in flower petals, which have been described as the main organs of expression in nonterpenoid floral scent genes in other plants such as *C. breweri* and *Antirrhinum majus*.[18–20] Whether this is due to a reduction of terpene emission as a consequence of the evolution of *A. thaliana* toward self-pollination remains to be determined.

The expression of terpene synthases in the stigma could be involved in protecting the moist surface area against fungal growth, since the monoterpenes produced have antimicrobial activity.[21] Similar expression patterns were found for a linalool synthase in the stigma of flowers from *C. breweri*.[22] Another potential function of terpenes in this tissue may be to protect against oxidative stress.[23] Expression of *At5g44630* occurs at the base of the *Arabidopsis* flower, an area in which sugar-producing nectaries are located.[24] The biosynthesis of several sesquiterpenes that have antimicrobial activity could therefore be important for defending this region against microbial infection. This might also be of significance in protecting the wound zone after abscission of the floral organs.

4.4.2 Insect Visitation to *A. thaliana* Flowers

Volatile terpenes are found in the aroma bouquet emitted by many insect-pollinated flowers.[25] A role in attracting insect pollinators was therefore a logical hypothesis for the emission of monoterpenes and sesquiterpenes from *A. thaliana* flowers. Although *A. thaliana*, unlike its close relative *Arabidopsis lyrata*, is a self-compatible species, and, at least in the laboratory, sets a copious number of seeds by self-pollination, several investigators have previously reported that *A. thaliana* flowers are sometimes visited in nature by insects such as hover flies and that a small amount of cross-pollination does occur.[26,27] These observations are consistent with further findings showing that natural *A. thaliana* populations exhibit polymorphisms at tested loci and contain heterozygous individuals at frequencies that cannot be accounted for solely by mutation rates.[28,29] Cross-pollination events could be of importance in wild *Arabidopsis* populations since the progeny arising from outcrossing often have greater reproductive fitness, thereby mitigating inbreeding depression.[30] This heterozygous advantage may have led to the retention of traits that promote outcrossing even in this mainly self-pollinating species. Indeed, the development of the *Arabidopsis* flower allows a short time window for cross-pollination, when the receptive stigma protrudes from the flower petals before the anthers mature. In addition, floral nectaries, located at the base of the stamens, provide sugars as rewards to visiting insects.[24]

We examined the visitation of insects to *A. thaliana* flowers in seminatural settings at the grounds of the botanical gardens in Halle, Germany, and Ann Arbor, Michigan. While a detailed accounting of these experiments will be given elsewhere, we observed a large number and different types of insects visiting the flowers. These included hover flies and other diptera, beetles, and thrips. The flowering plants of the German population were also frequently visited by solitary bees collecting and transferring flower pollen (Figure 4.5). Monitoring the frequency of these visits over the whole flowering season revealed regular daily visitation patterns that clearly corroborated the role of insects in cross-pollination events in wild *Arabidopsis* populations.[31]

It is not yet known whether the emission of terpenes from *A. thaliana* flowers is directly responsible for the attraction of these insects (as well as the efficacy of

FIGURE 4.5 Male solitary bee (*Andrena nigroaenea*) collecting pollen from *Arabidopsis* flowers in the Botanical Garden, Halle, Germany.

the insects in cross-pollinating the flowers). Such investigations should include GC-electroantennograms monitoring the antennal response to distinct terpene compounds of the volatile blend or wind tunnel experiments with insect species shown to have visited the *A. thaliana* flowers. In addition, it will be useful to determine the cross-pollination rates in synthetic populations of various *Arabidopsis* ecotypes and *TPS* mutant lines lacking or overproducing one or several floral terpene compounds.

4.5   CONCLUSION

We have shown that the *Arabidopsis* model system is as useful for the study of floral scent biosynthesis and emission as it is for so many other areas of plant biology. We have observed that *A. thaliana* flowers emit scent compounds, and that volatile monoterpenes and sesquiterpenes constitute 60% of the total floral scent output. The availability of the sequence of the entire *Arabidopsis* genome has allowed us to identify the complete *TPS* gene family, and specific *TPS* genes responsible for the synthesis of nearly all the terpenes emitted from flowers. With the modern tools available for experimentation in *Arabidopsis*, this model organism is the best system to elucidate the mechanisms regulating the processes of plant-pollinator interactions via volatiles.

ACKNOWLEDGMENTS

We thank Feng Chen for establishing the *At5g23960* and *At5g44630* expression constructs. We are grateful to Mathias Hoffmann for permission and support to examine insect visitations to *A. thaliana* flowers at the botanical garden in Halle, Germany. We are thankful to Katrin Heisse and Bettina Raguschke for excellent technical assistance and help with taking photos for Figure 4.4 and Figure 4.5. This work was supported by a National Science Foundation grant (IBN-0211697) to Eran Pichersky and by funds of the Max Planck Society to Jonathan Gershenzon.

REFERENCES

1. Dudareva, N. and Pichersky, E., Biochemical and molecular genetic aspects of floral scents, *Plant Physiol.* 122, 627, 2000.
2. Pichersky, E. and Gershenzon, J., The formation and function of plant volatiles: perfumes for pollinator attraction and defense, *Curr. Opin. Plant Biol.* 5, 237, 2002.
3. Davis, E.M. and Croteau, R., Cyclization enzymes in the biosynthesis of monoterpenes, sesquiterpenes, and diterpenes, *Top. Curr. Chem.* 209, 53, 2000.
4. Bohlmann, J., Meyer-Gauen, G., and Croteau, R., Plant terpenoid synthases: molecular biology and phylogenetic analysis, *Proc. Natl. Acad. Sci. USA* 95, 4126, 1998.
5. Aubourg, S., Lecharny, A., and Bohlmann, J., Genomic analysis of the terpenoid synthase (*AtTPS*) gene family of *Arabidopsis thaliana*, *Mol. Genet. Genomics* 267, 730, 2002.
6. Sun, T.-P. and Kamiya, Y., The *Arabidopsis* GA1 locus encodes the cyclase *ent*-kaurene synthetase A of gibberellin biosynthesis, *Plant Cell* 6, 1509, 1994.

7. Yamaguchi, S., Sun, T.-P., Kawaide, H., and Kamiya, Y., The GA2 locus of *Arabidopsis thaliana* encodes *ent*-kaurene synthase of gibberellin biosynthesis, *Plant Physiol.* 116, 1271, 1998.
8. Chen, F., Tholl, D., D'Auria, J.C., Farooq, A., Pichersky, E., and Gershenzon, J., Biosynthesis and emission of terpenoid volatiles from *Arabidopsis* flowers, *Plant Cell* 15, 481, 2003.
9. Donath, J. and Boland, W., Biosynthesis of acyclic homoterpenes: enzyme selectivity and absolute configuration of the nerolidol precursor, *Phytochemistry* 39, 785, 1995.
10. Schueller, S.K., Self-pollination in island and mainland populations of the introduced hummingbird-pollinated plant, *Nicotiana glauca* (Solanaceae), *Am. J. Bot.* 91, 672, 2004.
11. Bohlmann, J., Martin, D., Oldham, N.J., and Gershenzon, J., Terpenoid secondary metabolism in *Arabidopsis thaliana*: cDNA cloning, characterization, and functional expression of a myrcene/(*E*)-β-ocimene synthase, *Arch. Biochem. Biophys.* 375, 262, 2000.
12. Fäldt, J., Arimura, G., Gershenzon, J., Takabayashi, J., and Bohlmann, J., Functional identification of *AtTPS03* as (*E*)-beta-ocimene synthase: a new monoterpene synthase catalyzing jasmonate- and wound-induced volatile formation in *Arabidopsis thaliana*, *Planta* 216, 745, 2003.
13. Tholl, D., Chen, F., Petri, J., Gershenzon, J., and Pichersky, E., Two sesquiterpene synthases are responsible for the complex mixture of sesquiterpenes emitted from *Arabidopsis* flowers, *Plant J* 42, 757, 2005.
14. Wise, and Croteau. Monoterpene biosynthesis, in *Comprehensive Natural Products Chemistry: Isoprenoids*, Cane, D.E., Ed., Elsevier, Amsterdam, 1999, p. 97.
15. Cane, D.E., Sesquiterpene biosynthesis: cyclization mechanisms, in *Comprehensive Natural Products Chemistry: Isoprenoids*, Cane, D.E., Ed., Elsevier, Amsterdam, 1999, p. 155.
16. Bechtold, N., Ellis, J., and Pelletier, G., *In planta* Agrobacterium mediated gene-transfer by infiltration of adult *Arabidopsis thaliana* plants, *C. R. Acad. Sci. Paris Life Sci.* 316, 1194, 1993.
17. Jefferson, R.A., Kavanagh, T.A., and Bevan, M.W., Gus fusions: beta-glucuronidase as a sensitive and versatile gene fusion marker in higher plants, *EMBO J.* 6, 3901, 1987.
18. Wang, J., Dudareva, N., Bhakta, S., Raguso, R.A., and Pichersky E., Floral scent production in *Clarkia breweri* (Onagraceae) II. Localization and developmental modulation of the enzyme S-adenosyl-L-methionine:(iso)eugenol O-methyltransferase and phenylpropanoid emission, *Plant Physiol.* 114, 213, 1997.
19. Dudareva, N., D'Auria, J.C., Nam, K.H., Raguso, R.A., and Pichersky, E., Acetyl-CoA:benzylalcohol acetyltransferase: an enzyme involved in floral scent production in *Clarkia breweri*, *Plant J.* 14, 297, 1998.
20. Dudareva, N., Murfitt, L.M., Mann, C.J., Gorenstein, N., Kolosova, N., Kish, C.M., Bonham, C., and Wood, K., Developmental regulation of methyl benzoate biosynthesis and emission in snapdragon flowers, *Plant Cell* 12, 949, 2000.
21. Deans, S.G. and Waterman, P.G., Biological activity of volatile oils, in *Volatile Oil Crops: Their Biology, Biochemistry and Production*, Hay, R.K.M. and Waterman, P.G., Eds., Longman Scientific, Essex, 1993, p. 97.
22. Dudareva, N., Cseke, L., Blanc, V.M., and Pichersky, E., Evolution of floral scent in *Clarkia*: novel patterns of *S*-linalool synthase gene expression in the *C. breweri* flower, *Plant Cell* 8, 1137, 1996.

23. Loreto, F. and Velikova, V., Isoprene produced by leaves protects the photosynthetic apparatus against ozone damage, quenches ozone products, and reduces lipid peroxidation of cellular membranes, *Plant Physiol.* 127, 1781, 2001.

24. Davis, A.R., Pylatuik, J.D., Paradis, J.C., and Low, N.H., Nectar-carbohydrate production and composition vary in relation to nectary anatomy and location within individual flowers of several species of Brassicaceae, *Planta* 205, 305, 1998.

25. Knudsen, J.T., Tollsten, L., and Bergström, G., Floral scents: a checklist of volatile compounds isolated by head-space techniques, *Phytochemistry* 33, 253, 1993.

26. Jones, M.E., Population genetics of *Arabidopsis thaliana*. 1. Breeding system, *Heredity* 27, 39, 1971.

27. Snape, J.W. and Lawrence, M.J., Breeding system of *Arabidopsis thaliana*, *Heredity* 27, 299, 1971.

28. Loridon, K., Cournoyer, B., Goubely, C., Depeiges, A., and Picard, G., Length polymorphism and allele structure of trinucleotide microsatellites in natural accessions of *Arabidopsis thaliana*, *Theor. Appl. Genet.* 97, 591, 1998.

29. Abbott, R.J. and Gomes, M.F., Population genetic structure and outcrossing rate of *Arabidopsis thaliana* (L) Heynh., *Heredity* 62, 411, 1989.

30. Agren, J. and Schemske, D.W., Outcrossing rate and inbreeding depression in 2 annual monoecious herbs, *Begonia hirsuta* and *B. semiovata*, *Evolution* 47, 125, 1993.

31. Hoffmann, M.H., Bremer, M., Schneider, K., Burger, F., Stolle, E., Moritz, G., Flower visitors in a natural population of *Arabidopsis thaliana*, *Plant Biol.* 5, 491, 2003.

5 An Integrated Genomics Approach to Identifying Floral Scent Genes in Rose

Alexander Vainstein, Efraim Lewinsohn, and David Weiss

CONTENTS

5.1   INTRODUCTION

The ability of flowering plants to prosper throughout their long evolution has always been strongly dependent on the constant development of strategies to lure pollinators. This has led to the creation of elaborate perianth forms, splendid color patterns, and a broad spectrum of fragrances. Flower morphogenesis and pigmentation have been intensively studied in the past several decades and today the results of our deepened understanding of the underlying pathways have been harnessed for the improvement of such characters in some commercially important ornamentals.[1–3] In contrast, our knowledge of the biochemistry of fragrance production and of the mechanisms regulating its emission remain sketchy. This is due in part to the invisibility of scent as a character, to the shortcomings of humans' sense of smell, and to the highly variable nature of the trait (because of strong environmental influences, among other reasons). To date, no simple, efficient, and reliable methods of screening for genetic variation have been developed. Moreover, no convenient plant model systems that would enable biochemical or forward and reverse genetic studies of flower scent are available.

Unique floral scents are mixtures of low molecular weight volatile molecules,[4,5] many of which can now be identified by means of the sensitive analytical tools developed during the past few decades. Although all floral organs can emit fragrance compounds, petals are the main source of scent in most plants.[6] Some plants have developed highly specialized anatomical structures, termed scent glands, for fragrance production; in others, the nonspecialized floral epidermal cells are recruited for fragrance production and emission (for a review, see Dudareva and Pichersky[5]).

For centuries, rose has been the most important crop in the floriculture industry. The genus *Rosa* includes 200 species and more than 18,000 cultivars.[7,8] At an annual value of approximately $10 billion, roses are used as cut flowers and container and garden plants. Their economic importance also lies in the use of their petals as a source of natural fragrances and flavorings. The damask rose (*Rosa damascena*) is the most important species used to produce rose water, attar of rose, and essential oils in the perfume industry. However, despite the importance of scent, most modern rose cut-flower varieties lack a distinct fragrance, in contrast to some garden cultivars with their intense "rose" scent.[1] Because for many years selection for longevity was a major target, particularly in the "hybrid tea" used for cut flowers, and a concomitant loss of fragrance was observed, some negative correlation between these traits was suggested.[9] Nevertheless, the specific causes of the loss of fragrance during the course of breeding programs remain unknown.

There are a few studies in the literature characterizing rose floral scent volatiles.[10–17] In 1953 only 20 rose volatile compounds had been identified. Today more than 400 different volatile compounds are known to be potential contributors to the floral scent of roses.[18] The volatiles emitted by rose flowers can be classified into five main groups (Figure 5.1): alcohols, esters, monoterpenes, sesquiterpenes, and aromatic ethers.[13] Based on headspace analyses of the volatiles emitted by live flowers, Flament et al.[13] classified rose cultivars into five distinct groups:

Group 1. Hydrocarbon types, characterized by high (more than 40% of total) emissions of mono- and sesquiterpene hydrocarbons, but containing lower levels of other compounds such as monoterpenols, esters, and aromatic ethers. Cultivars such as Monica, Sonia Meilland, and Charles de Gaulle are in this group.

Group 2. Alcohol types without orcinol dimethylether. This group includes the classical "old" varieties that emit between 35% and 85% alcohols, such as phenylethyl alcohol, citronellol, geraniol, and nerol, and include *Rosa rugosa rubra*, *Rosa muscosa purpurea*, *Rosa gallica*, *R. damascena*, and *Rose a parfum de l'Hay*.

Group 3. Alcohol types containing orcinol dimethylether. This group includes roses with a deep scent, such as Margaret Merril, Westerland, and Chatelaine de Lullier, and contain up to 62% alcohols, but also volatile acetates, aromatic ethers, and other olfactorally important compounds, such as β-ionone.

Group 4. Ester types, emitting 30% to 90% esters, as well as moderate levels of alcohols and orcinol dimethylether, and including cultivars such as Baronne E. de Rothschild, Sutters Gold, Concorde 92, Susan Ann, and Nuage Parfumé (translated, Fragrant Cloud).

Group 5. Aromatic ether types, which emit more than 40% orcinol dimethylether, and include Sylvia, Gina Lollobrigida, and Papa Meiland.

Clearly, rose represents an excellent source for the identification of genes from different scent-related pathways.[19] However, the lack of an established genetic

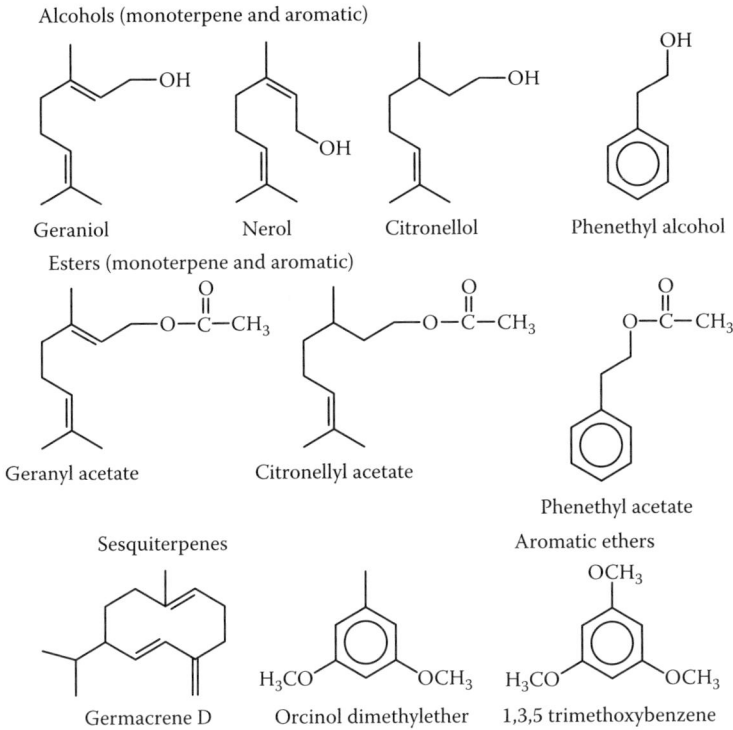

FIGURE 5.1 Chemical structures of the main rose volatiles.

framework, including characterized mutants and the recalcitrance of rose to genetic manipulation, makes the application of forward and reverse genetic approaches extremely cumbersome. The novel technologies of genomics, in contrast, enable quick access to plants with poor genetic characterization, enabling the choice of a model plant system based on the trait of interest (scent) rather than being limited to established model systems.[19,20] Indeed, several groups have recently used high-throughput technologies to identify new fragrance genes in fruits and vegetative tissues.[21–26] Aharoni et al.[21] were the first to combine expressed sequence tag (EST) database mining with metabolic profiling and microarray expression analyses to identify an aroma-related gene—alcohol acyltransferase—responsible for the production of volatile esters in strawberry fruit. Here we describe the use of genomics to study fragrance production in rose petals. A detailed chemical analysis of volatile composition, together with the identification of secondary metabolism-related genes whose expression coincides with scent production, has led to the discovery of several novel rose flower fragrance genes.

5.2   CHEMICAL ANALYSES OF ROSE FLORAL SCENT

Detailed analyses performed with 11 rose varieties revealed that the total volatile levels produced by their flowers correlate roughly with the scent characteristics of each cultivar.[17] Comparatively high levels of volatiles were detected in the very fragrant cultivars Fragrant Cloud (FC), Double Delight, Secret, and Dr. A. J. Verhage, while only low levels of volatiles were detected in the moderately and less fragrant cultivars. All the very fragrant cultivars produced a certain amount (7 to 60% of total volatiles) of alcoholic monoterpenes, including geraniol, citronellol, and nerol, all three of which have roselike fragrances (Figure 5.1).[27] Geraniol, citronellol, and nerol are in fact major constituents of rose oil, which also contains minor amounts of 2-phenylethyl alcohol.[27] In contrast to the very fragrant cultivars, the moderately and less fragrant cultivars were characterized by a typical absence of alcoholic monoterpenes, except for Bernstein-Rose, a moderately fragrant cultivar that contains low levels of geraniol. 2-Phenylethyl alcohol, a compound characterized by a mild rose odor[27] was detected in all 11 cultivars tested.[17] Benzyl alcohol, which has a weak, slightly sweet odor[27] was identified in 8 of the 11 cultivars. Among the very fragrant cultivars, volatile acetates were identified only in FC.

As a first step in the rose genomics study, headspace analyses of the aroma chemicals emitted by two rose cultivars during flower development, from a green closed bud through the corresponding intermediate stages and up to fully blooming flowers, were performed.[17,24] These cultivars were FC, a highly fragrant red-colored, modern hybrid tea, with a short vase life, and Golden Gate (GG), a long vase life yellow cultivar. Although GG is almost scentless to humans, some of the major volatiles emitted by the flowers can be easily detected by honeybees.[17] The volatiles emitted by FC flowers comprised almost exclusively short-chain alcohols and acetates, monoterpenes, and sesquiterpenes. The aromatic compound 2-phenylethyl alcohol was the most prominent alcohol (more than 39% of the total volatiles emitted by a fully open flower). High levels of the monoterpene alcohol citronellol (14% of the total volatiles at the fully open stage) were also detected in the FC headspace. Other volatile alcohols, such as benzyl alcohol, 2-ethyl hexanol, 1-hexanol (traces), linalool, geraniol, and 1,8-cineole, were detected at much lower levels (less than 1% of total volatiles). Low levels of 1,8-cineole were identified only during the first three flowering stages, 1-hexanol was detected only in the early stages of flower development, and geraniol was detected only at advanced stages.

The second major group of volatiles identified in the FC headspace was the acetate esters, with phenylethyl acetate (15% of the total volatiles) being the most abundant acetate. At anthesis, cis-3-hexenyl acetate, citronellyl acetate, geranyl acetate, 1-hexyl acetate, and trans-2 hexenyl acetate were also identified in the FC headspace. Neryl and benzyl acetate were also emitted, but at much lower levels (0.4% and 0.2% of total volatiles, respectively).

The third important group of volatiles in the FC headspace was the terpene hydrocarbons. The most abundant sesquiterpene in the FC headspace was germacrene D (8% of total volatiles). All the other sesquiterpenes, monoterpenes, and norisoprene β-ionone were present in very low concentrations, mostly at advanced stages of flower development, except for a negligible level of the sesquiterpene β-farnesene, which was identified only at early stages. In addition to the three aforementioned major groups of volatiles, negligible levels of phenolic derivatives were identified in the FC headspace; these included orcinol dimethylether, and methyl eugenol.[17]

In contrast, the headspace of GG flowers was largely comprised of orcinol dimethylether (60 to 98% of total volatiles at different stages of flower development). Aside from rose flowers, this volatile has only been found in a very few plant species, such as *Narcissus tazetta*, a plant native to central Europe and the Mediterranean,[11] and *Acnistus arborescens* (Solanaceae), indigenous to the southern Atlantic rainforest.[28] Orcinol dimethylether is almost undetectable to the human nose, but odor discrimination experiments with honeybees indicate that these insects readily detect this compound in rose petals.[17]

The GG flowers also emitted low levels of volatile hydrocarbon terpenes, including the monoterpenes α-pinene, β-pinene, limonene, and δ-carene, and the sesquiterpenes *trans*-caryophyllene and caryophyllene oxide. The GG headspace contained only minute levels of the short-chain alcohols *cis*-3-hexen alcohol and 2-ethyl hexanol. Volatile acetates were not detected. Interestingly, whereas the levels of volatiles in both analyzed varieties increased during flower development, the ratios of the major volatiles remained constant during the different stages, suggesting that emission of the different classes of major volatiles is regulated by similar mechanisms. Similar results have been reported for the monoterpene alcohols in flowers of *R. damascena*.[15]

5.3   GENOME AND PROTEOME

The National Center for Biotechnology Information (NCBI) stores 22,000 entries for Rosaceae nucleotides and 2000 protein records. Most of the DNA sequences were obtained for *Prunus* (http://www.genome.clemson.edu/projects/) from developing peach fruit mesocarp (14,000) and almond seeds (3500). In the past few years there has been a large increase in the number of rose petal sequences in the taxonomy browser of GenBank (http://www.ncbi.nlm.nih.gov/Taxonomy/Browser/wwwtax.cgi). *Rosa hybrida* cultivars are represented by about 3000 ESTs and about 30 protein sequences, while *Rosa chinensis* is represented by about 1800 ESTs and 5 protein sequences. Nearly 20% of these rose sequences have no homologues in the database and thus represent novel sequences that may encode rose-specific proteins. Indeed, more than a quarter of these novel genes have been identified in both *R. chinensis* and *R. hybrida*.

Almost all annotated *R. hybrida* ESTs come from two tetraploid varieties, FC and GG, with contrasting phenotypes. Analyses of the temporal regulation of volatile production in FC and GG flowers revealed that the levels of most scent compounds peak at advanced stages of flower development, just before anthesis. Hence flowers at this stage from both cultivars were used to construct complementary DNA (cDNA) libraries. Randomly chosen individual clones were sequenced and high-quality sequence information from 2873 individual clones—1834 from FC and 1039 from GG—was annotated and organized in contigs.[24] A very large number of unique genes (2139; 74% of the total ESTs) were represented in the database. The unique genes were classified into 18 functional groups. The largest group of petal sequences, containing about one third of all unique genes (32%), shows homology (E-value $< 1.0E^{-5}$) to genes coding for predicted proteins with unknown function. The second largest group (17%) contains unique sequences that have no homologues (E-value $\leq 1.0E^{-5}$). The high percentage of previously unknown sequences suggests rose petals as an intriguing source for novel genes. Another large group of genes in the rose database (15%) is represented by proteins putatively involved in metabolism. Among these, 233 unique genes are related to primary metabolism and 96 (about 5% of the total ESTs) are genes coding for enzymes putatively involved in secondary metabolism, that is, sesquiterpene synthases, monoterpene synthases, acetyl transferases, alcohol dehydrogenases, methyl transferases, and aldehyde dehydrogenases.

In a parallel study with the *R. chinensis* garden variety Old Blush, Channeliere et al.[29] generated a collection of 1794 ESTs from petals at different developmental stages, from young buds to senescing flowers. The high redundancy (65%) in this database was mainly due to the large number of transcripts coding for one stress-related gene and several lipid-transfer genes. The largest group of ESTs (about 25%) consisted of clones with no homologues in other plants. Transcripts related to metabolism represented about 9% of the ESTs.

To complement the genomic data, a rose (FC; same cultivar as that used for the generation of ESTs) petal proteome was analyzed. Using 2D-PAGE, we generated stage-specific (closed bud, mature flower, and flower at anthesis) petal protein maps with about 1000 unique protein spots. About 10% of these proteins were identified by mass spectrometry (MS), annotated, and classified into functional groups. This classification revealed energy, cell rescue, unknown function, and metabolism to be the largest classes, together comprising about 80% of all identified proteins. Despite the fact that secondary metabolism-related pathways are highly active in advanced stages of petal development,[30,31] only one identified protein, dihydroflavonol reductase, could be directly related to secondary metabolism (i.e., anthocyanin biosynthesis). Of particular interest is the fact that the relative sizes of the protein functional groups differed greatly from those of the corresponding functional groups constructed on the basis of an EST database (Figure 5.2).

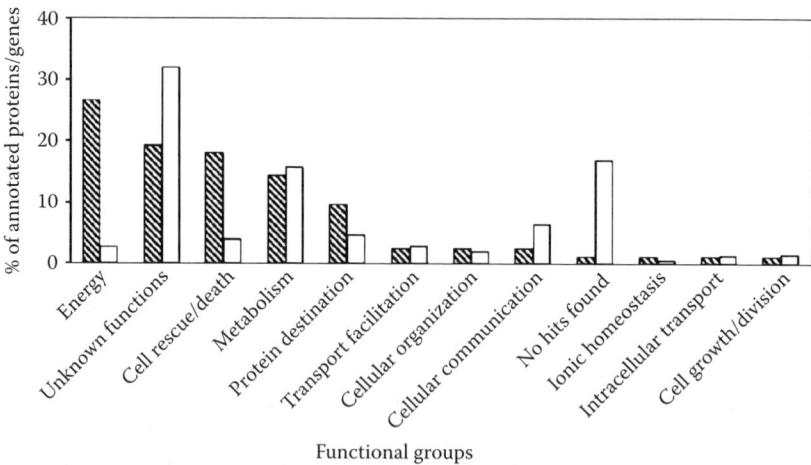

FIGURE 5.2 A comparison of the relative sizes of the functional groups deduced from protein and EST (unique genes) databases. Classification of both proteins (hatched bars) and genes (white bars) into functional groups was performed according to the Munich Information Center for Protein Sequences.

For example, in the EST database, genes with unknown function (including those with no homologues in the database) represent about 50% of the total unique genes, and cell rescue-related genes represent only about 4%. At the protein level, each of these functional groups consists of about 19% of the different identified proteins. The energy-related group is one of the smallest in the EST database (2.8% of the total unique genes), whereas it is the largest group (26% of the total identified proteins) in the protein database. This apparent discrepancy in the protein versus RNA (EST) database compositions is indicative of differences in gene expression at these two levels. Nevertheless, the annotated protein database still needs to be significantly increased for any unequivocal conclusions to be drawn.

5.4 IDENTIFICATION AND FUNCTIONAL CHARACTERIZATION OF NOVEL SCENT GENES

To create the infrastructure needed to explore the relationships between sequences and phenotypes in rose, expression profiling was performed using microarray analyses. Since most scent compounds are produced by mature petals but not by young green petals of FC, and not by mature GG petals, ESTs whose expressions are up-regulated during FC petal development were overlapped with those that show higher expression in mature FC versus mature GG petals. By using a twofold change in gene expression as the threshold in the microarray analyses, 80 genes with higher expression levels in FC petals as compared to GG petals were identified. Analyses of gene expression during FC petal development allowed identification of 88 genes that were up-regulated in mature petals. Overlapping the resultant sequence information from these comparisons led to the identification of about 40 genes differentially expressed in the two rose varieties examined.[24]

Since FC and GG petals differ in size and color, it was not surprising that genes putatively involved in these traits were represented among the 40 clones found in the overlap analysis. For example, genes related to cell growth included a putative xyloglucan endotransglycosylase[32] and a GAST-like gene.[33] In addition, genes potentially involved in anthocyanin production included a putative dihydroflavonol reductase and a gene similar to the transcriptional regulator myb26.[34] This MYB protein regulates phenylpropanoid biosynthesis and has a high affinity for promoters of phenylpropanoid genes. Fifteen of the 40 differentially expressed genes were classified into the metabolism functional group, seven with a putative role in scent production (e.g., decarboxylases, reductases, hydroxylases, and sesquiterpene synthase). To search for additional scent-related genes, we looked for ESTs whose expression increases during FC petal development, as revealed by the microarray analysis, regardless of their expression level in GG. Among these 88 unique genes, those with putative function in secondary metabolism included a putative alcohol acetyltransferase, two putative O-methyltransferases (OMT1 and OMT2),[35] and sequences with a strong similarity to known monoterpene synthases, decarboxylases, hydroxylases, aminotransferases, and aldehyde dehydrogenases. An alternative, candidate gene approach to the identification of rose scent genes was taken by Channeliere et al.[29] Based on the significant similarity between cv. Old Blush ESTs and sequences in databases, they identified genes from the isoprenoid pathway (i.e., sesquiterpene synthase) and from the phenylpropanoid pathway (i.e., O-methyltransferases) putatively involved in scent production.

To prove the function of the candidate fragrance genes, lysates from *Escherichia coli* expressing these putative scent genes or transgenic plants expressing the genes are used to test the associated enzymatic activities. To prove the function of a putative rose FC sesquiterpene synthase and to identify the type of sesquiterpene produced by this enzyme, the gene was cloned into an expression vector to obtain a catalytically active recombinant enzyme. *In vitro* assays using lysates from bacteria overproducing the gene product and the substrate farnesyl diphosphate resulted in the synthesis of a sesquiterpene, identified by gas chromatography-mass spectrometry (GC-MS) as germacrene D.[24] It is worth noting that the major sesquiterpene identified in the headspace of FC flowers was also germacrene D, constituting approximately 60% of the emitted sesquiterpenes. The molecular structure of germacrene D resembles that of its isomer germacrene C.[36] However, rose germacrene D synthase shows a greater similarity to cotton's (+)-δ-cadinene synthase, which catalyzes the synthesis of a sesquiterpene with a different skeleton, than to *Lycopersicon esculentum* germacrene C synthase. This suggests that the similarity between genes within this group does not necessarily determine the similarity between their products, and infers that specialization of the terpenoid genes, resulting in the production of discrete products, occurred several times during evolution, in a parallel fashion.[37,38]

Recently, transgenic tobacco and petunia expressing germacrene D synthase were generated.[39] Chemical (GC-MS) analyses of these transgenic plants showed that both produce the sesquiterpene germacrene D. The levels of germacrene D found in the headspace of transgenic petunia leaves were much higher than those found in that of tobacco leaves. Since nontransgenic tobacco leaves produce very high levels of sesquiterpenes, the amount of farnesyl pyrophosphate (FPP), the substrate for all sesquiterpenes, in this tissue may be low and therefore not accessible to the ectopically expressed germacrene D synthase enzyme. Petunia leaves, on the other hand, produce only low levels of sesquiterpenes and hence the pool of FPP substrate may be more accessible.

Similarly the function of the putative FC alcohol acetyltransferase (*RhAAT*) was confirmed *in vitro* by analyzing the catalytic activity of the recombinant gene product.[40] To test the substrate specificity of the *RhAAT1* gene product for potential alcoholic substrates and to determine its general catalytic and kinetic parameters, a sensitive radioassay as well as GC-MS analysis for the identification of products were employed. The *RhAAT1* protein was partially purified and a number of alcohols were tested as substrates. *RhAAT1* was most active with geraniol, catalyzing the production of geranyl acetate from geraniol and acetyl-CoA. The reaction depended on the presence of enzyme, the alcoholic substrate geraniol, and acetyl-CoA. Controls consisting of the same cells lacking a recombinant *RhAAT1* gene did not exhibit alcohol acetyltransferase activity. The enzyme also readily accepts citronellol (approximately 60% the rate obtained with geraniol), but is relatively inefficient at acetylating nerol, the *cis*-isomer of geraniol. The enzyme also exhibited lower levels of activity with *cis*-3-hexene 1-alcohol and 2-phenylethyl alcohol. Therefore it is likely that *cis*-3-hexenyl acetate and 2-phenylethyl acetate, the two main esters emitted from FC roses, are synthesized by other enzymes. The *RhAAT* gene was recently expressed in transgenic petunia, under regulation of the 35S promoter (Vainstein, A. and Weiss, D., unpublished). Chemical (GC-MS) analyses of headspace collected from flowers showed that the transgenic plants produce novel acetates (i.e., benzyl acetate and 2-phenylethyl acetate) not found in nontransgenic control plants. Since petunia flowers do not produce geraniol, but accumulate benzyl alcohol and 2-phenylethyl alcohol, it may be substrate availability rather than the enzyme substrate affinity level that determines the final product generated/emitted by the flower.

Using essentially the same approaches, several putative *R. hybrida* and *R. chinensis OMT* genes were functionally characterized. OMT catalyzes the transfer of a methyl group from *S*-adenosyl-L-Met (SAM) to different phenolic compounds.[41] Benzenoid 3,5-dimethoxytoluene, produced in this way, is one of the main scent compounds in flowers of many hybrid roses, contributing to their final fragrance. It derives from orcinol (3,5-dihydroxytoluene) in two successive methylation reactions. Functional analyses of two putative rose OMTs (OOMT1 and OOMT2) in *E. coli* revealed that both can carry out both reactions; however, OOMT1 showed a higher affinity to orcinol, whereas OOMT2 was more catalytically efficient with 3-methoxy,5-hydroxytoluene.[35,42] *R. chinensis* flowers emit a similar compound, 1,3,5-trimethoxybenzene, with three methoxyl groups, which is synthesized in consecutive reactions from 1,3,5,-trihydroxybenzene. While the identified OOMTs can methylate these intermediate steps in the production of 1,3,5-trimethoxybenzene, they cannot catalyze the methylation of 1,3,5-trihydroxybenzene to produce the first intermediate.

The rapid development of analytical tools for metabolite profiling, allowing the simultaneous identification of thousands of compounds, together with the ever-increasing phenomic, genomic, and EST, transcriptionomic, and proteomic databases, should be highly instrumental in deciphering the mechanisms regulating highly complex cellular networks. It is apparent that these advanced omics technology tools bypass the need for model systems. Instead, system biologists can now concentrate on nonmodel plants that are rich in the trait of interest. The molecular biochemistry of odor in general, and flower fragrance in particular, is still in its infancy. Nevertheless, as summarized here, major advances have been made in the past few years and we are now just beginning to grasp the intricate complexity of the molecular labyrinth of "That which we call a rose" (William Shakespeare; Romeo and Juliet, II, ii, 43).

ACKNOWLEDGMENTS

Work in the authors' laboratories has been supported by Research Grant Number US-3437-03 from BARD, the Israeli Ministry of Agriculture, the Israeli Ministry of Science, the Israel Science Foundation and the Hebrew University Intramural Research Fund Basic Project Awards.

REFERENCES

1. Zuker, A., Tzfira, T., and Vainstein, A., Cut-flower improvement using genetic engineering, *Biotech. Adv.* 16, 33, 1998.

2. Zuker, A., Tzfira, T., Ben-Meir, H., Ovadis, M., Shklarman, E., Itzhaki, H., Forkmann, G., Martens, S., Neta-Sharir, I., Weiss, D., and Vainstein, A., Suppression of anthocyanin synthesis by antisense *fht* enhances flower fragrance, *Mol. Breed.* 9, 33, 2002.

3. Ben-Meir, H., Zuker, A., Weiss, D., and Vainstein, A., Molecular control of floral pigmentation: anthocyanins, in *Breeding for Ornamentals: Classical and Molecular Approaches*, Vainstein, A., Ed., Kluwer Academic Press, Dordrecht, The Netherlands, 2002, p. 253.

4. Croteau, R. and Karp, F., Origin of natural odorants, in *Perfume: Art, Science and Technology*, Muller, P. and Lamparsky, D., Eds., Elsevier Applied Sciences, New York, 1991, p. 101.

5. Dudareva, N. and Pichersky, E., Biochemical and molecular genetic aspects of floral scents, *Plant Physiol.* 122, 627, 2000.

6. Pichersky, E., Raguso, R.A., Lewinsohn, E., and Croteau, R., Floral scent production in *Clarkia* (Onagraceae). I. Localization and developmental modulation of monoterpene emission and linalool synthase activity, *Plant Physiol.* 106, 1533, 1994.

7. Weiss, E.A., *Essential Oil Crops*, CAB International, Wallingford, Oxon, UK, 1997.

8. Gudin, S., Rose: genetics and breeding, *Plant Breed.* 17, 159, 1997.

9. Barletta, A., Scent makes a comeback, *Floraculture* 5, 23, 1995.

10. Ohloff, G. and Demole, E., Importance of the odoriferous principle of Bulgarian rose oil in flavor and fragrance chemistry, *J. Chromatogr.* 406, 181, 1987.

11. Mookherjee, B.D., Trenkle, R.W., and Wilson, R.A., Live vs. dead. Part II. A comparative analysis of the headspace volatiles of some important fragrance and flavor raw materials. *J. Essent. Oil Res.* 2, 85, 1989.

12. Dobson, H.E.M., Danielson, E.M., and van Wesep, I.D., Pollen odor chemicals as modulators of bumble bee foraging on *Rosa rugosa* Thunb. (Rosaceae). *Plant Spec. Biol.* 14, 153, 1999.

13. Flament, I., Debonneville, C., and Furrer, A., Volatile constituents of roses: characterization of cultivars based on the headspace analysis of living flower emissions, in *Bioactive Volatile Compounds from Plants*, Teranishi, R., Buttery, R.G., and Sugisawa, H., Eds., American Chemical Society, Washington, DC, 1993, p. 269.

14. Helsper, J.P.F., Davies, J.A., Bouwmeester, H.J., Krol, A.F., and van Kampen, M.H., Circadian rhythmicity in emission of volatile compounds by flowers of *Rosa hybrida* L. cv. Honesty, *Planta* 207, 88, 1998.

15. Oka, N., Ohishi, H., Hatano, T., Hornberger, M., Sakata, K., and Watanabe, N., Aroma evolution during flower opening in *Rosa damascena* Mill., *Z. Naturforsch Sect. C Biosci.* 54, 889, 1999.

16. Kim, H.J., Kim, K., Kim, N.S., and Lee, D.S., Determination of floral fragrances of *Rosa hybrida* using solid-phase trapping-solvent extraction and gas chromatography-mass spectrometry, *J. Chromatogr.* 902, 389, 2000.

17. Shalit, M. et al., Volatile compounds emitted by rose cultivars: fragrance perception by man and honey bees, *Isr. J. Plant Sci.* 52, 245, 2004.

18. Ohloff, G., Recent developments in the field of naturally-occurring aroma components, *Fortschr. Chem. Org. Naturst.* 35, 431, 1978.

19. Vainstein, A., Lewinsohn, E., Pichersky, E., and Weiss, D., Floral fragrance—new inroads into an old commodity, *Plant Physiol.* 127, 1383, 2001.

20. Fiehn, O., Kopka, J., Dormann, P., Altmann, T., Trethewey, R.N., and Willmitzer, L., Metabolite profiling for plant functional genomics, *Nat. Biotech.* 18, 157, 2000.

21. Aharoni, A., Keizer, L.C.P., Bouwmeester, H.J., Sun, Z., Alvarez-Huerta, M., Verhoeven, H.A., Blaas, J., van Houwelingen, A.M.M.L., De Vos, R.C.H., van der Voet, H., Jansen, R.C., Guis, M., Mol, J., Davis, R.W., Schena, M., van Tunen, A.J., and

O'Connell, A.P., Identification of the SAAT gene involved in strawberry flavor biogenesis by use of DNA microarrays, *Plant Cell* 12, 647, 2000.

22. Lange, B.M., Wildung, M.R., Stauber, E.J., Sanchez, C., Pouchnik, D., and Croteau, R., Probing essential oil biosynthesis and secretion by functional evaluation of expressed sequence tags from mint glandular trichomes, *Proc. Natl. Acad. Sci. USA* 97, 2934, 2000.

23. Gang, D., Wang, J., Dudareva, N., Nam, K.H., Simon, J.E., Lewinsohn, E., and Pichersky, E., An investigation of the storage and biosynthesis of phenylpropenes in sweet basil, *Plant Physiol.* 125, 539, 2001.

24. Guterman, I., Shalit, M., Menda, N., Piestun, D., Dafny-Yelin, M., Shalev, G., Bar, E., Davydov, O., Ovadis, M., Emanuel, M., Wang, J., Adam, Z., Pichersky, E., Lewinsohn, E., Zamir, D., Vainstein, A., and Weiss, D., Rose scent: genomic approach to discover novel floral fragrance-related genes, *Plant Cell* 14, 2325, 2002.

25. Chen, F., Tholl, D., D'Auria, J.C., Farooq, A., Pichersky, E., and Gershenzon, J., Biosynthesis and emission of terpenoid volatiles from Arabidopsis flowers, *Plant Cell* 15, 481, 2003.

26. Dudareva, N., Martin, D., Kish, C.M., Kolosova, N., Gorenstein, N., Faldt, J., Miller, B., and Bohlmann, J., (E)-{β}-ocimene and myrcene synthase genes of floral scent biosynthesis in snapdragon: function and expression of three terpene synthase genes of a new terpene synthase subfamily, *Plant Cell* 15, 1227, 2003.

27. Bauer, K., Garbe, D., and Surburg, H., *Common Fragrance and Flavor Materials*, Wiley-VCH, Weinheim, 2001.

28. Kaiser, R.D., Scent from rain forest, *Chimia* 54, 346, 2000.

29. Channeliere, S., Riviere, S., Scalliet, G., Jullien, F., Szecsi, J., Dolle, C., Vergne, P., Dumas, C., Bendahmane, M., Hugueney, P., and Cock, J.M., Analysis of gene expression in rose petals using expressed sequence tags, *FEBS Lett.* 515, 35, 2002.

30. Francis, M.J.O. and Allcock, O., Geraniol beta-D-glucoside: occurrence and synthesis in rose flowers, *Phytochemistry* 8, 1339, 1969.

31. Suzuki, K. et al., Molecular characterization of rose flavonoid biosynthesis genes and their application in petunia, *Biotechnol. Biotechnol. Equip.* 14, 56, 2000.

32. Schünmann, P.H.D., Smith, R.C., Lang, V., Matthews, P.R., and Chandler, P.M., Expression of XET-related genes and its correlation to elongation in leaves of barley (*Hordeum vulgare* L.), *Plant Cell Environ.* 20, 1439, 1997.

33. Shi, L., Gast, R.T., Golparaj, M., and Olszewski, N.E., Characterization of a shoot-specific, GA_3- and ABA-regulated gene from tomato, *Plant J.* 2, 623, 1992.

34. Uimari, A. and Strommer, J., Myb26: a MYB-like protein of pea flowers with affinity for promoters of phenylpropanoid genes, *Plant J.* 12, 1273, 1997.

35. Lavid, N., Wang, J., Shalit, M., Guterman, I., Bar, E., Beuerle, T., Menda, N., Shafir, S., Zamir, D., Adam, Z., Vainstein, A., Weiss, D., Pichersky, E., and Lewinsohn, E., *O*-methyltransferases involved in the biosynthesis of volatile phenolic derivatives in rose petals, *Plant Physiol.* 129, 1899, 2002.

36. Colby, S.M., Crock, J., Dowdle-Rizzo, B., Lemaux, P., and Croteau, R., Germacrene C synthase from *Lycopersicon esculentum* cv. VFNT cherry tomato: cDNA isolation, characterization, and bacterial expression of the multiple product sesquiterpene cyclase, *Proc. Natl. Acad. Sci. USA* 95, 2216, 1998.

37. Bohlman, J., Gershenzon, J., and Aubourg, S., Biochemical, molecular genetic and evolutionary aspects of defense-related terpenoid metabolism in conifers, in *Recent Advances in Phytochemistry*, vol. 34, *Evolution of Metabolic Pathways*, Romeo, J.T., Ibrahim, R., Varin L. and De Luca, V., Eds., Pergamon Press, New York, 2000, p. 109.

38. Pichersky, E. and Gang, D.R., Genetics and biochemistry of secondary metabolites in plants: an evolutionary perspective, *Trends Plant Sci.* 5, 439, 2000.
39. Vainstein, A., Adam, Z., Zamir, D., Weiss, D., Lewinsohn, E., and Pichersky, E., Rose fragrance: genomic approaches and metabolic engineering, *Acta Hort.* 612, 105, 2003.
40. Shalit, M., Guterman, I., Volpin, H., Bar, E., Tamari, T., Menda, N., Adam, Z., Zamir, D., Vainstein, A., Weiss, D., Pichersky, E., and Lewinsohn, E., Volatile ester formation in roses. Identification of an acetyl-coenzyme A. geraniol/citronellol acetyltransferase in developing rose petals, *Plant Physiol.* 131, 1868, 2003.
41. Ibrahim, R.K., Bruneau, A., and Bantignies, B., Plant O-methyltransferases: molecular analysis, common signature and classification, *Plant Mol. Biol.* 36, 1, 1998.
42. Scalliet, G., Journot, N., Jullien, F., Baudino, S., Magnard, J.L., Channeliere, S., Vergne, P., Durmas, C., Bendahmane, M., Cock, J.M., and Hugueney, P., Biosynthesis of the major scent components 3,5-dimethoxytoluene and 1,3,5-trimethoxybenzene by novel rose O-methyltransferases, *FEBS Lett.* 523, 113, 2002.

Section III

Cell Biology and Physiology of Floral Scent

6 Localization of the Synthesis and Emission of Scent Compounds within the Flower

Uta Effmert, Diana Buss, Diana Rohrbeck, and Birgit Piechulla

CONTENTS

6.1   INTRODUCTION

Previous chapters dealt with the types of compounds found in floral scent and how these compounds are synthesized inside the cell.[1] However, it has been known for many years that different parts of a flower can emit different mixtures of scent compounds or no scent at all. This chapter reviews our present knowledge regarding the spatial distribution of scent biosynthesis and emission within the flower, and the organization of certain structures that are sometimes associated with such localized scent production.

6.2 OSMOPHORES

More than 100 years ago, Arcangeli reported the observation that the release of scent can be connected with certain parts of an inflorescence.[2] He specified the upper part of the spadix of *Arum italicum* (Araceae) as the site of odor evaporation and called it *osmoforo*, the "odor bearer," whereas the lower part was called *antoforo*, the "flower bearer" (greek: *osmo*, odor; *antos*, flower; *pherein*, to bear). The relevance of this phenomenon for scent emission and pollination remained unvalued until Vogel[3] reintroduced the term "osmophore" to describe floral volatile production and emission via specialized defined tissue areas with glandular characteristics.

6.2.1 DEFINITION OF OSMOPHORES

Osmophores, also called floral scent glands, possess the ability to emit volatiles and comprise defined floral organs.[3,4] They can be found within the whole inflorescence as part of the perianth, bracts, appendices of peduncles, or anthers. Although osmophores might vary in shape and habitus, being plane-, whip-, brush-, club-, or palp-shaped, they have features in common. They usually face toward the adaxial side of the perianth and display a bullate, rugose, pileate, conical, or papillate epidermis (Figure 6.1).[3,5–10] Subjacent are several cell layers that form the glandular tissue, which merge into normal parenchyma cell layers.[4,6] Cells of the glandular tissue show enlarged nuclei compared to cells of nearby tissue and a dense cytoplasm.[4,6] A dispersed vacuome observed before and at anthesis, often turns into a large vacuole after anthesis.[6] Transmission electron microscopy revealed that cells of the glandular layers are supplied with an abundant rough or smooth endoplasmic reticulum (rER or sER, respectively), sometimes dictyosomes, many mitochondria, and lipoid droplets.[3–5] These droplets, probably lipid-protein mixtures, are surrounded by a monolayer of phospholipids and embedded in the cytosol.[11,12] They contain essential oils to be released and lipids like fatty acids and triacylglycerides.[3,4]

Hudak and Thompson[12] identified in cytosolic lipid-protein particle fractions of *Dianthus caryophyllus* petals (carnation, Caryophyllaceae) several volatile components of the carnation flower fragrance including hexanal, (*E*)-2-hexenal, nonanal, 2-hexanol, 3-hexen-1-yl benzoate, benzyl alcohol, and benzyl benzoate. As a result of pulse-labeling experiments with [^{14}C]-acetate, the authors suggest that the cytosolic lipid-protein bodies originate from membranes, indicating hydrophobic subcompartmentation. Most remarkable is the presence of enormous starch deposits in osmophores, which obviously ensures a carbon and energy supply for volatile production.[3,6] This is supported by the fact that the deposits disappear due to utilization while volatiles are emitted.[3] In the adaxial epidermis cells of osmophores, however, amyloplasts are usually absent. Based on these observations, Vogel distinguished the glandular cell layers as the site of volatile production and the osmophore epidermis as the site of emission.[3] As a consequence, inter- and intracellular transport toward the osmophore epidermis has to be postulated. Trafficking of lipoids is probably accomplished by close association of the rER with the plasma membrane, creating channels and multitubular structures closely related to lipoid bodies.[11,13] In addition, the transformation of homogeneous lipoid droplets into a multivesicular body just

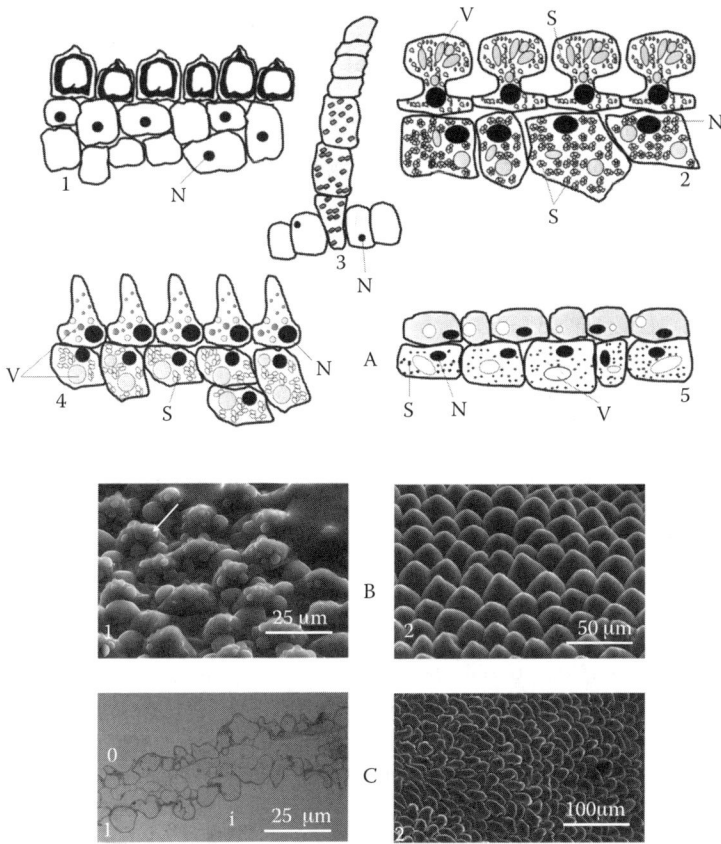

FIGURE 6.1 Epidermal surfaces of scent emitting flower tissue. (A) Schematic presentation of epidermal cells that are often observed in emitting floral tissue: (1) rugate to conical type; (2) pileate type; (3) trichome; (4) conical type; (5) flat to bullate type. N, nucleus; S, starch grains; V, vacuole. (B) Scanning electron microscope (SEM) picture of petals from scented and unscented rose (*Rosa hybrida*). (1) Adaxial epidermal cells with secretions (arrow) and (2) conical adaxial epidermis cells without secretions. (Courtesy of F. Ehrig.) (C) (1) Semithin cross-section (light microscope) of the petal lobe of *M. jalapa*. Both epidermata exhibit bullate appearance. Loose mesophyll cells and many intercellular spaces are visible. (2) Bullate adaxial epidermal cells of *M. jalapa* (environmental scanning electron microscope).

before fusion to the plasma membrane was reported.[14] Osmophores are often supplied with an extended vascular system and fragile intercellular net with phloematic bundles pervading an aerenchymatic tissue, and stomata are frequently found on the abaxial side of osmophores.[3,4,14]

6.2.2 Detection of Osmophores

In the middle of the past century, the human nose was the most important measure for determining osmophores in floral tissue.[3] A somewhat better, nonbiased indication for the presence of osmophores is based on neutral red staining of floral tissue.[3,15,16] Intact osmophore tissue is able to selectively take up and retain this vital stain. This phenomenon is attributed to the increased permeability of the cell wall in osmophores and the long-lasting storage ability of vacuoles. Applied nonionic neutral red molecules (aqueous solution, 0.01%, pH 8) penetrate the tonoplast and enter the vacuole. Since vacuoles act like an ion trap because of their slight acidic environment, a migration of neutral red cations is not possible. Therefore permeability of the cell wall and cuticle, as well as the presence of vacuoles are the prerequisites for the distinct staining of floral tissue with neutral red, which often correlates with the presence of osmophore tissue in the flower. Sudan black B, Sudan III, and Sudan IV are useful dyes for the detection of triacylglycerides and protein-bound lipids. Osmium tetroxide and Nile blue A will also stain lipids and NADI reagent indicates terpenoids.[16,17] In addition to staining, morphologic, and anatomic investigations, the objective evidence of volatile emission determined by headspace and gas chromatography-mass spectrometry (GC-MS) are indispensable for the identification of osmophore tissues.

6.2.3 EXAMPLES OF OSMOPHORES

Vogel[3] revived research interest in volatile emanation, and since then interest in the phenomenon of scent emission has increased. Species that have been investigated belong to the families of Alliaceae,[18] Aristolochiaceae,[3] Asclepiadaceae,[3] Fabaceae,[19] Rutaceae,[20] Solanaceae,[21] and the intensively studied Araceae[3,11,13,14,22] and Orchidaceae[3–8,15,16,23]. In the following paragraphs, some selected representative examples are described.

One of the most comprehensive investigations within the Araceae were performed with *Sauromatum guttatum* (voodoo lily). As typical for Araceae, the inflorescence of *S. guttatum* is made up of an unbranched spadix subtended by a colored bract called a spathe. Tiny female flowers are located at the base of the spadix and are crowned by small club-shaped organs, whereas male flowers are found in the upper part of the spadix, which develops a prominent apical appendix. GC-MS analysis proved that both the club-shaped organs and the appendix represent osmophores responsible for volatile emission.[11,13,14] The upper and lower half of the appendix produce 163 different volatile compounds, whereas 43 volatiles are solely assigned to the upper and 4 to the lower half of the appendix. A total of 105 volatiles are emitted by the club-shaped organs, and 29 of them are exclusively released by these extraordinary organs. The most prominent constituents are monoterpenes like α- and β-pinene, mainly released by the lower part of the appendix and α-phellandrene and limonene, mainly released by the club-shaped organs. α-Terpinolene and linalool are almost exclusively emitted by the latter.[14]

Papillose epidermis cells of the appendix contain an extremely fine, dispersed vacuome and large nuclei.[3] At the beginning of volatile emission, the formerly compact epidermal cell layer generates large intercellular channels. The same applies to the subjacent production layer.[11] The cells of this layer contain numerous mitochondria and amyloplasts. Electron microscopic investigations of the appendix reveal the presence of rER associated with Golgi dictyosomes present before volatile emission. The rER and Golgi network seems to play a key role in the production, accumulation, and secretion of osmiophilic lipoid droplets.[11,13,22] Lipoid droplets are also observed in the cells of the glandular layers of the club-shaped organs, but are not present in epidermis cells.[14]

Flowers of the Orchidaceae show high morphologic differentiation, thus this family has the most polymorphic osmophore structures. Two striking examples are presented. The *Restrepia* inflorescence is single-flowered, with one slender dorsal and two fused lateral sepals, two antennae-forming petals with broadened apices, and a narrow, short column. The osmophore of *Restrepia antennifera* was originally identified as a palp-shaped adaxially thickened segment that separates the basal and distal parts of the dorsal sepal.[3] Detached dorsal sepals retrieve the foul odor of the entire inflorescence, while the other flower parts are scentless to the human nose. More recent investigations of *Restrepia hemsleyana*, *Restrepia muscifera*, and *Restrepia shuttleworthii*, however, assigned the osmophore character to the adaxial apex of the dorsal sepal and to the adaxial and abaxial apices of the petals.[15] The osmophore epidermis expresses very conspicuous pilei-forming cells (Figure 6.1). The cell head contains dense cytoplasm with a polylocular vacuome and some small starch grains, whereas the neck contains an enlarged nucleus. In older buds, epidermis cells undergo increasing vacuolation and some starch grains are associated with plastoglobuli. At anthesis, small lipoid droplets and osmiophilic aggregates appear inside the tonoplasts. These observations could be supplemented by electron microscopic investigations.[15] Cuticular pores of irregular size, shape, and arrangement are found in all adaxial osmophore areas, whereas the abaxial side of the dorsal sepal and all other perianth areas lack these pores. Cells of the subjacent glandular layers contain a smaller vacuome, and toward the epidermis, enlarged nuclei embedded in a dense cytoplasm. These cells develop rER and sER close to the plasmalemma and possess large numbers of amyloplasts. The starch agglomerations fade away during volatile evaporation, accompanied by vacuole enlargement and cytoplasm depletion. Only a thin peripheral layer of cytoplasm with very few amyloplasts, mitochondria, and rER/sER remain. Some large lipoid osmiophilic droplets can still be observed.

Osmophores are also found in *Stanhopea* sp. (Orchidaceae). The inflorescence consists of narrow petals, a lip divided into a basal hypochile, the horn-bearing mesochile, a distal epichile, and an elongated column flexed toward the lip. The osmophore lines the pouch of the hypochile and shows species-dependent epidermal structures. A flat to bullate surface is typical of *Stanhopea pulla*, a papillate surface is found in *Stanhopea candida*, a papillate to rugate surface is found in *Stanhopea tigrina*, a papillate surface with unicellular trichomes is found in *Stanhopea saccata*, and papillae with short, unicellular trichomes are present in *Stanhopea martiana*.[8] The ultrastructures of epidermal and subepidermal cells resemble mainly those of *Restrepia*.[3,6–8] Myrcene, α-pinene, β-pinene, α-terpineol, 1,8-cineol, methyl salicylate, phenylethyl acetate, and indol are frequently found scent compounds in *Stanhopea*.[23,24]

An osmophore might also be present in *Boronia megastigma* (brown boronia, Rutaceae), which is native to southwestern Australia. The tissues of various flower organs contain lysigenous oil glands, stigma, and androecium lack these characteristic glands. Instead, the adaxial surface of the stigma and the tips of the stamina show a papillate epidermis. Subjacent cell layers of all three organs are characterized by a dense cytoplasm and intracellular spaces. Fifty percent of the total scent is emitted by the stigma and the androecium. The compounds are almost exclusively β-ionone, dodecanol acetate, and *cis*-n-heptadecene, while the other flower parts emit mostly monoterpenes.[20,25,26]

Finally, a recently described example of a putative osmophore structure is found in *Gilliesia graminea* (Alliaceae). The striking similarity between flowers of this species and sexually deceitful orchids (e.g., *Ophrys*) caused closer investigation of this inflorescence, which also mimics insect shapes. Two thick structures extending abaxial from the staminal column (appendage 1) imitate the insect abdomen, and several narrow structures surrounding the staminal column appear as legs (appendage 2). The latter are thought to possess osmophores. Both appendage types display an epidermis with papillose cells (Figure 6.1).[18]

6.3 EPIDERMAL EMISSION

Osmophores with characteristic epidermal cells and subjacent glandular cell layers are found in a few, often highly evolved genera. Although typical osmophores are predestined for volatile production and emission, they are not always needed for scent emission. For example, the conspicuous absence of typical osmophores and papillate or rugose epidermis was observed for the well-scented *Clarkia breweri* (Onagraceae) and *Stephanotis floribunda* (Asclepiadaceae)[27] (Figure 6.2). Consequently, the question arises, how does a flower that apparently lacks a typical osmophore emit volatiles? The most striking difference is that the epidermis cells are involved in both biosynthesis and the emission of volatiles. Thus the osmophores' characteristic starch agglomerations and lipoid droplets are primarily found in epidermis cells and not in the underlying cell layers.[3,28] The adaxial epidermis often displays delicate bullate, conical, or papillose cells in order to facilitate volatile emission, but the multilayered secretory production and releasing tissues present in typical osmophores are missing.[29–31] Therefore emitting floral tissues of this type are often very fragile, comprising only a very few loosely connected mesophyll cells and layers that are dispersed by an extensive intercellular system (Figure 6.1).[29,31,32]

In this context, two methods of volatile emanation are discussed. Lipoid secretions emerge on the adaxial surface of the emitting tissue and subsequently evaporate

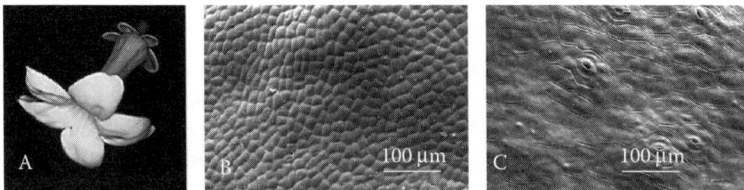

FIGURE 6.2 Petal epidermis of *S. floribunda*. (A) An individual *S. floribunda* flower. (B) The adaxial epidermis of the lobe exhibiting flat to slightly bullate cells (SEM). (C) The abaxial side of the lobe with a flat appearance of the epidermis. Some stomata are observed (SEM).

into the atmosphere (Figure 6.1B). Such secretions have been observed in *Rosa hybrida*. Often, an apparent secretion cannot be observed (e.g., *C. breweri*, *Antirrhinum majus* [snapdragon, Scrophulariaceae], and *Mirabilis jalapa* [four o'clock, Nyctaginaceae]) (Figure 6.1C), which indicates that evaporation is an immediate process.[29,31,33] Further details on the localization of scent synthesis in these species are presented in Section 6.5.

6.4   ORGAN- AND TISSUE-SPECIFIC EMISSION

More recent investigations have demonstrated that all flower organs are not equally employed in scent emission. Spatial differences within a flower (perianth, gynoecium, androecium) are quite common. Although petals are often the main source, and decisive for the whole flower bouquet, the stamen, pistil, and sepals contribute or are dominantly or even solely responsible for the emission of certain compounds. Examination of spatial emission patterns of *C. breweri* revealed that petals are mostly responsible for S-linalool, methyl eugenol, and methyl isoeugenol emission, whereas linalool oxide is released from the pistil. Although the benzenoid esters benzyl acetate, benzyl benzoate, and methyl salicylate are released from all flower parts, petals are responsible for most of the benzyl acetate/methyl salicylate emission, while the pistil is the primary source of benzyl benzoate release.[34,35,36] In *Chrysanthemum coronarium* (garland, Asteraceae), the compounds camphor and *cis*-chrysanthenyl acetate are primarily emitted by tubular and ligulate florets. The involucral bracts are responsible for sesquiterpene release, such as *trans*-α- and *trans*-β-farnesene, but also emit considerable amounts of myrcene and *cis*-ocimene.[37] Sesquiterpenes are also only found in sepal and gynoecium samples of *Rosa rugosa* (hedgehog rose, Rosaceae), whereas the constituents of the petal scent resemble the dominating compounds of the rose bouquet (citronellol, nerol, geraniol, 2-phenylethanol). In part, these petal volatiles are also retrieved in other flower parts, but only geranial and citronellol emission seems to be restricted to the petals.[38] Comparative analysis of volatiles emitted from different flower organs of *Ranunculus acris* (buttercup, Ranunculaceae) showed that petals, stamens, sepals, and gynoecium comprise identical volatiles, although they contribute different amounts; petals and stamens contribute the most.[39]

Remarkable also is the observation that emitting tissue sometimes comprises only certain parts of a floral organ. The basal (nectariferous) region of petals of *R. acris* is characterized by a higher emission of α-farnesene and 2-phenylethanol and an increased diversity of volatiles than in the apical region.[39] *M. jalapa*, a tropical plant that primarily emits *trans*-β-ocimene and small amounts of β-myrcene, *cis*-ocimene, *trans*-epoxy-ocimene, and benzyl benzoate, exhibits a perianth that consists of a tube and a five-lobed limb.[31,40,41] This corolla like calyx is divided into four sections (tube, transition zone between tube and limb, petaloid lobes, and a starlike center of the limb). The segments were separately examined by GC-MS analysis for scent emanation, revealing that the petaloid lobes are the region of highest emission of *trans*-β-ocimene, which correlates with a bullate epidermis on the adaxial side (Figure 6.1C and Figure 6.3).[31]

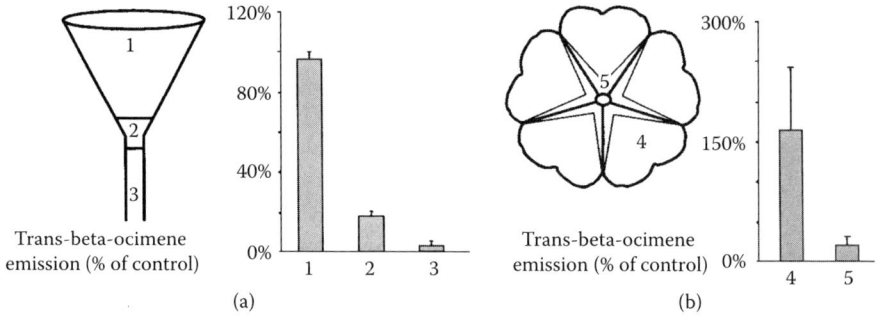

FIGURE 6.3 Localization of scent emission from *M. jalapa* flowers. (a) Schematic of the *M. jalapa* flower into (1) limb, (2) transition zone, and (3) tube. Relative emission of trans-β-ocimene in each floral part is presented. (b) Schematic of (4) petaloid lobe and (5) star-shaped center. Relative emission of trans-β-ocimene in each part is presented. The petaloid lobes are the primary source of ocimene emission.

Scents released by defined areas of flower organs are described, for example, for the wings of *Spartium junceum* (Spanish broom, Fabaceae), the vexillum of *Lupinus cruckshanksii* (Lupine, Fabaceae), or the paracorolla of *Narcissus jonquilla* (jonquil, Amaryllidaceae).[3,4] These areas are also stainable with neutral red and often such fragrance-emitting areas are congruent with ultraviolet (UV) light absorbing (visible) nectar guides, which represent an important cue to direct pollinators toward nectaries.[3,4] Lex[42] established the term "odor guide" for these fragrance-emitting areas and showed in her study that nectar guides were always correlated with odor guides.

Volatiles emitted by pollen are often remarkably different from those emitted by the other flower parts.[43] Pollen volatiles are significant constituents, but sometimes account for only a small part of the whole flower fragrance.[38] This phenomenon is reported for *R. acris*[39] as well as for *R. rugosa*,[38] *C. coronarium*,[37] *Filipendula vulgaris* (dropwort, Rosaceae), and *Lupinus polyphyllus* (lupine, Fabaceae).[43] Pollen volatiles comprise the same major classes of compounds known for floral scent, but most pollen odors are dominated by a few specific volatiles.[44] Protoanemonin is one of the typical compounds almost exclusively detected in pollen.[39] Furthermore, carbonylic compounds and long-chained linear hydrocarbons are present.[37,38,43]

6.5 GLANDULAR TRICHOMES

Volatile production and emission from vegetative tissue is linked to compartments such as glandular trichomes, oil glands, oil ducts, and cavities. Glandular and nonglandular trichomes have been well described for vegetative tissues.[45,47] These structural barriers and chemical weapons are important components of resistance to herbivores for plants in general.[48–51] Various types of glandular trichomes have been described (e.g., peltate, capitate, conoidal, and digitiform) that can be present on the same plant organ or tissue. Many plant species respond to insect damage by increasing the density or number of trichomes on new leaves.[52–60] Recent studies demonstrate that a jasmonic acid-dependent pathway also regulates the systemic increase in physical defenses such as trichomes in *Arabidopsis thaliana* or *Lycopersicon esculentum*.[61,62] Exogenously applied jasmonic acid up-regulates and salicylic acid down-regulates trichome formation on new leaves, while gibberellin and jasmonic acid exhibit a synergistic effect on trichome production. The *jai1* tomato mutant (homolog to the F-box protein coronative-insensitive *COI* from *A. thaliana*), which is defective in jasmonic acid signaling, shows several defense-related phenotypes, including the inability to express jasmonic acid-responsive genes, severely compromised resistance to two-spotted spider mites, and abnormal development of glandular trichomes.

Since floral organs are metamorphogenized leaves, it can be assumed that floral glandular trichomes exist that harbor volatile secondary metabolites that may function in defense as well as attraction. However, it turns out that the yield of essential oils and fragrances is generally low in floral tissue (e.g., rose, 0.075% w/v; acacia, 0.084% w/v; and jasmine, 0.04% w/v), which may be the result of a different primary emanation process in flowers compared to vegetative tissue.[45] Presently it remains unclear how much trichome volatiles contribute to floral fragrance compositions. In contrast to our knowledge of glandular and nonglandular trichomes in vegetative tissues, the literature is limited about the presence, absence, and distribution of glandular trichomes/glands in floral organs and tissues. Microscopic studies of the adaxial and abaxial epidermis of various plant species were performed. The well-scented *S. floribunda*, which emits at least 27 different compounds, with methyl benzoate, linalool, α-farnesene, benzyl benzoate, and 1-nitro-2-phenylethane as major volatiles, does not exhibit trichomes on either the adaxial or the abaxial petal epidermis (Figure 6.2).[63] Capitate hairs, but no peltate glands, are present on the abaxial, but not the adaxial petal site of *Nicotiana alata* and *Nicotiana suaveolens* (Figure 6.4 and Figure 6.5). Both Solanaceae species emit terpenoid-rich (e.g., sabinene, β-myrcene, limonene, *trans*-β-ocimene, 1,8-cineole) and benzenoid-rich (e.g., methyl benzoate, methyl salicylate) scents nocturnally.[64] Headspace scent collection of separated petal lobes and petal tubes followed by GC-MS analysis revealed that fragrances are only emitted from lobe tissue (Piechulla B. et al., unpublished results). Since the same type of trichomes are present on lobe and tube tissue, it seems very likely that they are not involved in the synthesis or emission of fragrance compounds.

Mattern and Vogel[65] showed that many Laminaceae have glandular trichomes on the corolla. *Plectranthus ornatus* (Laminaceae), cultivated as an ornamental or as a source of essential oils, shows an unusual conoidal trichome with long unicellular conical heads on the calyx and the corolla.[17,66] On stamens and carpels, peltate trichomes are numerous. On the calyx, which is two-lipped (an upper and lower lip with four small teeth) capitate, digitiform, and conodial trichomes are clustered on the abaxial calyx surface, while peltate trichomes are scarce and are restricted to the periphery of the lips. Digitiform and conoidal trichomes are the most conspicuous trichome types on the adaxial calyx surface. Histochemical studies reveal the presence of hyaline, a slightly viscous or orange-brown secretion in these glandular trichomes. The capitate and conoidal trichomes stain positively for lipophilic and hydrophilic substances, the digitiform types give positive reactions for hydrophilic substances, and peltate trichomes give positive reactions for lipophilic and terpenoid

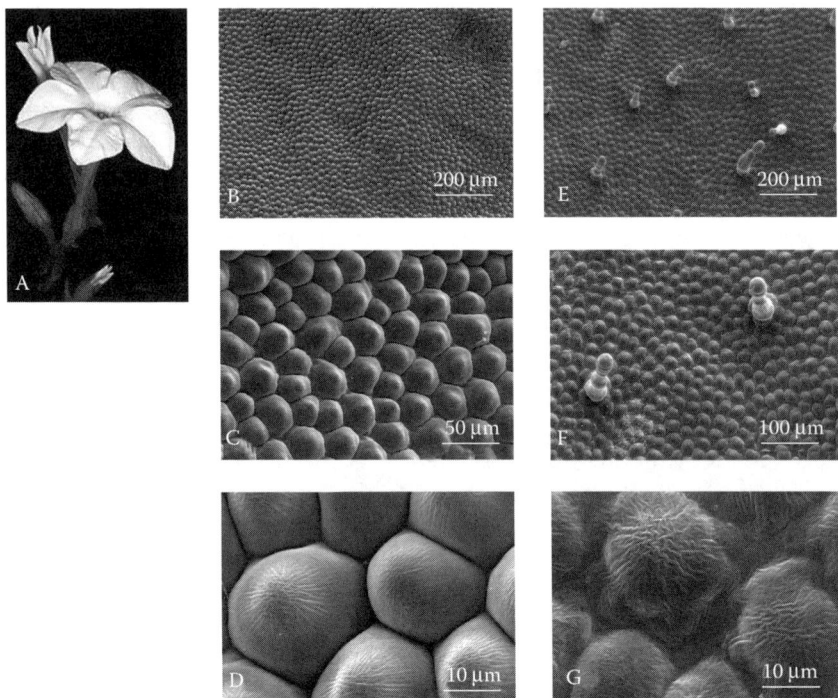

FIGURE 6.4 Petal epidermis of *N. alata*. (A) An individual *N. alata* flower. (B–D) The adaxial epidermis of the lobe exhibiting conical cells with a wrinkled cuticle at the tip (SEM, increasing magnification). (E–G) The abaxial epidermis of the lobe exhibiting bullate epidermis cells and several capitate trichomes. Higher magnifications show the wrinkled cuticle in the center of the epidermis cells.

compounds. Thus the trichomes of *P. ornatus* produce various amounts of secretory materials; however, volatiles have yet not been reported.

A large number of variable glandular hairs are found on the reproductive organs of *Salvia* species (Laminaceae).[67] *Salvia dominica* peltate hairs, almost exclusively on the abaxial site of the calyx, are responsible for the secretion of neryl acetate, α-terpineol, and α-terpinyl acetate as major compounds, while myrcene, 1,8-cineole, and β-pinene are minor compounds. The stalky hairs are present on both sides of the corolla and produce relatively large amounts of linalyl acetate. Linalool and linalyl acetate appear in large amounts in abaxial hairs on the calyces, bracts, and peduncles of *Salvia sclarea*. The different morphological structures of the prominent scent-secreting hairs on the inflorescence rather than the peltate hairs on the vegetative tissue led Werker et al.[67] to suggest that they may have a different function (e.g., in luring specific pollinators).

The adaxial epidermis of *Phragmopedilum grande* (Orchidaceae, hybrid: *Phragmopedilum caudatum* × *Phragmopedilum longifolium*) of the distal part of two extremely elongated and twisted petals is supplied with unique uniseriate trichomes.[3] The trichomes consist of seven to eight cells and were found to be the source of the foul-smelling scent. The trichomes, present on completely unrolled petals, often

FIGURE 6.5 Petal epidermis of *N. suaveolens*. (A) An individual *N. suaveolens* flower. (B) The adaxial epidermis of the lobe exhibiting bullate epidermis cells (SEM). (C) The abaxial epidermis with flat epidermis cells and several capitate trichomes (SEM). (D) Higher magnification of the capitate trichome (SEM).

exhibit one or two dead brown cells at the tip, while the other cells remain turgescent and vital. The latter contain a large nucleus, a dispersed vacuome, and small starch grains. During flowering, when the apical cells become necrotic, terpenoid secretions can be observed. This is one of the few floral trichomes that have excretion ability and a typical osmophore function.

Trichomes on the osmophore of *Stanhopea saccata* (Orchidaceae) have also been observed; however, their functions remain unknown.[8] Secretory glands with oil-filled subcuticular space are observed on florets and developing ovules of *Chamaemelum nobile* (chamomile, Compositae), and endogenous oil glands are present in petals of *Syzygium aromaticum* (clove, Myrtaceae).[45] Vegetative excretory structures integrated into floral tissue are found in *Dictamnus albus* (burning bush, Rutaceae), having oil glands at the filaments, and in *Boronia megastigma* (Rutaceae), having lysigenous oil glands on the petal epidermis and in the mesophyll of bracts, sepals, petals, the receptacle, ovary, and nectary.[3,20,25,26] The latter glands are the site of α- and β-pinene and limonene production, monoterpenes which are found in the volatile blend of *B. megastigma*.

In the flower tube of *Antirrhinum majus*, two stripes of yellow hairs are located on the boundaries between the ventral and lateral petals. These hairs are on the adaxial epidermal surface. Environmental scanning and light microscopy revealed that these hairs are unicellular, consisting of a long stalk and a head. Cross-sections incubated with antibodies against an enzyme involved in scent production (benzoic acid methyltransferase [BAMT]) stain the head of these hairs, indicating the presence of this biosynthetic enzyme.[29]

Microscopic studies of *M. jalapa* showed that a large number of uniseriate and multicellular capitate trichomes are present on the abaxial site of the petaloid lobe (Figure 6.6A).[31] Scanning electron microscopic investigations implied a glandular character of the trichome because of the enlarged headlike cell. After electron beam disruption, organic matter leaking from the trichome was observed. Trichomes from *M. jalapa* were collected and analyzed by GC-MS. Surprisingly, not *trans*-β-ocimene as expected, but β-farnesene was found, a substance that is well known as a chemical defense compound.[68–70]

Anthers, which present the pollen, are important for the reproductive success of the plant. To limit destruction of the pollen or stamen by herbivores and pathogens, it is therefore not surprising that glandular trichomes are frequently found on the anthers. The anthers of the four fertile stamens of *Leonorus sibiricus* bear small glandular scales that rupture at the slightest touch and release a sticky substance. Their location strongly suggests a role in the production of adhesive substances, but release of volatiles has not been shown.[71] A few other reports also demonstrate that such anther glands provide an accessory pollenkitt (*Cyclanthera*, Curcubitaceae[72]; *Hedyhinum*, Zingiberaceae[73]; *Drymonia*, Gesneriaceae[74]). Anther glands present on the connectives found in *Cyphomandra* (tamarillo, Solanaceae) produce a reward for euglossine bees. Sazima et al.[21] analyzed the volatiles and the flower morphology and showed that the dorsal papillate epidermis is attached to glandular mesophyll cell layers that are responsible for production and secretion of 1,8-cineol, *trans*-β-ocimene, and germacrene D, and resembling typical osmophore characteristics. Anther glands are also quite common in Mimosoideae. A comprehensive survey of anther glands in this tribe showed that among four gland types, the unusual conical type on the ventral side of the gland just above the anther sacs was only found in *Pentaclethra macroloba* (Parkieae, Mimosoieae, Fabaceae).[19] Although the function is still unknown, the authors speculate that this structure may be involved in volatile production and emission.

Glandular trichomes present on floral organs may be a source of floral volatiles. Many investigations have shown a correlation between trichome appearance on floral tissue and scent emission; however, to our knowledge, a definite volatile presence in floral trichomes has only been demonstrated with *M. jalapa*. Since the *Mirabilis* floral trichomes contain a compound with a biological defense function, the idea is put forward that typical floral pollinator-attracting volatiles are emitted differently compared to the characteristic glandular trichome-based vegetative defense compounds. Further investigations are needed to support or reject this hypothesis.

6.6 CELLULAR AND SUBCELLULAR LOCALIZATION OF SCENT BIOSYNTHESIS

Osmophores, epidermis, and different floral organs are the structures that are the sources of fragrance emission. Recent progress in elucidating the biosynthetic pathways of volatiles has provided further tools for detailed localization of scent synthesis and emission. In the past 10 years, several enzymes involved in volatile biosynthesis

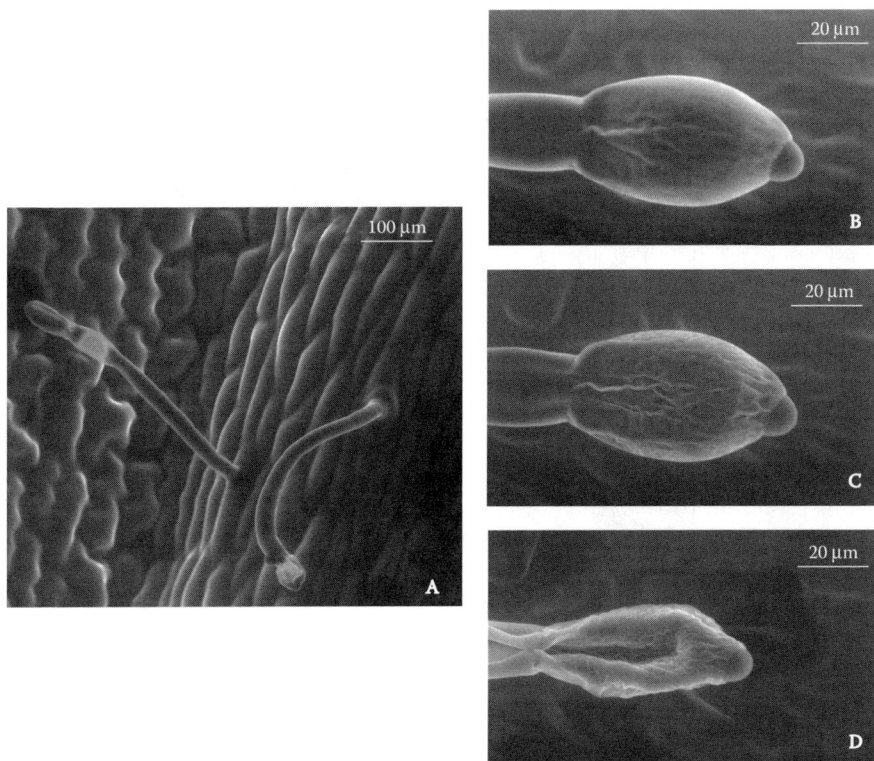

FIGURE 6.6 Trichomes of *M. jalapa*. (A) Trichomes of the abaxial epidermis of *M. jalapa*. Trichomes are localized on the veins (environmental scanning electron microscope [ESEM]). (B) Higher magnification of the trichome head (ESEM). (C,D) Collapsing trichome head after electron beam disruption (ESEM).

of terpenoids, benzenoids, and fatty acid derivatives have been isolated and characterized. In most cases, the respective genes have been cloned and analyzed, and now are available as molecular tools.

The presence of biosynthetic enzymes or their respective transcripts in particular floral organs and tissues is presently the best indicator for volatile synthesis. Northern blots performed with a probe against the benzenoid carboxyl methyltransferase (BSMT) from *N. suaveolens* show that the respective messenger RNAs (mRNAs) primarily accumulate in petal tissue and only to very small extent in other floral parts (Figure 6.7). Such organ-specific accumulation patterns are also found in snapdragon (*A. majus*) and petunia.[75–77] The expression of acetyl-CoA:benzyl alcohol acetyltransferase (BEAT) and linalool synthase (LIS) from *C. breweri* is not so organ specific, since transcripts are detectable in petals, sepals, stamen, stigma, and style to different degrees.[33,78] Separation of the perianth of snapdragon revealed significantly higher transcript levels of the ocimene synthase in lower lobe tissue.[76] In addition to the transcript appearance in extracts of floral tissue, the presence of enzyme activities in flower extracts is taken as an indicator of localized scent synthesis (e.g., in snapdragon and *S. floribunda*).[63,75] Furthermore, the molecular tools can be

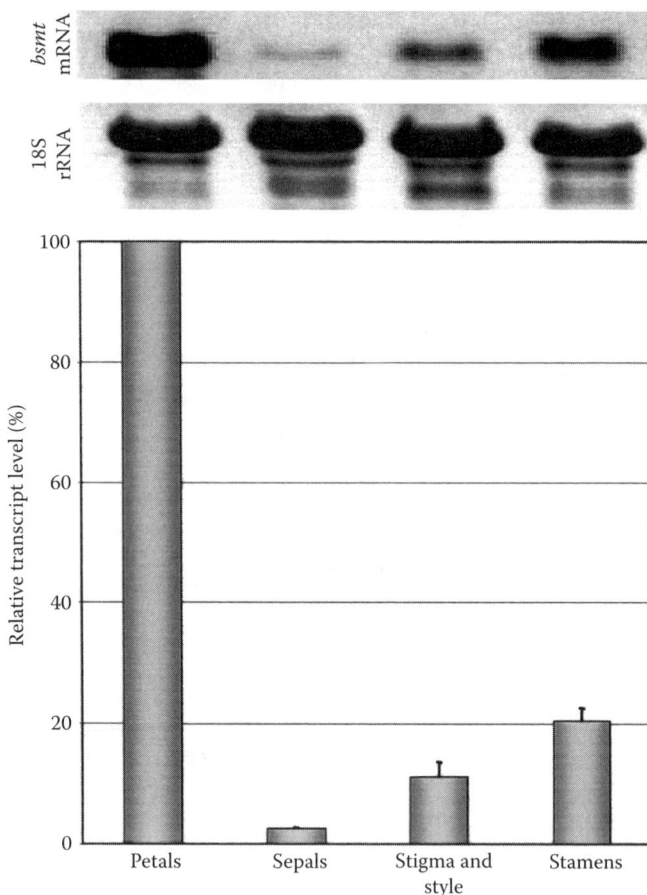

FIGURE 6.7 Expression of benzoic/salicylic acid methyltransferase (BSMT) in floral organs of *N. suaveolens*. Total RNA isolated from different floral organs was hybridized with a *BSMT* gene-specific probe. Differential mRNA accumulation is observed in the different tissues relative to 18S rRNA levels. Relative transcript levels are calculated. The highest level in petals was 100%.

used to trace respective transcripts and enzymes at the cellular and subcellular level. The spatial distribution of the LIS transcripts in *C. breweri* buds, pistils, and petals was performed by *in situ* hybridization using sense and antisense RNA probes.[33] In cross-sections of the flower, it can be seen that LIS mRNA transcripts are mainly concentrated in the secretory zone of the four-lobed stigma and also in the epidermal layers of the petals. Since up to 70% of the total LIS activity of *C. breweri* flowers is found in the petals, they are regarded as the major source of emitted linalool.[33,34,79] The petal epidermal cells produce water-insoluble linalool on the surface, from which it can most easily escape into the atmosphere. Thus it appears that *C. breweri* has evolved its ability to emit large amounts of linalool simply by highly expressing LIS

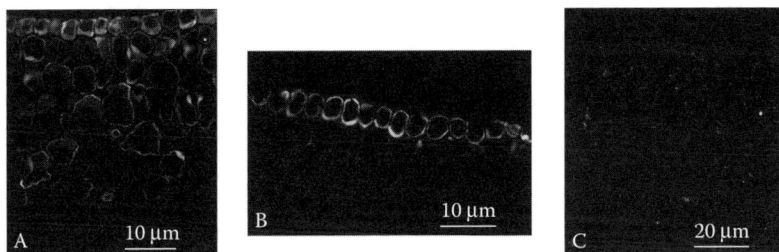

FIGURE 6.8 Immunofluorescence localization of SAMT in *S. floribunda* petals. Cross-sections of the (A) *S. floribunda* lobe, (B) transition zone (upper part of the tube), and (C) lower part of the tube were incubated with antibodies against the salicylic acid methyltransferase (SAMT) and FITC-labeled secondary antibodies.

in the epidermal cells of the petals, without the concomitant development of specialized scent glands.[33]

The distribution of volatile synthesizing enzymes in floral tissue has only been investigated with *S. floribunda* (Figure 6.8) and *A. majus*.[29] Salicylic acid methyltransferase (SAMT), the enzyme that methylates salicylic acid using *S*-adenosyl-L-methionine as a methyl donor, was found in the petals of *S. floribunda* when specific antibodies against the SAMT enzyme and fluorescein-labeled (FITC) secondary antibodies were incubated with thin sections from various petal regions (petal lobe, upper part of the tube, lower part of the tube). The SAMT enzyme is primarily present in the epidermal cells of the petal lobe, but underlying cell layers also stain to some extent (Figure 6.8A). Since epidermis as well as subjacent cell layers express the SAMT enzyme, it is likely that a typical osmophore, in the sense of Vogel's definition, exists in *S. floribunda*, although only parts of the petals are involved. Interestingly, in the upper part of the flower tube, the SAMT is exclusively restricted to the adaxial epidermal cells (Figure 6.8B), and no enzyme could be detected in the lower part of the petal tube (Figure 6.8C). These experiments clearly define the petal lobes of the *Stephanotis* flower as the area where methyl salicylate synthesis occurs, which correlates well with GC-MS analysis (Piechulla B. et al., unpublished results). A similar enzyme from snapdragon flowers to the one mentioned above, benzoic acid methyltransferase (BAMT), turns out to be epidermis specific. Both the adaxial and abaxial epidermal petal cells are differentially involved in scent biosynthesis.[29] Methyl benzoate is predominantly produced in the conical cells of the adaxial epidermis of the lower lobe and tube. Apparently the cells between both epidermata do not contain much BAMT enzyme. They form a very loose structure with large intercellular spaces. BAMT expression was also found in the yellow hairs within the tube located on the bee's way to the nectar. Subcellular localization with immunogold-labeled antibodies localizes the BAMT in the cytoplasm of the epidermal cells, adjacent to the primary cell wall.[29]

As more enzymes and genes of the biosynthetic pathways of floral scent compounds become available, the more *in situ* hybridization and immunofluorescence experiments can be performed that will help to clarify our understanding of floral scent synthesis and emission on the cellular level.

6.7   CONCLUSION

It is imprecise to identify the flower as the source of scent emission. For many insects, defined guidance and orientation within the flower is absolutely necessary. Therefore it is not surprising that many investigations clearly demonstrate that not only flower organs (e.g., classical osmophores), but also certain areas or parts of a floral organ emit distinct scents or scent compositions. Modern techniques allow us to obtain detailed information about which floral organs and tissues, and in which cells or cell layers scent synthesizing enzymes or transcripts are present. The precise localization of scent synthesis in many plant species will provide helpful information to further understand transport and transport mechanisms, as well as the process of scent emission in floral tissues. Furthermore, such investigations might support and clarify the present view that glandular trichomes on vegetative as well as floral tissue produce defensive volatiles, while the pollinator attracting volatiles are synthesized and emitted from classical osmophores or from other distinct floral tissues.

ACKNOWLEDGMENTS

The authors thank Claudia Dinse (Rostock University, Rostock, Germany) for technical assistance and Bettina Hause (Institute for Plant Biochemistry, Halle, Germany) for technical support. Figure 6.1B was provided by Fred Ehrig (Federal Center for Breeding Research on Cultivated Plants, Institute of Resistance Research and Pathogen Diagnostics, Aschersleben, Germany) and Figure 6.7 was from Saudra Sascheubrecher (Max Planck Institute for Biochemistry, Martinsried, Germany). Financial support was provided by the Deutsche Forschungsgemeinschaft (German Research Foundation) to B.P.

REFERENCES

1. Knudsen, J.T., Tollsten, L., and Bergström, G., Floral scents: a checklist of volatile compounds isolated by head-space techniques, *Phytochemistry* 33, 253, 1993.
2. Arcangeli, D.I.G., Osservazioni sull'impollinazione in alcune aracee, *Nuovo Giorn. Bot. Ital.* 7, 72, 1883.
3. Vogel, S., Duftdrüsen im Dienste der Bestäubung, *Akad. Wiss. Lit. Mainz Math.-Nat. Klasse* 10, 600, 1962.
4. Vogel, S., *The Role of Scent Glands in Pollination: On the Structure and Function of Osmophores*, Amerind, New Delhi, India, 1990.
5. Pridgeon, A.M. and Stern, W.L., Osmophores of *Scaphosepalum* (Orchidaceae), *Bot. Gaz.* 146, 115, 1985.
6. Stern, W.L., Curry, K.J., and Pridgeon, A.M., Osmophores of *Stanhopea* (Orchidaceae), *Am. J. Bot.* 74, 1323, 1987.
7. Curry, K.J., Initiation of terpenoid synthesis in osmophores of *Stanhopea anfracta* (Orchidaceae): a cytochemical study, *Am. J. Bot.* 74, 1332, 1987.
8. Curry, K.J., Stern, W.L., and McDowell, L.M., Osmophore development in *Stanhopea anfracta* and *S. pulla* (Orchidaceae), *Lindleyana* 3, 212, 1988.
9. Curry, K.J., McDowell, L.M., Judd, W.S., and Stern, W.L., Osmophores, floral features, and systematics of *Stanhopea* (Orchidaceae), *Am. J. Bot.* 78, 610, 1991.

10. Davies, K.L. and Turner, M.P., Morphology of floral papillae in *Maxillaria Ruiz &* *Pav.* (Orchidaceae), *Ann. Bot.* 93, 75, 2004.
11. Skubatz, H., Kunkel, D.D., Patt, J.M., Howald, W.N., Hartman, T.G, and Meeuse, B.J.D., Pathway of terpene excretion by the appendix of *Sauromatum guttatum*, *Proc. Natl. Acad. Sci. USA* 92, 10084, 1995.
12. Hudak, K.A. and Thompson, J.E., Subcellular localization of secondary lipid metabolites including fragrance volatiles in carnation petals, *Plant Physiol.* 114, 705, 1997.
13. Skubatz, H., Kunkel, D.D., Howald, W.N., Trenkle, R., and Mookherjee, B., The *Sauromatum guttatum* appendix as an osmophore: excretory pathways, composition of volatiles and attractiveness to insects, *New Phytol.* 134, 631, 1996.
14. Hadacek, F. and Weber, M., Club-shaped organs as additional osmophores within the *Sauromatum inflorescence*: odour analysis, ultrastructural changes and pollination aspects, *Plant Biol.* 4, 367, 2002.
15. Pridgeon, A.M. and Stern, W.L., Ultrastructure of osmophores in *Restrepia* (Orchidaceae), *Am. J. Bot.* 70, 1233, 1983.
16. Stern, W.L., Curry, K.J., and Whitten, W.M., Staining fragrance glands in orchid flowers, *Bull. Torrey Bot. Club* 113, 288, 1986.
17. Ascensão, L., Mota, L., and de M. Casto, M., Glandular trichomes on the leaves and flowers of *Plectranthus ornatus*: morphology, distribution and histochemistry, *Ann. Bot.* 84, 437, 1999.
18. Rudall, P.J., Bateman, R.M., Fay, M.F., and Eastman, A., Floral anatomy and systematics of Alliaceae with particular reference to *Gilliesia*, a presumed insect mimic with strongly zygomorphic flowers, *Am. J. Bot.* 89, 1867, 2002.
19. Luckow, M. and Grimes, J., A survey of anther glands in the mimosoid legume tribes Parkieae and Mimoseae, *Am. J. Bot.* 84, 285, 1997.
20. Bussell, B.M., Considine, J.A., and Spadek, Z.E., Flower and volatile oil ontogeny in *Boronia megastigma*, *Ann. Bot.* 76, 457, 1995.
21. Sazima, M., Vogel, S., Cocucci, A.A., and Hausner, G., The perfume flowers of *Cyphomandra* (Solanaceae): pollination by euglossine bees, bellows mechanism, osmophores, and volatiles, *Plant Syst. Evol.* 187, 51, 1993.
22. Skubatz, H. and Kunkel, D.D., Further studies of the glandular tissue of the *Sauromatum guttatum* (Araceae) appendix, *Am. J. Bot.* 86, 841, 1999.
23. Whitten, W.M. and Williams, N.H., Floral fragrance of *Stanhopea* (Orchidaceae), *Lindleyana* 7, 130, 1992.
24. MacTavish, H.S. and Menary, R.C., Volatiles in different floral organs, and effect of floral characteristics on yield of extract from *Boronia megastigma* (Nees), *Ann. Bot.* 80, 305, 1997.
25. MacTavish, H.S., Davies, N.W., and Menary, R.C., Emission of volatiles from brown *Boronia* flowers: some comparative observations, *Botany* 86, 347, 2000.
26. Raguso, R.A. and Pichersky, E., A day in the life of a linalool molecule: chemical communication in a plant-pollinator system. Part 1: Linalool biosynthesis in flowering plants, *Plant Spec. Biol.* 14, 95, 1999.
27. Mazurkiewicz, W., Über die Verteilung des ätherischen Oeles im Blütenparenchym und über seine Lokalisation im Zellplasma, *Zeitschr. Allgem. österr. Apotheker-Vereins* 23, 805, 1913.
28. Kolosova, N., Sherman, D., Karlson, D., and Dudareva, N., Cellular and subcellular localization of S-adenosyl-L-methionine: benzoic acid carboxyl methyltransferase, the enzyme responsible for biosynthesis of the volatile ester methylbenzoate in snapdragon flowers, *Plant Physiol.* 126, 956, 2001.

29. Lopez, H.A. and Galetto, L., Flower structure and reproductive biology of *Bougainvillea stipitata* (Nyctaginaceae), *Plant Biol.* 4, 508, 2002.
30. Effmert, U., Große, J., Röse, U., Ehrig, F., Kägi, R., and Piechulla, B., Volatile composition, emission pattern and localization of floral scent emission in *Mirabilis jalapa* (Nyctaginaceae), *Am. J. Bot.* 92, 2, 2005.
31. Goodwin, S.M., Kolosova, N., Kish, C.M., Wood, K.V., Dudareva, N., and Jenks, M.A., Cuticle characteristics and volatile emission of petals in *Antirrhinum majus*, *Physiol. Plant.* 117, 435, 2003.
32. Dudareva, N., Cseke, L., Blanc, V.M., and Pichersky, E., Evolution of floral scent in *Clarkia*: novel patterns of S-linalool synthase gene expression in the *C. breweri* flower, *Plant Cell* 8, 1137, 1996.
33. Pichersky, E., Raguso, R.A., Lewinsohn, E., and Croteau, R., Floral scent production in *Clarkia* (Onagraceae). I. Localization and developmental modulation of monoterpene emission and linalool synthase activity, *Plant Physiol.* 106, 1533, 1994.
34. Wang, J., Dudareva, N., Bhakta, S., Raguso, R.A., and Pichersky, E., Floral scent production in *Clarkia breweri* (Onagraceae). II. Localization and developmental modulation of the enzyme S'-adenosyl-L-methionine:(iso)eugenol O-methyltransferase and phenylpropanoid emission, *Plant Physiol.* 114, 213, 1997.
35. Dudareva, N., Raguso, R.A., Wang, J., Ross, J.R., and Pichersky, E., Floral scent production in *Clarkia breweri*. III. Enzymatic synthesis and emission of benzenoid esters, *Plant Physiol.* 116, 599, 1998.
36. Flamini, G., Cioni, P.L., and Morelli, I., Differences in the fragrances of pollen, leaves, and floral parts of garland (*Chrysanthemum coronarium*) and composition in the essential oils from flower heads and leaves, *J Agric. Food. Chem.* 51, 2267, 2003.
37. Dobson, H.E.M., Bergström, G., and Groth, I., Differences in fragrance chemistry between flower parts of *Rosa rugosa* Thunb. (Rosaceae), *Isr. J. Bot.* 39, 143, 1990.
38. Bergström, G., Dobson, H.E.M., and Groth, I., Spatial fragrance patterns within the flowers of *Ranunculus acris* (Ranunculaceae), *Plant Syst. Evol.* 195, 221, 1995.
39. Heath, R.R. and Manukian, A., An automated system for use in collecting volatile chemicals released from plants, *J. Chem. Ecol.* 20, 593, 1994.
40. Levin, R.A., McDade, L.A., and Raguso, L.A., The systematic utility of floral and vegetative fragrance in two genera of Nyctaginaceae, *Syst. Biol.* 52, 334, 2003.
41. Lex, T., Duftmale an Blüten, *Zeitschr. Vergl. Physiol.* 36, 212, 1954.
42. Dobson, H.E.M., Groth, I., and Bergström, G., Pollen advertisement: chemical contrasts between whole-flower and pollen odors, *Am. J. Bot.* 83, 877, 1996.
43. Dobson, H.E.M. and Bergström, G., The ecology and evolution of pollen odors, *Plant Syst. Evol.* 222, 63, 2000.
44. Svoboda, K. and Svoboda, T., *Secretory Structures of Aromatic and Medicinal Plants: A Review and Atlas of Micrographs*, Microscopix Publications, Knighton 2000, p. 3.
45. Gang, D.R., Wang, J., Dudareva, N., Nam, K.H., Simon, J.E., Lewinsohn, E., and Pichersky, E., An investigation of the storage and biosynthesis of phenylpropanes in sweet basil, *Plant. Physiol.* 125, 539, 2001.
46. Pichersky, E. and Gershenzon, J., The formation and function of plant volatiles: perfumes for pollinator attraction and defense, *Curr. Opin. Biol.* 5, 237, 2002.
47. Levin, D.A., The role of trichomes in plant defense, *Q. Rev. Biol.* 48, 3, 1973.
48. Agren, J. and Schemske, D.W., Evolution of trichome number in a naturalized population of *Brassica rapa*, *Am. Nat.* 143, 1, 1994.
49. Fernandes, G.W., Plant mechanical defenses against insect herbivory, *Rev. Bras. Entomol.* 38, 421, 1994.

50. Mauricio, R. and Rausher, M.D., Experimental manipulation of putative selective agents provides evidence for the role of natural enemies in the evolution of plant defense, *Evolution* 51, 1435, 1997.

51. Myers, J.H. and Bazely, D.R., Thorns, spines, prickles, and hairs: are they stimulated by herbivory and do they deter herbivores, in *Phytochemical Induction by Herbivores*, Raupp, M.J. and Tallamy, D.W., Eds., Wiley, New York, 1991, p. 325.

52. Agrawal, A.A., Induced responses to herbivory and increased plant performance, *Science* 279, 1201, 1998.

53. Agrawal, A.A., Induced responses to herbivory in wild radish: effects on several herbivores and plant fitness, *Ecology* 80, 1713, 1999.

54. Agrawal, A.A., Benefits and costs of induced plant defense for *Lepidium virginicum* (Brassicaceae), *Ecology* 81, 1804, 2000.

55. Pullin, A.S. and Gilbert, J.E., The stinging nettle, *Urtica dioica*, increases trichome density after herbivore and mechanical damage, *Oikos* 54, 275, 1989.

56. Baur, R., Binder, S., and Benz, G., Nonglandular leaf trichomes as short-term inducible defense of the grey alder, *Alnus incana* (L.), against the chrysomelid beetle, *Agelastica alni.* L., *Oecologia* 87, 219, 1991.

57. Traw, M.B., Is induction response negatively correlated with constitutive resistance in black mustard?, *Evolution* 56, 2196, 2002.

58. Traw, M.B. and Dawson, T.E., Differential induction of trichomes by three herbivores of black mustard, *Oecologia* 131, 526, 2002.

59. Traw, M.B. and Dawson, T.E., Reduced performance of two specialist herbivores (Lepidoptera: Pieridae, Coleoptera: Chrysomelidae) on new leaves of damaged black mustard plants, *Environ. Entomol.* 31, 714, 2002.

60. Traw, M.B. and Bergelson, J., Interactive effects of jasmonic acid, salicylic acid, and gibberellin on induction of trichomes in *Arabidopsis*, *Plant Physiol.* 133, 1367, 2003.

61. Li, L., Zhao, Y., McCaig, B.C., Wingerd, B.A., Wang, J., Whalon, M.E., Pichersky, E., and Howe, G.A., The tomato homolog of CORONATINE-INSENSITIVE1 is required for the maternal control of seed maturation, jasmonate-signaled defense responses, and glandular trichome development, *Plant Cell* 16, 126, 2004.

62. Pott, M.B., Pichersky, E., and Piechulla, B., Evening specific oscillations of scent emission, SAMT enzyme activity, and SAMT mRNA in flowers of *Stephanotis floribunda*, *J. Plant. Physiol.* 159, 925, 2002.

63. Raguso, R.A., Levin, R.A., Foose, S.E., Holmberg, M.W., and McDade, L.A., Fragrance chemistry nocturnal rhythms and pollination "syndromes" in *Nicotiana*, *Phytochemistry* 63, 265, 2003.

64. Mattern, G. and Vogel, S., Lamiaceen-Blüten duften mit dem Kelch – Prüfung einer Hypothese. I. Anatomische Untersuchungen. Vergleich der Laub- und Kelchdrüsen, *Beitr. Biol. Pflanzen* 68, 125, 1994.

65. Ascensão, L., Figueiredo, A.C., Barroso, J.G., Pedro, L.G., Schripsema, J., Deans, S.G., and Scheffer, J.C.J., *Plectranthus madagascariensis*: morphology of the glandular trichomes, essential oil composition, and its biological activity, *Int. J. Plant Sci.* 159, 31, 1998.

66. Werker, E., Ravid, U., and Putievsky, E., Glandular hairs and their secretions in the vegetative and reproductive organs of *Salvia sclarea* and *S. dominica*, *Isr. J. Bot.* 34, 239, 1985.

67. Gibson, R.W. and Pickett, J.A., Wild potato repels aphids by release of aphid alarm pheromone, *Nature* 302, 608, 1983.

68. Ave, D.A., Gregory, P., and Tingey, W.M., Aphid repellent sesquiterpenes in glandular trichomes of *Solanum berthaultii* and *S. tuberosum*, *Entomol. Exp. Appl.* 44, 131, 1987.
69. Mondor, B., Baird, S., Slessor, K., and Roitberg, B., Ontogeny of alarm pheromone secretion in pea aphid, *Acyrthosiphon pisum*, *J. Chem. Ecol.* 26, 2875, 2000.
70. Moyano, F., Cocucci, A.A., and Sérsic, A.N., Accessory pollen adhesive from glandular trichomes on the anthers of *Leonurus sibiricus* L. (Lamiaceae), *Plant Biol.* 5, 411, 2003.
71. Vogel, S. Die Klebstoffhaare an den Antheren von *Cyclanthera pedata* (Curcurbitaceae), *Plant Syst. Evol.* 137: 291, 1981.
72. Vogel, S. Blütensekrete als akzessorischer Pollenkitt, *Mitteil. Botaniker-Tagung Wien* 123, 1984.
73. Steiner, K.E., The role of nectar and oil in the pollination of *Drymonia serrulata* (Gesneriaceae) by Epicharis bees (Anthophoridea) in Panama, *Biotropica* 17, 217, 1985.
74. Dudareva, N., D'Auria, J.C., Nam, K.H., Raguso, R.A., and Pichersky, E., Developmental regulation of methyl benzoate biosynthesis and emission in snapdragon flowers, *Plant Cell* 12, 949, 2000.
75. Dudareva, N., Martin, D., Kish, C.M., Kolosova, N., Gorenstein, N., Fäldt, J., Miller, B., and Bohlmann, J., (*E*)-β-ocimene and myrcene synthase genes of floral scent biosynthesis in snapdragon: function and expression of three terpene synthase genes of a new terpene synthase subfamily, *Plant Cell* 15, 1227, 2003.
76. Negre, F., Kish, C., Boatright, J., Underwood, B., Shibuya, K., Wagner, C., Clark, D., and Dundareva, N., Regulation of methylbenzoate emission after pollination in snapdragon and petunia flowers, *Plant Cell* 15, 1, 2003.
77. Dudareva, N., D'Auria, J.C., Nam, K.H., Raguso, R.A., and Pichersky, E., Acetyl-CoA:benzylalcohol acetyltransferase: an enzyme involved in floral scent production in *Clarkia breweri*, *Plant J.* 14, 297, 1998.
78. Raguso, R.A. and Pichersky, E., Floral volatiles from *Clarkia breweri* and *C. concinna* (Onagraceae): recent evolution of floral scent and moth pollination, *Plant Syst. Evol.* 194, 55, 1995.

7 Examination of the Processes Involved in the Emission of Scent Volatiles from Flowers

Reinhard Jetter

CONTENTS

7.1   INTRODUCTION

In earlier chapters of this book, two major approaches were summarized that have contributed greatly to our current knowledge about floral scent formation. On one hand, diverse chemical studies have documented the accumulation of volatile molecules in the headspace around flowers, in many cases detailing scent composition as a function of floral development, circadian rhythm, or environmental conditions. On the other hand, much new insight has been gained into the molecular processes that are relevant for the biosynthesis of floral volatiles, allowing the first inferences on the subcellular localization of the enzymes involved. These two very successful approaches focused on two distinct steps in scent production, separated in space and time. Information from both approaches can now be combined to assess the transport of scent molecules between both locations, that is, between the site of biosynthesis inside the cells of flower tissues or specialized structures such as osmophores or glands (Chapter 6; see also Vogel[1]) and the surface of the floral organ, where they evaporate into the surrounding airspace. The transport steps occurring between biosynthesis and evaporation will be the topic of this chapter.

Diverse methods have been employed to show in which floral organs scent molecules are produced in various plant species. All studies taken together show how temporally and spatially regulated expression of scent biosynthesis genes causes the selective formation of individual scent constituents only in certain floral organs, or even only in restricted parts of these organs.

Further molecular genetic investigations, mostly performed on *Clarkia breweri* and *Antirrhinum majus*, showed that genes relevant for scent formation are expressed uniformly and almost exclusively in cells of the epidermal layer of various floral parts.[2-4] Either idioblastic epidermal cells or groups of cells, or the majority of epidermal pavement cells can be responsible for floral scent production, depending on the species and the specific scent components being biosynthesized. As those epidermal areas involved in floral scent formation are lacking stomatal pores, the volatile molecules must be emitted directly from the surface of the epidermis cell in which they have been synthesized. The epidermal cells of floral organs are covered by a cuticle, a thin, continuous layer of lipids deposited onto the periclinal cell walls (Figure 7.1). Hence scent emission occurs directly from the cuticle surface and only short-distance transport of scent molecules between their intracellular site of biosynthesis and this surface is required.

The transport mechanisms must play an important role in the overall emission of floral scents. Unfortunately, they are relatively difficult to investigate and the individual processes can hardly be singled out for separate quantification. As a consequence, transport of scent molecules has long been neglected in the literature and direct evidence is only scarcely available to date. This chapter can therefore only attempt to first describe a model for the export of floral scent compounds, then

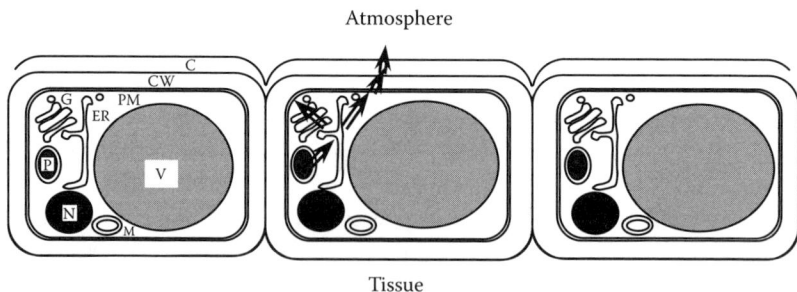

FIGURE 7.1 Schematic cross-section of the plant epidermis. Selected organelles are shown to illustrate the subcellular organization of scent biosynthesis. Trafficking of pathway intermediates and lipophilic products must occur between the plastids and the ER toward the plasma membrane. Further transport steps include transfer across the epidermal cell wall and the cuticle, as well as evaporation from the cuticle surface into the atmosphere. C, cuticle; CW, cell wall; ER, endoplasmic reticulum; G, Golgi apparatus; M, mitochondria; N, nucleus; P, plastids; PM, plasma membrane.

compare the few available literature data, and finally point to the various open questions in this field. Based on general considerations, our model will use chemical, physiological, morphological, biochemical, and molecular arguments to describe the export of volatiles from floral scent-producing cells. Four major steps are likely to be involved in floral volatile emission: (1) trafficking within the epidermal cell, (2) export from the plasma membrane into the epidermal apoplast and subsequent transport across the cell wall, (3) permeation of the cuticle, and (4) evaporation at the surface of the cuticle. These steps will be described consecutively in four sections (Section 7.2 to Section 7.5) of this chapter. Then the overall flux and barrier properties will be discussed based on experimental data for emission and internal pools (Section 7.6).

7.2   INTRACELLULAR TRAFFICKING

Our understanding of the major biosynthetic pathways leading to plant volatile products has greatly improved over the past three decades. Selected enzymes were characterized in model plant species, most importantly *C. breweri* and *A. majus*, where they catalyze the formation of volatiles in specialized cells of vegetative organs. Based on the subcellular localization of these enzymes, the spatial organization of partial and entire pathways has started to emerge. Similar studies on other plant species, although typically addressing only small portions of the pathways, have since helped to generalize the compartmentalization models for vegetative tissues of diverse species. In recent years, experimental evidence, mostly for *A. majus*, has also been provided for the subcellular localization of key enzymes in the biosynthesis of floral scent compounds.[4] The results are in good accordance with those for vegetative organs, further confirming the general model for compartmentalization of volatile biosynthetic pathways. Based on this model, the subcellular transport steps involved in scent formation can be postulated and will be described in the following paragraphs for the three major scent constituent classes of monoterpenoids, sesquiterpenoids, and benzenoids.

Monoterpenoid biosynthesis proceeds via geranyl pyrophosphate (GPP) as a common precursor, by stereospecific cyclization and deprotonation (or hydration) into various carbon structures, and the modification of their functional groups.[5] In diverse plant taxa, specialized cells of vegetative organs are responsible for monoterpenoid formation, where they originate from the plastids.[6] There is strong evidence showing compartmentalized biosynthesis of GPP inside plastids via the D-glyceraldehyde-3-phosphate/pyruvate pathway.[7] Biochemical studies and immunogold localization showed that the following cyclization step is also restricted to plastids (e.g., to leucoplasts of oil gland secretory cells of peppermint leaves)[8] and to nonphotosynthetic plastids in *Quercus ilex* leaves.[9] In accordance with these findings, all the known monoterpene synthase sequences apparently possess plastid-targeting sequences.

In floral tissue, monoterpenoid biosynthesis was first indirectly localized in the plastids by showing that isolated chromoplasts from daffodil petals are capable of catalyzing the production of limonene, myrcene, ocimene, and linalool.[10] Further evidence came from the presence of N-terminal signal peptides on the proteins involved

in monoterpenoid formation in petals of *A. majus*. Both the small and large subunits
of GPP synthase and the terpene synthases catalyzing the cyclization of GPP into
myrcene and (*E*)-β-ocimene possess N-terminal sequences targeting the enzymes to
plastids.[11] Accordingly, substantial specific immunogold labeling with antibodies
made against the small subunit of *A. majus* GPP synthase was found within the
nonpigmented plastids (leucoplasts) of the conical epidermal cells, which are extensively involved in scent production.[12] In summary, all the evidence available to date
consistently shows that monoterpenoid constituents of floral scents originate from
epidermal plastids.

Kleinig[6] proposed that monoterpenoids are transported from the plastids to the
cytosol for further modification (e.g., hydroxylation by cytochrome P450 oxidases
bound to the endoplasmic reticulum [ER]), but there are few direct studies localizing
these reactions. In a similar way, linalool oxides emitted from epidermal cells of
C. breweri pistils might be formed by modification of linalool at the ER. This
precursor is likely to be biosynthesized inside the plastids of the pistil epidermal
cells via GPP and its cyclization. Hence the pistils and petals of *C. breweri* would
share the monoterpenoid biosynthetic steps located inside the plastids, while the ER-
based modification would occur only in pistils. The cellular details of linalool and
linalool oxide biosynthesis in *Clarkia* flowers have not yet been explored experimentally.

Sesquiterpenoid biosynthesis proceeds via farnesyl pyrophosphate as the last
common precursor, formed through the distinct mevalonate pathway.[13] There is
ample evidence from vegetative tissues of diverse plant species that this pathway is
located at the ER of all plant cells. Similarly the ensuing stereospecific cyclization
and deprotonation (or hydration) of farnesyl pyrophosphate into various carbon
structures also take place at the ER. Sesquiterpenoid biosynthesis in vegetative plant
tissues thus does not necessitate trafficking of precursors and intermediates between
subcellular compartments. The localization of sesquiterpenoid biosynthesis within
epidermal cells of floral tissues has not been investigated to date, but it seems plausible
that it proceeds at the ER of respective epidermal cells.

Benzenoids are typical scent components, accounting for large portions of the
volatiles emitted by flowers in diverse species such as *Rosa damascena*,[14]
snapdragon, and petunia.[15] Although their biosynthesis has been investigated to a
certain extent in floral tissues, very little information on the subcellular localization
of respective enzymes and the compartmentalization of pathways is available. The
only exception is benzoic acid carboxyl methyltransferase (BAMT), an enzyme that
catalyzes the transfer of the methyl group of *S*-adenosyl-L-methionine to the carboxyl
group of benzoic acid. The resulting volatile ester, methyl benzoate, is one of the
most abundant scent compounds of snapdragon. Substantial immunogold labeling
was found within the cytoplasm of epidermal cells, suggesting an association of the
BAMT protein with the ER.[4] The localization of BAMT within the cytosol near the
plasma membrane and cell wall has been explained by the presence of large vacuoles,
which would leave little room for the cytosol, thereby compacting the immunogold
labeling.

In summary, the current understanding is that various floral scent components
are either formed inside the epidermal plastids and exported to the cytosol, are formed

in association with the ER, or are formed in plastids and further modified at the ER. In all cases, the products end up in the cytosol and, as lipophilic molecules, are likely associated with membrane systems of the ER. To date, no evidence is available for the mechanisms that traffic these compounds toward the plasma membrane. Involvement of the Golgi apparatus is plausible, and glycosylation of hydroxylated scent components has to be taken into account. It could be involved in trafficking of compounds or their storage in the vacuole. Alternatively, direct vesicular transport or protein-mediated molecular movement across the aqueous environment cannot be excluded at this point. In the case of snapdragon flowers, where benzenoid biosynthesis occurs at the ER close to the plasma membrane, direct diffusion over the membrane gap into the plasma membrane also seems feasible.

Finally, direct contact between membranes of the ER and the plasma membrane could create a lipophilic pathway for intracellular trafficking of floral scent molecules. Evidence supporting this mechanism comes only from the highly specialized scent-producing cells in the *Sauromatum guttatum* appendix, where Skubatz et al.[16] described membrane channels created by fusion of the rough ER to the plasma membrane. The authors interpreted this microscopic finding as the transport mechanism en route to emission of the sesquiterpenoids copaene and caryophyllene. Unfortunately, biochemical evidence to corroborate this conclusion is missing. It is unclear whether these mechanisms are applicable to other volatile classes, including monoterpenoids and aromatics, and to other plant species. It is also unclear whether similar mechanisms occur in flowers of species that produce and emit scent volatiles from nonspecialized epidermal cells.

7.3 EXPORT OF SCENT MOLECULES FROM THE PLASMA MEMBRANE AND TRANSPORT ACROSS THE EPIDERMAL CELL WALL

Export from the plasma membrane into the periclinal cell wall involves transfer of the relatively unpolar scent molecules from a lipophilic into an aqueous compartment. The unpolar nature of scent constituents can be described using their octanol-water partition coefficients, a parameter comparing their relative solubilities in lipophilic and aqueous environments. Small polar metabolites have octanol-water partition coefficients in the range of $\log K_{OW} = 0–2$,[17] nonionic surfactants like polyethoxylates have a range of $\log K_{OW} = 3–7$,[18] and highly lipophilic chemicals like alkyl phenols have a range of $\log K_{OW} = 7–10$.[19] The corresponding values for monoterpenoids range from $\log K_{OW} = 2$ for the alcohols to $\log K_{OW} = 5$ for the hydrocarbons.[20] Accordingly, the water solubilities for monoterpenoids between less than 35 ppm for hydrocarbons and several thousand parts per million for their oxygenated derivatives have been reported.[21]

This illustrates the very low solubility of scent molecules in any aqueous environment, which will consequently substantially hamper their transport across the cell wall. Obviously both physicochemical processes, detachment from the plasma membrane, and transport across the cell wall have to be facilitated energetically, probably by direct or indirect action of proteins. The biochemical principles acting in this respect have not been investigated, so this step remains the second unknown in the overall scent export process. We can only speculate about the mechanisms involved, in part based on comparisons with export of other lipophilic molecules across epidermal plant cell walls.

As one possibility, parts of the plasma membrane could detach in a process comparable to exocytosis, to form vesicles of amphiphilic lipids. Vesicular transport across the cell wall might be directed by gradients of bilayer constituents or scent molecules. Vesicle motion could be kept up by diffusion or enhanced by proteins. As an alternative possibility, specialized proteins might be involved both in export from the plasma membrane and transport across the cell wall. The adenosine triphosphate (ATP) binding cassette (ABC) transporters are one class of candidate proteins, as they have been shown to bind various lipophilic (and hydrophilic) substrates and transport them across membranes. It has been shown that they are frequently involved in the transfer of xenobiotics across plasma membranes (or tonoplasts) of plants cells.[22] More recently, evidence was provided that an ABC transporter is involved in the export of cuticular waxes from epidermal cells of *Arabidopsis thaliana* stems.[23] The *CER5* gene codes for an ABC half transporter that is an essential element in transporting waxes during growth of the tissue, and hence during surface formation, from the biosynthesizing epidermal cells toward the cuticle. It is located in the plasma membrane and indirect evidence showed that it participates in the nonspecific transport of all classes of wax constituents. It cannot be excluded that it also transports other substrates, such as volatile molecules. Conversely, other ABC transporters might be present that could be involved in specific or unspecific transport of lipophilic volatiles, both in vegetative and floral epidermis. To date, only one ABC transporter has been shown to be specifically expressed in the floral tissue of *Nicotiana tabacum*.[24] Although the corresponding mutant phenotype was not characterized, it was concluded that this protein functions in the export of stigma lipid exudate rather than in the export of scent molecules.

Alternatively, lipid transfer proteins (LTPs) could be involved in volatile transport across the epidermal cell wall. LTPs are known to reversibly bind lipids, in many cases nonspecifically, thus increasing their solubility in aqueous solution.[25] The LTP-lipid complexes are thought to be transported along gradients of the bound molecules, effectively shuttling them from points of relatively high concentration toward sinks. LTPs have been detected in the epidermal apoplast of barley leaves,[26] suggesting a role in the export of cuticular lipids, but this function has not been proven experimentally. Finally, other lipid-binding proteins might be involved in the transport of scent molecules across the epidermal cell wall.

7.4   TRANSPORT OF SCENT MOLECULES ACROSS THE PETAL CUTICLE

Aboveground tissues of land plants, as long as they are in a primary state of development, have an intact epidermis consisting of a single layer of tightly packed living cells. These epidermal cells biosynthesize and export the components of the plant cuticle, which they deposit as a continuous lipid membrane outside their periclinal cell walls. Over the past few decades, numerous investigations have addressed the structure, composition, and function of the cuticular membranes of leaves, fruits, and stems of selected plant species.[27] In contrast, relatively little is known about the cuticles of floral organs, and it has even been questioned whether flower surfaces are covered by a cuticle.[28] Recent chemical analyses of petal cuticles have shown that their composition is similar to that of vegetative organs, thus providing experimental evidence for the presence of cuticles on floral organs.

The following paragraphs summarize these chemical data and compare them to typical leaf and fruit cuticle compositions. Cuticles on vegetative organs consist of two major components: an insoluble polymer matrix of cutin, comprising between 40% and 80% of cuticular weight,[29] that is associated with soluble cuticular waxes. The comparison between cuticles of floral and vegetative organs will then describe the physical structure of petal cuticles and discuss their barrier properties for the transport of lipophilic molecules.

7.4.1 Petal Cutin Composition

The cutin of vegetative organs consists mainly of C_{16} and C_{18} fatty acids with a primary hydroxyl group on the ω-carbon, through which they form polyester chains. Additional secondary alcohol functions are likely involved in connecting the chains into a three-dimensional polymer matrix.[30] As ester hydrolysis does not suffice to entirely depolymerize the cuticular matrix from the leaves of many plant species, some of these connections are probably formed by ether bonds.[31,32] Evidence supporting this model for the molecular structure of cutin on vegetative organs came from gas chromatography-mass spectrometry (GC-MS) studies performed after depolymerizing the cutin matrix, as well as from x-ray diffraction, Fourier transform infrared spectroscopy (FTIR), and solid-state nuclear magnetic resonance (NMR) investigations.[33-36]

To date, there are no published reports on the cutin composition of floral tissues that can be compared with this general model for the cutin structure of vegetative organs. Only recently, cuticular membranes were isolated for the first time from petal tissue of *Cosmos bipinnatus*.[37] The cuticular material was then extracted to remove soluble waxes, leaving a colorless solid material. As transesterification with BF_3/methanol did not leave a residue, the material was a polyester lacking ether cross-linkages. The depolymerization yielded a mixture of C_{16} and C_{18} fatty acid derivatives that are typical cutin monomers of vegetative tissues (Figure 7.2). They were identified using MS of the trimethylsilyl (TMSi) derivatives of the fatty acid methyl esters, and quantified using a gas chromatography-flame ionization detector (GC-FID). 10,16-Dihydroxyhexadecanoic acid was the predominant monomer, accompanied by smaller amounts of 10,18-dihydroxyoctadecanoic acid and 6- and 7-hydroxyhexadecanoic acids. In addition, homologous series of n-alkanoic acids, 2-hydroxyalkanoic acids, and primary alcohols with chain lengths between C_{18} and C_{26} were present. These results clearly show that petals of *C. bipinnatus* have cutin that is very similar in composition to the polymer matrix of cuticles from vegetative tissues.

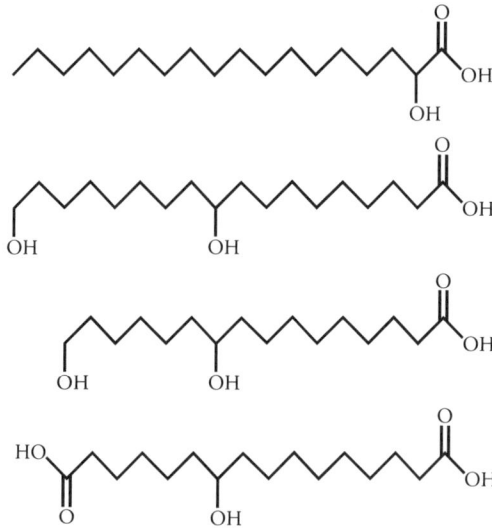

FIGURE 7.2 Structures of selected cutin monomers identified in the cuticular membrane on petals of *C. bipinnatus*.

7.4.2 PETAL WAX COMPOSITION

In the cuticles of vegetative organs, the polymer matrix of cutin is associated with cuticular waxes (i.e., soluble lipophilic compounds). These waxes are complex mixtures of very long chain fatty acid (VLCFA) derivatives, usually containing one primary functional group.[38] Homologous series of fatty acids, aldehydes, alcohols, alkanes, and alkyl esters are found in a specific composition characteristic for the plant species, organ, and state of development.[39] In leaf, stem, and fruit wax analyses, it was found that the vast majority, if not all of the compounds are fully saturated and have straight carbon chains. In addition to these ubiquitous constituents, cuticular wax mixtures may also contain triterpenoids and minor secondary metabolites, such as sterols and flavonoids. The cuticular waxes are arranged into an intracuticular layer in close association with the cutin matrix, as well as an epicuticular film exterior to this, which may include crystals.[40]

Over the past four decades, reports on the composition of cuticular waxes from flower parts of various plant species have been published. However, this covers only a relatively small selection of species, and large numbers of plant families have not yet been analyzed. The selection of species investigated is strongly biased; for example, there are a disproportionate number of reports on *Cistus* species[41,42] and *Rosa* species due to the horticultural interest in this genus.[43,44] In many cases only partial analyses have been performed, so the picture might be misleading and incomplete. For example, early investigations frequently focused on hydrocarbon composition and did not identify or quantify compounds in other wax fractions.[45,46]

All those classes of compounds that are usually found in cuticular wax mixtures of vegetative plant parts (i.e., alkanes, primary alcohols, aldehydes, fatty acids, and alkyl esters) have been identified in the waxes of floral surfaces. In most cases, chain length ranges and distributions were found that are typical for waxes; for example, C_{20} to C_{30} acids, aldehydes, and primary alcohols, with a strong predominance of even-numbered carbon chains, and C_{21} to C_{35} alkanes with a predominance of odd-numbered chains. These findings are in accordance with the generally accepted scheme for wax biosynthesis, including fatty acid elongation, and either reductive modification of oxygen functionalities or decarbonylation leading to alkanes with one less carbon.[38] All the molecules thus described have unbranched carbon backbones and contain only C-C single bonds (i.e., they are fully saturated).

In addition, in the flower waxes of some species, specialty compounds were identified that had previously been described for leaf waxes of only a few species. These include diverse triterpenoids (e.g., triterpene alcohols, acids, and esters thereof),[47,48] as well as other cyclic components like tocopherols.[47] Most interestingly, a large variety of hydrocarbons other than the ubiquitous n-alkanes have been reported for petal waxes. These included series of iso-alkanes, anteiso-alkanes, and alkanes with single methyl branches farther away from the chain terminus (Figure 7.3). These compounds had rarely been described as trace components of waxes from vegetative tissues. It is noteworthy that they also are ubiquitous constituents of insect cuticular waxes. Finally, unsaturated hydrocarbons have repeatedly been identified in floral wax mixtures, mostly alkenes with Z double bonds, in many cases Δ9.[41]

A number of reports have described novel compounds in floral waxes. These include very long chain aliphatics with secondary functional groups (e.g., secondary alcohols,[49] ketones,[44] alkanediols,[50] and γ-lactones[51]). Finally, special esters have been identified, including secondary alcohols,[52] phytyl esters, and cinnamyl esters.[53]

For *A. majus*, one of the model species for which most of the biochemical and molecular evidence regarding scent biosynthesis is available, a detailed petal wax analysis has recently been published.[52] The floral cuticular wax mixtures were dominated by n-alkanes with chain lengths between C_{27} and C_{33}, while branched alkane isomers and free fatty acids were detected at lower percentages. Similar petal wax characteristics were recently reported by Hager[37] in a comparative analysis for a number of species (Figure 7.4). This confirmed previous literature findings on a vast array of other plant species, which also had floral wax compositions dominated by

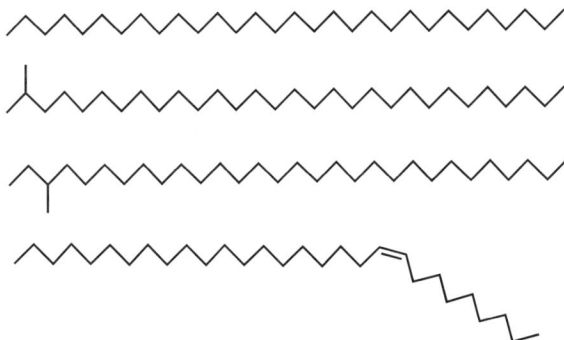

FIGURE 7.3 Structures of selected hydrocarbons reported in the cuticular wax mixtures on petals of various plant species. One representative each from the homologous series of n-alkanes, iso-alkanes, anteiso-alkanes, and alkenes is shown.

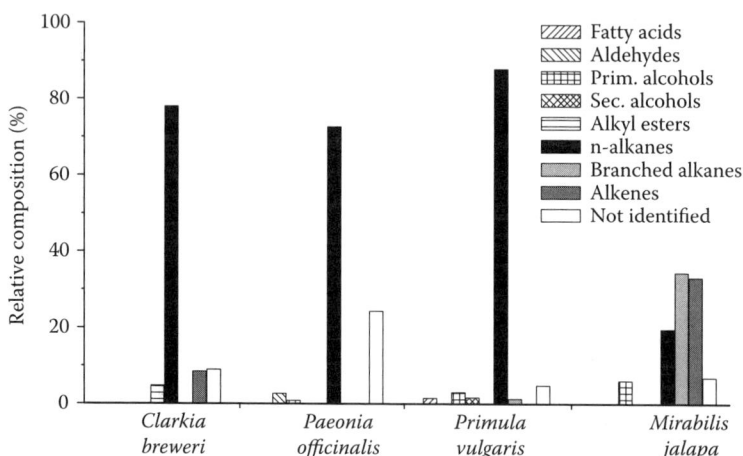

FIGURE 7.4 Relative composition of the wax mixtures from petals of four plant species.[37]

hydrocarbons. In many cases n-alkanes alone contributed more than half of the wax mixture. Even the petal waxes of distantly related species frequently differed only in small quantitative details of alkane homologue patterns, while conversely, wax compositions of flowers and leaves of the same species in some cases differed drastically.

Some plant species were further characterized by significant percentages of other hydrocarbon classes. For example, branched alkanes contributed approximately 26% to the petal waxes of A. majus[52] and accounted for 34% of the petal wax of *Mirabilis jalapa* (Figure 7.4). On the other hand, unsaturated hydrocarbons accounted for 5%, 8%, and 11% of the petal wax mixtures of *Cistus ladanifer*, *Cistus incanus*, and *Cistus albidus*, respectively,[41] and 8% and 33% in *C. breweri* and *M. jalapa*, respectively (Figure 7.4). Finally, individual species had floral wax mixtures dominated by compound classes other than hydrocarbons. For example, triterpenoids were the prevalent class of compounds in *Cichorium intybus*, while alkyl esters predominated in raspberry flowers.[47]

Relatively few literature reports give data on the wax coverage of floral organs, either as wax mass per unit area or tissue weight. Values vary between 8 and 14 mg/g petal dry weight for A. majus,[52] corresponding to approximately 45 μg/cm^2 of surface area. Hager[37] reported coverages of 1.3 μg/cm^2 for *C. breweri*, 4.0 μg/cm^2 for *M. jalapa*, 4.5 μg/cm^2 for *Primula vulgaris*, and 5.2 μg/cm^2 for *Paeonia officinalis*. Thus the amount of wax found on petal surfaces is typically less than for vegetative tissues. Also, for species in which waxes on both floral and vegetative organs have been investigated, the petals tend to have less wax coverage than the corresponding leaves or fruits.

7.4.3 BARRIER PROPERTIES OF PETAL CUTICLES

The transport of molecules across the cuticle of leaves and fruits of selected plant species has been studied using various model compounds. Based on the resulting physical characteristics, models for the structures and barrier properties of cuticular membranes have been established. Two fundamental mechanisms are postulated for the movement of molecules across this membrane: a nonpolar pathway for the transport of lipophilic compounds and water,[54] and a polar pathway important for the transport of larger hydrophilic compounds.[55] Although the transport of scent constituents across cuticles has not been investigated, it is very likely that these lipid molecules will move exclusively along the nonpolar pathway.

It has been shown that the major barrier imposed on the nonpolar pathway is localized in the waxes,[56] and that transport of lipids through the cuticle (of vegetative organs) occurs across continuous layers of these waxes. It would be very important to know how the chemical characteristics of the cuticular wax mixture cause resistance against transport of compounds. But unfortunately a correlation between the barrier properties and either the wax composition or the wax amount has not been found. It has even been shown that the barrier quality is generally independent of cuticle thickness.[57] This means that the thick cuticles of some plant species are not necessarily good permeation barriers, and accordingly should not be described as an adaptation against drought without experimental evidence.

It has been hypothesized that, due to differences in the molecular geometry of the various constituents, a number of physical phases coexist within the cuticular wax mixture.[58] According to this model, the hydrocarbon tails of the aliphatic compounds are packed side by side, creating monolayer rafts with crystalline order, while the terminal functional groups are exposed on both sides of the sheets. Due to thermal movements and variations in chain lengths between the crystalline sheets, there will be amorphous zones with lower density. While small lipophilic molecules like scent constituents are excluded from the crystalline domains, they can dissolve in and diffuse across those amorphous areas. Transport across the cuticle will thus be limited by the length of the diffusion pathways through the amorphous zones, which are in turn influenced by the size, shape, and orientation of the crystalline rafts. In conclusion, this model states that the cuticular barrier is mainly determined by the crystallinity of the waxes (i.e., by the aliphatic chain length distribution), whereas wax amounts and relative amounts of compound classes are less important. Although this hypothesis has been developed for cuticular membranes of vegetative organs, it can also be applied to floral tissues.

Based on the somewhat limited information available to date, it can be assumed that petal cuticles have the same general composition and structure as those of vegetative organs (see Chapter 8, Sections 7.4.1 and 7.4.2). The most notable characteristics of petal waxes were the relatively small amounts of waxes, the large proportions of alkanes, and the admixture of branched and unsaturated hydrocarbons. The latter are of special interest, as both the methyl side groups and the double bonds disturb the linear geometry of the molecules. It has been suggested that these compounds, when packed together with other alkyl chains, prevent the crystalline arrangement of molecules. They effectively increase the amorphous domains, and thus shorten the transport pathways and increase the permeability of the membrane. Alternatively, portions of these relatively bulky molecules might be excluded from the crystalline domains, thereby adding to the volume of amorphous zones and weakening the transport barrier. In both cases, it must be concluded that the special composition of petal waxes should cause relatively poor barrier properties. Interestingly, this hypothesis implies that (at least some) petal cuticles do not impose an important resistance against transport of scent molecules, but at the same time are also a relatively poor barrier against water loss.

Currently there are no experimental data to support this model for transport of compounds across petal waxes, and it would be interesting to compare the overall crystallinities of wax mixtures on vegetative and floral organs. FTIR showed that the leaf wax mixtures of *Citrus aurantium*, *Fagus sylvatica*, and *Hordeum vulgare*, with relatively broad chain length distributions in the first two cases and a very strong predominance of a single chain length in the latter, had crystallinities of 20%, 30%, and 52% at room temperature, respectively.[59,60] Comparable investigations could be carried out on petal waxes, of species with high percentages of n-alkanes and of species with varying portions of n- and iso-alkanes. As petal waxes in many cases consist of relatively few components, they provide an opportunity to test the influence of single compound classes on the structure of the wax mixture. These data on composition-structure relationships could be expanded to structure-function relationships by adding data on the barrier properties of respective petal cuticles. The permeability of test compounds would have to be determined (in meters per second) under controlled conditions, similar to investigations on leaf cuticles. Unfortunately respective experiments have not been reported to date, probably due in part to the small size and fragile nature of floral organs.

7.5 EVAPORATION OF SCENT MOLECULES: SURFACE MORPHOLOGY

In a hallmark investigation covering a wide range of taxa, Vogel[1] described floral scent glands with a wide variety of shapes (see also Chapter 9). The surfaces of these osmophores were further characterized in subsequent histologic and electron microscopic studies.[61–63] Trichomes, glandular hairs, and scent-producing idioblasts have been described for a number of species, but a discussion of their structure and function is beyond the scope of this review. In many other species, the entire scent-emitting surface is made up of epidermal pavement cells (which are also the site of volatile biosynthesis). Scanning electron microscopy revealed that in some of these cases the surface is structured only due to the slightly convex shape of the cells, while the cuticle surface on them is smooth. *C. breweri* may serve as an example to illustrate this type of surface, as its petals showed no evidence for glandular structures that would contain specialized cells mainly dedicated to scent biosynthesis or for subcellular structures that would increase surface area.[3]

For many other species, scent-producing floral organs have epidermis cells with pronounced conical shapes. The conical shape of the epidermal cells obviously increases the scent-emitting surface of the cells when compared with flat epidermal cells typical for vegetative organs.[64] The characteristic dome-shaped surface of individual cells sometimes develops only during floral development, probably synchronized with the onset of scent biosynthesis.[52,65] It therefore has been speculated that the conical cell shape is an adaptation with functions in scent emission or in the interaction of floral surfaces with pollinators. Petal epidermal cells with a conical shape tend to have more elaborate three-dimensional structures, but on a smaller scale (Figure 7.5). Their surfaces frequently show ridges that are a few micrometers wide, run in parallel, and have radial orientation patterns with respect to the cell outline. For *C. bipinnatus*, it has been shown that these ridges are cuticular structures consisting of cutin and waxes, rather than cell wall extensions.[37]

The specific occurrence of a conical shape associated with cuticular ridges on petal surfaces has triggered hypotheses that these structures play a part in the pollination syndrome. One frequently stated interpretation is that these structures facilitate emission of scent molecules by increasing the surface area of the epidermal cells. Although this idea is convincing at first, it should be viewed critically as experimental evidence to prove it is missing. It should also be noted that this inter-pretation is based on the assumption that surface evaporation is a rate-limiting step in the overall emission of scent molecules. Neither the mechanisms of other transport steps nor the involved permeability barriers can be assessed to date, but it seems plausible that transport both across the cell wall and through the cuticle will be relatively slow. Therefore surface evaporation is likely not the (only) slow step in the process, and an increase in surface area should have only limited impact on emission rates. Instead of only enhancing scent emission, the microstructure of petal surfaces might also function in other aspects of petal-insect interactions. It is likely beneficial if the same surfaces that emit attractive volatiles also modify the visual appearance of the flowers by diffuse reflection of light, improve surface adhesion of insects walking on the flower surfaces, and provide tactile stimuli for surface recognition by pollinators.

Many leaf surfaces are covered with epicuticular wax crystals (i.e., nanometer structures protruding from the cuticle surface into the air).[40] Similar features have been described for the floral surfaces of only a few species, where they were usually not localized on the scent-producing epidermal cells.[66] It has been hypothesized that these epicuticular crystals create slippery surfaces that help to catch insects in trap-pollination systems. In contrast, the epidermal pavement cells involved in scent biosynthesis, both in species with normal cell shapes and in species with conical cells and cuticular ridges, typically do not carry epicuticular wax crystals, but instead are covered by a smooth cuticle.[3,67]

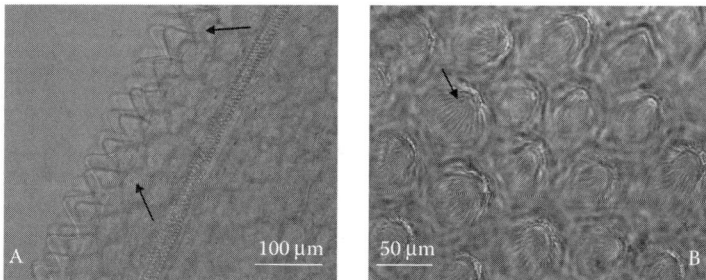

FIGURE 7.5 Light micrographs of the petal surface of *C. bipinnatus*. Epidermal cells have a conical shape and the cell surface is structured by cuticular ridges (arrows).

7.6   INTERPRETATION OF TRANSPORT DATA

Previous sections of this chapter have described a mechanistic model for the transport of scent molecules between the site of biosynthesis and the site of evaporation into the headspace. Unfortunately very little direct experimental evidence has been published to date that can corroborate the steps detailed in this model. It would be particularly interesting to separately assess the permeability of the epidermal plasma membrane, cell wall, and cuticle for selected model compounds. The results could be used to calculate the resistance imposed by each transport segment. Once respective results become available, they will have to be compared with data on the overall resistance of the transport process in order to assess the contribution each step makes to the total transport barrier. The overall resistance in turn can be inferred from the mass flow of compounds, the surface area across which this flow occurs, and the concentration gradient driving transport. These quantities can be estimated based on literature data comparing the headspace accumulation of products with the biosynthetic capacities and the precursor pools for model plant species.

The headspace concentration of scent components has been reported for diverse plant species and is described in detail in Section 1.1. It should be noted that results are usually given as amounts per flower and time, or as amounts per tissue weight and time, but the permeability of the transport barriers can only be assessed using amounts per area and time. Thus the surface area per flower or per tissue weight would be necessary to transform the headspace results. For *A. majus*, the only species for which these data are available,[52] it can be concluded that the petals emit up to 75 µg/mm²/day.

Next, the gradient between scent constituent concentrations inside and outside the tissue has to be taken into account. While the headspace amounts of these compounds have been reported for many plant species, relatively little is known about their internal concentrations. Essential oils were analyzed for diverse plant species, but in many cases only qualitative data were given and for only some species (and floral organs) have the amounts of essential oil constituents per flower (or surface area) been reported. Based on these results, the concentration of scent molecules inside epidermal cells can be inferred, assuming that only these cells biosynthesize and thus contain essential oils. Accordingly, the petal epidermis of *A. majus* should, for example, contain roughly 5 µg/mm³ of methyl benzoate (using published data[52] and assuming an average epidermis thickness of 50 µm, as well as even distribution of methyl benzoate in all petal areas). This calculation demonstrates that scent components do accumulate inside the cell to substantial concentrations, and local concentrations in subcellular compartments must be even higher. As this internal concentration greatly exceeds the concentration of scent in the headspace surrounding the tissue, a gradient is established that acts as a driving force for the transport of the compounds toward the surface.

In the same study on *A. majus* flowers, Goodwin et al.[52] also monitored the dynamics of essential oil quantities as a function of flower development. Their results have three important implications: (1) transport phenomena are relatively fast, occurring in time scales of hours; (2) significant transport barriers exist that cause transient buildup of internal material; and (3) these barriers are not rate limiting for emission of scent compounds.

Snapdragon petals represent the first system in which all the data have been acquired that are necessary to quantitatively describe the transport of scent compounds, assessing the overall flow rate, permeability, and resistance. Based on a flow rate for methyl benzoate of 0.7 μg/s/flower, surface areas of 750 mm^2/flower, and concentration gradients between internal and external pools of 5 μg/mm^3, an overall permeability on the order of 10^7 m/s can be predicted. This corresponds to an overall resistance of 10^{-7} s/m imposed on transport. Although these values represent only relatively rough estimates, calculated from various parameters using a number of assumptions, they give a quantitative description of the transport barrier. Hence this successful approach should be repeated by similar investigations on other plant species, allowing quantitative comparisons between the barriers involved in transport of scent compounds.

It should be noted that our efforts to describe the export and emission of scent molecules in quantitative terms might be complicated by (at least) one other factor. The absence of both appreciable internal pools of scent volatiles and the biosynthetic enzymes of such compounds in scent-emitting floral tissue of *Jasminum* species prompted Watanabe et al.[68] to discover that fragrance components are stored as nonvolatile glycosides. However, Loughrin et al.[69] found that the level of glycosidically bound volatiles in *Nicotiana* species was not correlated with the emission levels of such volatiles, but with the age of the flower; older, senescing flowers had higher levels of stored glycosides. It has been speculated that glycosides are either precursors, storage forms, or detoxification products of scent compounds. The role of these derivatives, as well as their amounts, may depend on plant species and physiology. Hence the (reversible?) formation of glycosides must be regarded side branch to the flux of scent material between biosynthesis and emission, and their amounts would accordingly have to be taken into account when quantifying transport phenomena and barrier properties. Unfortunately the presence and amounts of these derivatives have not been investigated in most of the studies on essential oils of flowers.

To further refine our understanding of transport mechanisms involved in scent emission, a number of approaches seem feasible. Among them, two appear especially promising: (1) The concentration of scent compounds within the mixture of cuticular waxes should be quantified, if possible as a function of flower development or diurnal rhythms. Respective data would help to distinguish between internal pools of scent compounds, showing where exactly they (transiently) accumulate and hence pointing to the rate-limiting step in transport. (2) Model species could be used to generate transgenic plants with altered capacity for scent biosynthesis. In one experiment, a volatile biosynthetic gene was introduced into petunia and the protein levels and enzyme activity of monoterpene synthase were detected.[70] Nonetheless, the amounts of the target compound, linalool, in the floral scent bouquet were found to be only slightly changed. This result could be due to limited precursor availability, restricted capacities of pathway enzymes, mismatches in intracellular localization and trafficking, as well as rate-limiting barriers in epidermal cell export. Hence the transgenic plants are very interesting tools to investigate various aspects of scent production and test a number of related hypotheses, so it would be very interesting to characterize them in detail.

7.7 CONCLUSION

This chapter outlined a mechanistic model for the export and emission of scent molecules from floral surfaces. It is based on the finding that in most plant species investigated to date, the amount of volatiles emitted is limited by the capacity of biosynthetic pathways. Temporal and spatial patterns of biosynthesis are tightly controlled, mostly through gene expression levels limiting enzyme levels and activities, and less by substrate availability. Biosynthesis is in many species restricted to epidermal cells of certain regions of floral organs. Most of the steps involved in the export of scent products require energy, including intracellular trafficking, export from the plasma membrane, and transport through the epidermal cell wall. Consequently these steps impose transport barriers and small amounts of products build up in corresponding compartments. A critical concentration is built up, resulting in a gradient from inside to outside. This gradient drives transport across the cell wall and cuticle. The transport may be facilitated by proteins, especially in the transfer of the lipophilic molecules across the aqueous environment of the cell wall. In case higher internal concentrations accumulate, the compounds might be toxic to the synthesizing cell. Therefore they are either metabolized (no evidence so far) or stored as glycosylates.

REFERENCES

1. Vogel, S., Duftdrüsen im Dienste der Bestäubung. Über Bau und Funktion der Osmophoren, *Abhandlungen der Math.-Naturw. Klasse* 10, 46, 1962.
2. Pichersky, E., Raguso, R.A., Lewinsohn, E., and Croteau, R., Floral scent production in *Clarkia* (Onagraceae). I. Localization and developmental modulation of monoterpene emission and linalool synthase activity, *Plant Physiol.* 106, 1533, 1994.
3. Raguso, R.A. and Pichersky, E., A day in the life of a linalool molecule: chemical communication in a plant-pollinator system. I. Linalool biosynthesis in flowering plants. *Plant Spec. Biol.* 14, 95, 1999.
4. Kolosova, N., Sherman, D., Karlson, D., and Dudareva, N., Cellular and subcellular localization of S-adenosyl-L-methionine:benzoic acid carboxyl methyltransferase, the enzyme responsible for biosynthesis of the volatile ester methylbenzoate in snapdragon flowers, *Plant Physiol.* 126, 956, 2001.
5. Wise, M.L. and Croteau, R., Monoterpene biosynthesis, in *Comprehensive Natural Products Chemistry: Isoprenoids*, vol. 2, Cane, D.E., Ed., Elsevier Science, London, 1999, p. 97.
6. Kleinig, H., The role of plastids in isoprenoid biosynthesis, *Annu. Rev. Plant Physiol. Plant Mol. Biol.* 40, 39, 1989.
7. Lichtenthaler, H.K., The 1-deoxy-D-xylulose-5-phosphate pathway of isoprenoid biosynthesis in plants, *Annu. Rev. Plant Physiol. Plant Mol. Biol.* 50, 47, 1999.
8. Turner, G., Gershenzon, J., Nielson, E.E., Froehlich, J.E., and Croteau, R., Limonene synthase, the enzyme responsible for monoterpene biosynthesis in peppermint, is localized to leucoplasts of oil gland secretory cells, *Plant Physiol.* 120, 9966, 1999.
9. Loreto, F., Ciccioli, P., Brancaleoni, E., Cecinato, A., Frattoni, M., and Sharkey, T.D., Different sources of reduced carbon contribute to form three classes of terpenoid emitted by *Quercus ilex* L. leaves, *Proc. Natl. Acad. Sci. USA* 93, 4126, 1996.

10. Mettal, U., Boland, W., Beyer, P., and Kleinig, H., Biosynthesis of monoterpene hydrocarbons by isolated chromoplasts from daffodil flowers, *Eur. J. Biochem.* 170, 613, 1988.

11. Dudareva, N., Martin, D., Kish, C.M., Kolosova, N., Gorenstein, N., Faldt, J., Miller, B., and Bohlmann, J., (*E*)-β-Ocimene and myrcene synthase genes of floral scent biosynthesis in snapdragon: function and expression of three terpene synthase genes of a new terpene synthase subfamily, *Plant Cell* 15, 1227, 2003.

12. Tholl, D., Kish, C.M., Orlova, I., Sherman, D., Gershenzon, J., Pichersky, E., and Dudareva, N., Formation of monoterpenes in *Antirrhinum majus* and *Clarkia breweri* flowers involves heterodimeric geranyl diphosphate synthases, *Plant Cell* 16, 977, 2004.

13. Bohlmann, J., Meyer-Gauen, G., and Croteau, R., Plant terpenoid synthases: molecular biology and phylogenetic analysis, *Proc. Natl. Acad. Sci. USA* 95, 4126, 1998.

14. Picone, J.M., Clery, R.A., Watanabe, N., MacTavish, H.S., and Turnbull, C.G., Rhythmic emission of floral volatiles from *Rosa damascena semperflorens* cv. "Quatre Saisons," *Planta* 219, 468, 2004.

15. Negre, F., Kish, C.M., Boatright, J., Underwood, B., Shibuya, K., Wagner, C., Clark, D.G., and Dudareva, N., Regulation of methylbenzoate emission after pollination in snapdragon and petunia flowers, *Plant Cell* 15, 2992, 2003.

16. Skubatz, H., Kunkel, D.D., Patt, J.M., Howald, W.N., Hartman, T.G., and Meeuse, B.J., Pathway of terpene excretion by the appendix of *Sauromatum guttatum*, *Proc. Natl. Acad. Sci. USA* 92, 10084, 1995.

17. Leo, A., Hansch, C., and Elkins, D., Partition coefficients and their uses, *Chem. Rev.* 71, 525, 1971.

18. Müller, M.T., Zehner, A.J.B., and Escher, B.I., Liposome-water and octanol-water partitioning of alcohol ethoxylates, *Environ. Toxicol. Chem.* 18, 2191, 1999.

19. Sherblom, P.M., Gschwend, P.M., and Eganhouse, R.P., Aqueous solubilities, vapor pressures, and 1-octanol-water partition coefficients for C9–C14 linear alkylbenzenes, *J. Chem. Eng. Data* 37, 394, 1992.

20. Griffin, S., Wyllie, S.G., and Markham, J., Determination of octanol-water partition coefficient for terpenoids using reversed-phase high-performance liquid chromatography, *J. Chromatogr. A* 864, 221, 1999.

21. Weidenhamer, J.D., Macias, F.A., Fischer, N.H., and Williamson, G.B., Just how insoluble are monoterpenes, *J. Chem. Ecol.* 19, 1799, 1993.

22. Theodoulou, F.L., Plant ABC transporters, *Biochim. Biophys. Acta* 1465, 79, 2000.

23. Pighin, J.A., Zheng, H., Balakshin, L.J., Goodman, I.P., Western, T.L., Jetter, R., Kunst, L., and Samuels, A.L., Plant cuticular lipid export requires an ABC transporter, *Science* 306, 702, 2004.

24. Otsu, C.T., da Silva, I., de Molfetta, J.B., da Silva, L.R., de Almeida-Engler, J., Torraca, P.C., Goldman, G.H., and Goldman, M.H., NtWBC1, an ABC transporter gene specifically expressed in tobacco reproductive organs, *J. Exp. Bot.* 55, 1643, 2004.

25. Kader, J.-C., Lipid-transfer proteins in plants, *Annu. Rev. Plant Physiol. Plant Mol. Biol.* 47, 627, 1996.

26. Hollenbach, B., Schreiber, L., Hartung, W., and Dietz, K.J., Cadmium leads to stimulated expression of the lipid transfer protein genes in barley: implications for the involvement of lipid transfer proteins in wax assembly, *Planta* 203, 9, 1997.

27. Kerstiens, G., *Plant Cuticles*, 1st ed., BIOS Scientific, Oxford, 1996.

28. Galen, C., Sherry, R.A., and Carroll, A.B., Are flowers physiological sinks or faucets? Costs and correlates of water use by flowers of *Polemonium viscosum*, *Oecologia* 118, 461, 1999.

29. Holloway, P.J., The chemical constitution of plant cutins, in *The Plant Cuticle*, Linnean Society Symposium Series, vol. 10, Cutler, D.F., Alvin, K.L., and Price, C.E., Eds., Academic Press, London, 1982, p. 45.

30. Kolattukudy, P.E., Biochemistry and function of cutin and suberin, *Can. J. Bot.* 62, 2918, 1984.

31. Schmidt, H.W. and Schönherr, J., Development of plant cuticles: occurrence and role of non-ester bonds in cutin of *Clivia miniata* Reg. leaves, *Planta* 156, 380, 1982.

32. Heredia, A., Biophysical and biochemical characteristics of cutin, a plant barrier biopolymer, *Biochim. Biophys. Acta* 1620, 1, 2003.

33. Stark, R.E., Zlotnik-Mazori, T., Ferrantello, L.M., and Garbow, J.R., Molecular structure and dynamics of intact plant polyesters: solid-state NMR studies, in Lewis, N.G. and Paice, M.G., Eds., *Plant Cell Wall Polymers: Biogenesis and Biodegradation*, ACS Symposium Series, vol. 399, American Chemical Society, Washington, DC, 1989, p. 214.

34. Lai, S., Lai, A., Stange, R.R., Jr., McCollum, T.G., and Schirra, M., Characterization of the wound-induced material in *Citrus paradisi* fruit peel by carbon-13 CP-MAS solid state NMR spectroscopy, *Phytochemistry* 63, 177, 2003.

35. Villena, J.F., Dominguez, E., Stewart, D., and Heredia, A., Characterization and biosynthesis of non-degradable polymers in plant cuticles, *Planta* 208, 181, 1999.

36. Deshmukh, A.P., Simpson, A.J., and Hatcher, P.G., Evidence for cross-linking in tomato cutin using HR-MAS NMR spectroscopy, *Phytochemistry* 64, 1163, 2003.

37. Hager, D. Zusammensetzung und Funktion der Kutikyla von Bluetenblaettern, Diploma Thesis, University of Wuerzburg, 2003.

38. Kunst, L. and Samuels, A.L., Biosynthesis and secretion of plant cuticular wax, *Prog. Lipid Res.* 42, 51, 2003.

39. Baker, E.A., Chemistry and morphology of plant epicuticular waxes, in *The Plant Cuticle*, Linnean Society Symposium Series, vol. 10, Cutler, D.F., Alvin, K.L., and Price, C.E., Eds., Academic Press, London, 1982, p. 139.

40. Jeffree, C.E. Structure and ontogeny of plant cuticles, in *Plant Cuticles: An Integrated Functional Approach*, Kerstiens, G., Ed., BIOS Scientific, Oxford, 1996, p. 33.

41. Gülz, P.-G. et al., Untersuchungen über die Kutikularwachse in der Gattung *Cistus* L. (Cistaceae). VI. Die Zusammensetzung der Alkene aus *Cistus*-Petalen und die Lage der Doppelbindung in diesen Alkenen am Beispiel von 3 *Cistus*-Arten, *Z. Pflanzenphysiol.* 94, 35, 1979.

42. Gülz, P.-G., Proksch, P., and Schwarz, D., Untersuchungen über die Kutikularwachse in der Gattung *Cistus* L. (Cistaceae). V. Zusammenfassende Darstellung über die Kohlenwasserstoffe und Wachsester aus den Epikutikularwachsen der Blätter und Petalen, *Z. Pflanzenphysiol.* 92, 341, 1979.

43. Mladenova, K., Stoianova-Ivanova, B., and Daskalov, R.M., Trialkyltrioxanes in flower wax of some decorative roses, *Phytochemistry* 15, 419, 1976.

44. Stoianova-Ivanova, B., Mladenova, K., and Popow, S., The composition and structure of ketones from rose bud and rose flower waxes, *Phytochemistry* 10, 1391, 1971.

45. Stransky, K. and Streibl, M., On natural waxes 12. Composition of hydrocarbons in morphologically different plant parts, *Coll. Czech.* 34, 103, 1969.

46. Salasoo, I., Alkane distribution in epicuticular wax of Epacridaceae, *Phytochemistry* 22, 937, 1983.

47. Griffiths, D.W., Robertson, G.W., Shepherd, T., Birch, A.N., Gordon, S.C., and Woodford, J.A., A comparison of the composition of epicuticular wax from red raspberry (*Rubus idaeus* L.) and hawthorn (*Crataegus monogyna* Jacq.) flowers, *Phytochemistry* 55, 111, 2000.

48. Hennig, S., Gülz, P.-G., and Hangst, K., Organ specific composition of epicuticular waxes of *Cistus albidus* L. Cistaceae, *Z. Naturforsch. C* 43, 806, 1988.
49. Wollrab, V., Secondary alcohols and paraffins in the plant waxes of the family Rosaceae, *Phytochemistry* 8, 623, 1969.
50. Stoianova-Ivanova, B., Hadjieva, P., and Tamas, J., Nonacosane-5,8-diol: a new component of plant waxes, *Phytochemistry* 13, 1523, 1974.
51. Hadjieva, P., Stoianova-Ivanova, B., and Danieli, B., Composition and structure of lactones from rose flower wax, *Chem. Phys. Lipids* 12, 60, 1974.
52. Goodwin, S.M., Kolosova, N., Kish, C.M., Wood, K.V., Dudareva, N., and Jenks, M.A., Cuticle characteristics and volatile emissions of petals in *Antirrhinum majus* L., *Physiol. Plant.* 117, 435, 2002.
53. Griffiths, D.W., Robertson, G.W., Shepherd, T., and Ramsay, G., Epicuticular waxes and volatiles from faba bean (*Vicia faba*) flowers, *Phytochemistry* 52, 607, 1999.
54. Schreiber, L., Kirsch, T., and Riederer, M., Diffusion through cuticles: principles and models, in *Plant Cuticles: An Integrated Functional Approach*, Kerstiens, G., Ed., BIOS Scientific, Oxford, 1996, p. 109.
55. Schönherr, J. and Schreiber, L., Size selectivity of aqueous pores in astomatous cuticular membranes isolated from *Populus canescens* (Aiton) Sm. leaves, *Planta* 219, 405, 2004.
56. Schönherr, J., Water permeability of isolated cuticular membranes: the effect of cuticular waxes on diffusion of water, *Planta* 131, 159, 1976.
57. Riederer, M. and Schreiber, L., Protecting against water loss: analysis of the barrier properties of plant cuticles, *J. Exp. Bot.* 52, 2023, 2001.
58. Riederer, M. and Schreiber, L., Waxes: the transport barriers of plant cuticles, in *Waxes: Chemistry, Molecular Biology and Functions*, Hamilton, R.J., Ed., The Oily Press, Dundee, 1995, 131.
59. Reynhardt, E.C. and Riederer, M., Structure and molecular dynamics of the cuticular wax from leaves of *Citrus aurantium*, *J. Phys. D Appl. Phys.* 24, 478, 1991.
60. Reynhardt, E.C. and Riederer, M., Structures and molecular dynamics of plant waxes II. Cuticular waxes from leaves of *Fagus sylvatica* L. and *Hordeum vulgare* L., *Eur. Biophys. J.* 23, 59, 1994.
61. Williams, N.H., Floral fragrances as cues in animal behavior, in *Handbook of Experimental Pollination*, Jones, C.E. and Little, R.J., Eds., Van Nostrand-Reinhold, New York, 1983, p. 50.
62. Stern, W.L., Curry, K.J., and Pridgeon, A.M., Osmophores of *Stanhopea* (Orchidaceae), *Am. J. Bot.* 74, 1323, 1987.
63. Curry, K.J., Initiation of terpenoid synthesis in osmophores of *Stanhopea anfracta* (Orchidaceae): a cytochemical study, *Am. J. Bot.* 74, 1332, 1987.
64. Noda, K., Glover, B.J., Linstead, P., and Martin, C., Flower color intensity depends on specialized cell shape controlled by a Myb-related transcription factor, *Nature* 369, 661, 1994.
65. Vincent, C.A. and Coen, E.S., A temporal and morphological framework for flower development in *Antirrhinum majus*, *Can. J. Bot.* 82, 681, 2004.
66. Vogel, S., Remarkable nectaries: structure, ecology, organophyletic perspectives. 2. Nectarioles, *Flora* 193, 1, 1998.
67. Eveling, D.W., Examination of the cuticular surfaces of fresh delicate leaves and petals with a scanning electron microscope: a control for artifacts, *N. Phytol.* 96, 220, 1984.

68. Watanabe, N., Watanabe, S., Nakajima, R., Moon, J.H., Shimokihira, K., Inagaki, J., Etoh, H., Asai, T., Sakata, K., and Ina, K., Formation of flower fragrance compounds from their precursors by enzymatic action during flower opening, *Biosci. Biotech. Biochem.* 57, 1101, 1993.
69. Loughrin, J., Hamilton-Kemp, T.R., Burton, H.R., Andersen, R.A., and Hildebrand, D.F., Glycosidically bound volatile components of *Nicotiana sylvestris* and *N. suaveolens* flowers, *Phytochemistry* 31, 1537, 1992.
70. Lücker, J., Bouwmeester, H.J., Schwab, W., Blaas, J., van Der Plas, L.H., and Verhoeven, H.A., Expression of *Clarkia S*-linalool synthase in transgenic petunia plants results in the accumulation of *S*-linalyl-β-D-glucopyranoside, *Plant J.* 27, 315, 2001.

Section IV

Plant-Insect Interactions and Pollination Ecology

8 Relationship between Floral Fragrance Composition and Type of Pollinator

Heidi E.M. Dobson

CONTENTS

8.1   INTRODUCTION

Flowers attract pollinator animals through a concert of visual and olfactory stimuli, and the role of floral scent volatiles in attracting as well as eliciting landing, feeding, and in some cases mating and ovipositing behaviors on the flower varies with each flower-animal association.[1–3] Flower-animal interactions include a broad diversity of both invertebrates (insects) and vertebrates that function as pollinators,[4,5] and the relative importance of floral scent in pollination depends on both the purpose of the animal's visit to the flower and features of the animal's biology. Flowers reward pollinators with food used in direct consumption or to provision nests (e.g., nectar, pollen, oils), or with materials used in nest building or sexual reproduction (e.g., resins, fragrance chemicals). Some flowers are deceitful in attracting animals, whereby they mimic oviposition sites, mates, or food sources of pollinator animals; other flowers have a nursery pollination system where they provide a breeding site for their pollinators. In each plant species, flower visitation is restricted to only a subset of the local flower-visiting animals by the phenology of bloom (e.g., seasonal, day versus night); the type, quality, and quantity of food rewards; the floral morphology, which limits access to the rewards; and the floral advertisement stimuli, such as colors, shapes, and scents, which variously stimulate animals to seek, locate, and land on the flowers. Most flowers are visited by a diversity of animal species, but only a few may actually effect pollination, as revealed in careful field studies (see, e.g., Kandori[6]). Furthermore, the assemblage of animal species that pollinate may differ in their relative importance in the plant's pollination and in the selective pressures they exert on floral traits, including floral scent chemistry.[7–9] Therefore identifying pollinators among visitors and identifying the most important and floral trait selecting of the pollinators is essential in any attempt to determine if floral scent chemistry is associated with and being selected by the type of pollinator.

    The goal of this chapter is to provide a general survey of the literature of floral fragrance chemistry in relation to pollinator type, and represents a tentative effort to uncover possible associations between floral scent and pollinators, whereas the two following chapters provide more detailed studies on the overall behavior of two specific types of pollinators. For the purpose of the general survey in this chapter, pollinator animals are grouped into taxonomic units or, where applicable and possible, into "functional groups" as suggested by Fenster et al.[9] The literature on floral scent chemistry has vastly increased in the past decade since a somewhat similar review of the role of floral volatiles in insect biology,[1] and greater emphasis is directed here to the phylogenetic, ecological, and ethological aspects of floral scent chemistry in pollination. This review focuses on the floral scents of plant species for which the pollinators are known or relatively well characterized; however, in most species, comparative information on the importance of different pollinators is not available, and plants are therefore grouped by the types of pollinators as given in the literature. It must be kept in mind that information used here to classify plants by pollinator type is subject to change with more detailed studies of the pollination and reproductive biology of each plant species, and plants placed under one pollinator group may in fact belong under a different group. Furthermore, the distinctions between the groups of pollinator types are not clear-cut, with some boundaries being more blurry than others. Plant species studied with respect to floral scent chemistry, but for which no pollination information or only visitor lists were located, have generally been omitted, except when the floral scent is particularly distinctive (e.g., some species with sapromyophilous flowers).

Attempts to group plants according to their pollination biology into pollination syndromes, which are based on combinations of floral characteristics typically associated with particular classes of pollinators, date back more than 100 years.[4,9,10] Fragrance types have been added to the floral traits, but they are subjectively defined (to the human nose) and not based on a foundation of chemical studies, making them particularly fuzzy characteristics.[2] Early studies aimed at determining if floral scent chemistry does in fact covary with distinct pollinator guilds, and if there has been convergent evolution in floral volatile compositions in plants pollinated by particular types of animals, were made by the laboratory of Gunnar Bergström in Sweden, where Jette Knudsen, Lars Tollsten, and Susanna Andersson undertook chemical surveys of floral scents in species pollinated by moths, bats, butterflies, and hummingbirds.[11–14] Other researchers have followed, and the amount of data currently available provides deeper insight into possible trends, or lack thereof, in floral scent chemistry vis-à-vis pollination by different groups of animals. Among the most informative floral scent studies are those of taxonomic plant groups that either display a range of pollination syndromes or represent primarily one syndrome while having a few species pollinated by a different group of animals; these studies provide excellent opportunities to determine if the shifts in pollinators are correlated with parallel shifts in floral scent chemistry (e.g., Nyctaginaceae,[15] *Nicotiana* in Solanaceae,[16] Caryophyllaceae,[17–19] *Tritoniopsis* in Iridaceae[20]). Most importantly, any effort to uncover relationships between floral scent chemistry and pollinator types requires an integration of methodological approaches encompassing chemical scent analyses and behavioral studies (to uncover which volatiles play a role in attracting pollinators and eliciting behavioral responses necessary for pollination), as well as electrophysiological investigations (electroantennographic methods to determine the sensitivity of an animal's olfactory sensory organs to floral volatiles). Information on these different facets are only occasionally available, and are included here wherever possible.

The classification of fragrances is highly challenging, given the combinatorial potential of the wide diversity of floral scent volatiles and the variation in their relative abundance.[2] Accordingly, flowers can present flower-visiting animals with a virtually unlimited universe of species-specific odor blends that can be learned and recognized during foraging. The diversity of volatiles emitted from flowers can be grouped into the main chemical classes of fatty-acid derivatives, terpenoids, and benzenoids, as well as nitrogen- and sulfur-containing compounds.[21] Some volatiles are nearly ubiquitous in floral scents, others are found in only certain species. However, not all fragrance compounds occur independently, given that metabolites of the same biosynthetic pathway branches are often emitted together.[22] In evaluating the chemical profile of a floral scent, it must be remembered that reports on fragrance chemistry are never final: different methodological approaches may yield different scent compositions in the same plant species, especially in the relative proportions of individual volatiles. In addition, floral scents of any given species display natural variation in the number and relative abundance of constituent volatiles between plant populations, within populations, within a plant, and within a flower, both temporally and spatially. These features are mentioned here where appropriate. As was very aptly summarized by Raguso,[2] floral scent is best described as a mosaic product of biosynthetic pathway dynamics, phylogenetic constraints, and balancing selection due to pollinator and florivore attraction. And all conclusions drawn herein must be viewed in this light.

8.2   INVERTEBRATE POLLINATORS: INSECTS

8.2.1   Generalist Pollination (Diverse Insects)

Many plant species fall under a generalist pollination syndrome, where the flowers are pollinated by a diversity of insects (beetles, flies, bees, butterflies) that feed on the exposed nectar and pollen, presented in open and typically bowl-shaped flowers.[4,5,10] Although various insects are thought to be pollinators, some may be more important and therefore drive selection in scent chemistry more strongly than others; more detailed pollination studies are necessary to uncover the relative effectiveness of the different pollinators.[6,19,23,24] Typical examples of families with species displaying this pollination syndrome include Rosaceae, Apiaceae, Arecaceae, and Ranunculaceae (Figure 8.1).

The floral scents in general do not show any particular unifying pattern other than that they contain the three major groups of volatiles (fatty acid derivatives, terpenoids, and benzenoids), with one often predominating over the others. The most commonly encountered dominant scent components in the examples below are the terpenoids β-ocimene and pinene. Most of the scent compounds are common floral volatiles, which may potentially make the flowers attractive to a range of insect pollinators.[25] In *Conopodium majus* (Apiaceae), where the chemical scent profiles differed among populations and mainly in terms of the dominant compounds, the authors attribute this variation to possible adaptations to differences in local pollinator fauna.[26] The presence of nitrogen-bearing volatiles in several species, especially in the Apiaceae, might reflect an adaptation to attract muscoid flies, and these species may belong more appropriately to the pollination category of myophily.

FIGURE 8.1 (See color insert following page 178.) Example of generalist pollination: *R. acris* (Ranunculaceae) has bowl-shaped flowers with pollen and nectar readily available to various insects; the flowers are visited and pollinated by flies, beetles, micropterigid moths, and solitary bees. Shown here is the bee *Chelostoma florisomne* (Megachilidae), which collects pollen only on *Ranunculus* species. Photo by Dr. Heidi E.M. Dobson.

Species with generalist pollination that display scents in which terpenoids clearly dominate include *Rubus* (Rosaceae),[27,28] *Ranunculus* (Ranunculaceae),[29] and *Chamaedora linearis* (Arecaceae), which is visited by drosophilid flies, nitidulid beetles, and small bees,[30] each with primarily *trans*-β-ocimene. *Listera ovata* (Orchidaceae) contains predominantly linalool and *trans*-β-ocimene;[31] *Ceratonia siliqua* (Leguminosae) predominantly linalool and its oxides;[32] the palm *Geonoma macrostachys* (Arecaceae), which is visited by a diversity of insects (beetles, flies, bees) while being possibly pollinated mainly by male euglossine bees and syrphids, shows a dominance of *trans*-farnesene.[30] Several Apiaceae are dominated by mixtures of terpenes, namely *Carum* with *trans*-β-farnesene, β-caryophyllene, and limonene, *Laserpitium* with α- and β-pinene and myrcene,[33] and species that also have nitrogen-containing volatiles, namely species of *Conopodium* with α-pinene, sabinene, and caryophyllene,[26] *Aegopodium* with limonene and β-pinene, *Anthriscus* with sabinene, myrcene, and α-pinene, and *Heracleum* and *Pastinaca* with *cis*- and *trans*-ocimene.[33]

Floral scents dominated by benzenoids include those of two willows (*Salix*, Salicaceae), with benzenoid ethers (1,4-dimethoxy benzene) accompanied by low representations of various terpenoids,[34] and *Filipendula vulgaris* (Rosaceae), with 2-phenylethanol, along with some green-leaf volatiles.[35]

Some species have scents in which the predominant volatiles represent different chemical classes, such as *Trimenia* (Trimeniaceae), with 2-phenylethanol followed by 8-heptadecene.[36] Those that display mixtures of terpenoids and benzenoids are exemplified by *Angelica* (Apiaceae), with myrcene, limonene, linalool, β-ocimene, and the benzenoids 2-hydroxybenzaldehyde and phenylacetaldehyde,[25] and *Daphne mezereum* (Thymelaeaceae), with linalool and *trans*-β-ocimene accompanied by some benzenoids (e.g., benzaldehyde, benzyl alcohol).[37] In *Platanthera stricta* (Orchidaceae), which is pollinated by empidid flies, bees, and geometrid moths, the scent consists mostly of monoterpenoids (especially lilac aldehydes and alcohols, and α-pinene) and benzenoids, but the strong presence of lilac compounds suggests that Lepidoptera might be the more important players in the selection of scent chemistry.[38,39] A special case is presented by parasitic rust fungi that infect species of Brassicaceae and form structures that mimic flowers in appearance, nectar reward, and scent, and attract a variety of insects (bees, butterflies, flies).[40] The scents of these "pseudoflowers," which vary with both the fungus and host plant, are composed mainly of benzenoids (e.g., 2-phenylethanol, phenylacetaldehyde) and fatty acid derivatives (especially green-leaf volatiles), and are generally reminiscent of the scent of flowers pollinated by settling moths.

8.2.2 COLEOPTERA

Beetles visit flowers to feed on pollen, various floral tissues, and other floral exudates,[10] and sometimes to mate and lay eggs, and flowers pollinated by beetles are generally placed under the syndrome of cantharophily.[4] The flowers are described as having no particular shape or visual attraction, being usually dull in color (cream, greenish, or brownish), with the exception of brightly colored (typically red) flowers pollinated by hopliine beetles; they display morphologies that provide easy access to rewards (pollen, nectar, or other flower parts) and have exposed sexual parts.[4,10] Although the floral odors are often characterized as strong, they range widely: they may be fecal, musky, honey-like, fruity, fermented, or undetectable.[41] Beetle-pollinated plants show a range of pollination systems (reviewed by Bernhardt[41]); included is pollination by deceit, where flowers depend on beetles associated with carrion or dung to transfer pollen, and this is discussed under flowers pollinated by insects associated with decaying matter. The number of beetle species that serve as pollinators can be subdivided into as many as 85 genera in 34 families, but the overwhelmingly best represented are members of the Scarabaeidae, and many plant species are pollinated by a combination of species. Beetle pollination overall is more common in families with woody habits and has been documented most extensively within tropical moist and Mediterranean or warm temperate regions, but the majority of floral scent studies have been conducted on tropical species. Some distinction can be made between beetle pollination in the warmer tropical areas and in the temperate regions.

8.2.2.1 Tropical Dynastine Scarabaeidae

Flowers pollinated by scarab beetles (Scarabaeidae) of the subfamily Dynastinae and tribe Cyclocephalini are found in the tropics, belong to various families, and show convergent evolution in typically having chamber blossoms in which thermogenesis is common, anthesis is nocturnal, and floral scents are strong.[10,41,42] The beetles are attracted to the flowers on the first evening of anthesis, when the flowers are in the female stage, and remain there until the second evening, when the flowers enter the male stage and beetles are released, covered with pollen. This mode of pollination has been rather extensively documented and floral scents analyzed in several species.

   The scent chemistry of some species show similarities in containing methoxylated benzenoids, which may be present as nonmajor compounds (*Nymphaea*, Nymphaeaceae),[43] as codominants with esterified fatty acid derivatives (*Nymphaea*)[44] or with jasmone (*Montrichardia*, Araceae),[45] or as dominant constituents (*Phytelephas*, Arecaceae).[46] Other species contain predominantly esterified fatty acid derivatives (*Victoria*, Nymphaeaceae),[47] methoxylated pyrazine (*Phytelephas*),[46] or oxygenated terpenoids (*Cyclanthus*, Cyclanthaceae;[48] and *Magnolia*, Magnoliaceae[49–51]).

8.2.2.2   Other Tropical Beetles

In the tropical and subtropical regions, flowers pollinated by nondynastine beetles typically have either small floral chambers or fall into the brush or bilabiate flower forms.[41] Flowers with small floral chambers contrast with the flowers pollinated by dynastine beetles in being primarily diurnal in anthesis.[42] Brush blossoms usually consist of many small flowers with reduced perianths and massed together, as is common in some palms, and small-bodied beetles are the major pollinators; the bilabiate flower form is confined principally to the orchids.[41] The beetle pollinators encompass members of many families, including the Nitidulidae, Scraptiidae, Curculionidae, Staphylinidae, and various combinations thereof, as found also in those species for which floral scents have been analyzed.

There seems to be no particular pattern in floral scent chemistry that unifies these plants. Esters are common, both of fatty acid and benzenoid derivation, but scents dominated by terpenoids are also common, making any generalization about scent chemistry untenable. The scarab beetle *Pachnoda marginata*, which is native to equatorial Africa and destructively feeds on flowers, displays a wide breadth of attraction to floral scent volatiles and may not be much different in its olfactory orientation from pollinator beetle species. It is behaviorally most strongly attracted to esters, as well as aldehydes, acids, terpenoids, and nitrogen-containing compounds (e.g., methyl salicylate, methyl benzoate, cinnamic aldehyde, anethole, eugenol, isovaleric acid, butyl butyrate, linalool, and methyl anthranilate).[52]

The Annonaceae and Magnoliaceae both have a large representation of species falling into this pollination type, and in each family a wide variation in floral scent chemistry is evident. Species belonging to four genera within the Annonaceae and pollinated by similar insects have scents that may be variously dominated by fatty acid-derived esters, benzenoid esters, terpenoids, or naphthalene, depending on the genus.[53] Within the genus *Magnolia* (Magnoliaceae) alone, the scent may be composed mainly of benzenoids, terpenoids, or fatty acid-derived methyl esters,[50,54] or in the case of *Magnolia grandiflora*, monoterpenes and fatty acid-derived methyl esters.[54,55] In *Magnolia kobus*, pollinated by Nitidulidae and Scraptiidae, high intraspecific scent variation has been documented among both individuals and populations in terms of the relative proportion of major volatiles, which include linalool and its oxides, other monoterpenes, benzenoids, and nitrogen-containing compounds.[56]

Greater uniformity in scent chemistry is observed in curculionid-pollinated *Eupomatia* species (Eupomatiaceae), which have floral scents containing mainly esters of fatty acid derivatives.[57] Similarly, species of *Zygogynum* and *Exospermum* (Winteraceae) that are pollinated by weevils emit scent consisting almost entirely of fatty acid-derived esters and their alcohols.[58,59]

With respect to other plant taxa, the floral scents of three *Clusia* species (Clusiaceae) pollinated by undetermined beetles are dominated by benzenoids (methoxylated *p*-anisaldehyde in one species and a mixture in another) or fatty acid-derived hydrocarbons, and may also include nitrogen-containing compounds.[60] In the gymnosperm cycads (Cycadaceae), one species of *Macrozamia* seemingly pollinated by Tenebrionidae beetles has a cone scent dominated by monoterpenes.[61] Other cycads pollinated by Curculionidae emit scents dominated by linalool or other monoterpenes in the case of *Macrozamia*,[61,62] or blends of either aliphatic hydrocarbons (e.g., 1,3-octadiene) mixed with some terpenoids (linalool) or benzenoids (methyl salicylate) accompanied by various aliphatic hydrocarbons in the genera *Zamia* and *Encephalartos*.[63]

In the Arecaceae, palm species pollinated by beetles have floral scents that are generally characterized by large amounts of one or a few dominant compounds. In *Ceroxylon*, *Mauritia*, and *Geonoma* species pollinated by Nitidulidae and Chrysomelidae beetles, the principal volatiles are fatty acid derivatives, especially aliphatic hydrocarbons (C_{13}–C_{15}).[30] Similarly in *Wettinia* palms (Arecaceae), which are thermogenic and appear to be pollinated mainly by curculionids and nitidulids (although visited also by bees and Hemiptera), the floral scents contain a very high predominance of fatty acid derivatives, mainly aliphatic hydrocarbons (e.g., dodecane) or 3-pentanone.[30] In the phytelephantoid palms that are pollinated by various small beetles in addition to some dynastines, the scents not only are dominated by mainly methoxylated benzenoids and pyrazines, but also contain other benzenoids, terpenoids, and fatty acid derivatives,[46] and *Elaeis guineensis*, which is pollinated by curculionids, has a scent also dominated a methoxylated benzenoid.[64] A special situation is presented by the dwarf palm *Chamaerops humilis*, which displays a nursery pollination system with curculionids (the beetles feed, oviposit, and complete larval development in the flowers): this species is exceptional in that the "floral" volatiles attracting pollinators during bloom are emitted by the leaf instead of flowers; the scent consists mainly of terpenoids, particularly two monoterpenes (*trans*-β-ocimene, *trans*-β-farnesene) and one unidentified sesquiterpene.[65–67]

8.2.2.3   Beetles of Temperate Regions

Outside the tropics, beetle-pollinated flowers typically have open, bowl-shaped morphologies or are in flat-topped inflorescences (similar to the brush flower form), where pollen and nectar are easily accessible, and do not have floral chambers to trap insects. Regions with Mediterranean climates have an additional beetle pollination mode referred to as the "painted bowl," where the often reddish perianth forms a deep cup with exposed sexual parts; floral scents appear to be very weak or undetectable to humans[41] and the beetles are mainly hopliine Scarabaeidae and Glaphyridae. Beetle-pollinated flowers in the temperate regions are associated with various families that include among others the Nitidulidae, Cantharidae, Staphylinidae, Chrysomelidae, Byturidae, Malachiidae, Dasytidae, Bruchidae, Melyridae, and Scarabaeidae.[5,68] Many plant species visited and pollinated by beetles fall into the generalist pollination syndrome, but some species appear to rely more heavily on beetles than other insects and are included in this section.

The floral scent chemistry generally varies widely among species. In *Actaea* (Ranunculaceae) species pollinated by byturids and other beetles, the floral scents consist mainly of monoterpenoids, with some fatty acid derivatives and only trace benzenoids.[69,70,71] In the two species of *Hypecoum* (Papaveraceae) that are documented to be pollinated mainly by various beetles (Malachidae, Bruchidae, Scarabaeidae, Dasytidae), the scents are mostly benzenoids (e.g., benzaldehyde and ethers), including various nitrogen-containing compounds (e.g., benzeneamines, *o*-nitroanisole, nicotinaldehyde).[72] *Arum creticum* (Araceae), which emits a freesia-like odor, attracts scarab, staphylinid, and bruchid beetles, but is pollinated by bruchids, has a floral scent strongly dominated by benzyl alcohol accompanied by other benzenoids (methyl salicylate, methyl benzoate, benzaldehyde, eugenol), fatty acid-derived alcohols and esters, terpenoids (e.g., geraniol, citronellol), and some nitrogen-bearing volatiles (e.g., indole).[73] In *Cyathostegia mathewsii* (Leguminosae), the flowers appear to be pollinated by scarab beetles (subfamily Melolonthinae) and emit a scent consisting mainly of fatty acid-derived alkenes (especially heptadecene), accompanied by linalool and its oxides, and lesser amounts of other terpenoids (e.g., *trans*-β-farnesene), various benzenoids (including methoxylated compounds), and *cis*-jasmone.[74] The scarab beetle *Phyllopertha horticola* (subfamily Rutelinae), which feeds destructively on flowers, is strongly attracted to geraniol, eugenol, and methylanthranilate, which may be important in orienting this and other beetles to host flowers.[75]

8.2.3 Diptera

Flies are an important order of flower-pollinating insects worldwide, and most families have members associated with flowers.[76] Families especially prominent are the Syrphidae, Bombyliidae, Muscidae, Tachinidae, and Anthomyiidae. Flies form a major portion of the pollinators at high elevations and latitudes,[77,78] where they replace the small bees that are most prevalent at lower altitudes, but Bombyliidae show decreases at higher elevations in parallel with bees. In general, flies are generalists in their associations with flowers, visiting a variety of flowers to feed on pollen and nectar; some flies are attracted to flowers as sites for mating and oviposition, and these relationships are more specialized.[4]

Pollination by flies encompasses a broad range of flowers in terms of both floral morphology and scent, reflecting the wide diversity of feeding, mating, and oviposition sites used by this large group of insects.[4,5] Short-tongued flies visiting flowers for food are restricted to flowers that have an open, bowl-shaped morphology with readily accessible nectar and exposed sexual parts (including pollen), such as the Apiaceae,[4] and many of the plant species visited are also pollinated by a variety of insects, including bees and beetles, and can be placed in the category of generalist pollination. In contrast, long-tongued flies (e.g., Bombyliidae) visit flowers that produce nectar in floral tubes of variable lengths,[79,80] such as flowers frequented by bees and Lepidoptera. Flies pollinating flowers that mimic oviposition sites may have various morphologies and scents. A single fly species may visit distinct flowers for different purposes (i.e., food versus oviposition) and therefore pollinate flowers falling under different pollination syndromes. This can lead to complex behavioral responses of flies to flowers, as demonstrated by newly emerged flies of Calliphoridae and Sarcophagidae, where attraction to flowers was determined by an interplay of floral scents and colors: the flies preferred yellow in the presence of sweet scents, which signal food sources, and brown-purple in the presence of the odor of excrement, which is typical of egg-laying sites.[81] Floral scent is certainly important in eliciting food-seeking behavior in flies that otherwise breed on decaying animal material, as demonstrated when several species of calliphorid, sarcophagid, and anthomyiid flies resting on a wall were aroused when visually hidden nectar plants of *Cicuta* (Apiaceae) were placed in their vicinity, and eventually located the flowers.[82]

8.2.3.1 "Typical" Myophily

Among the broad group of plant species visited by food-seeking flies, there appear to be ones that rely more heavily on flies as pollinators and form a syndrome of myophily. In myophilous plants, the floral scents are described as generally sweet, being either heavy or light.[5] Within the typically sweet scent, there tends also to be a slight hint of urine or sweat, which may be caused by the presence of various acids or nitrogen-containing compounds; this is readily evident in flowers of hawthorn, and many Apiaceae.

One species fitting the myophilous syndrome is edelweiss (*Leontopodium alpinum*, Asteraceae).[83] Although the floral scent has the green-leaf volatile *cis*-3-hexenyl acetate as its most prevalent compound, it has a heavy representation of a variety of acids, especially acetic acid and 3-methyl-2-pentenoic acid, as well as benzaldehyde, several monoterpenes (e.g., α-pinene, limonene), and the sesquiterpene germacra-1(10),5-dien-4-ol. Another species is *Gypsophila paniculata* (Caryophyllaceae), where the scent contains major amounts of methylbutyric acids and smaller representations of ethanol along with the dominant volatile ocimene.[84] Several fly-visited species of *Theophrastra* and *Deherainia* (Theophrastaceae), for which the fly pollinators have not been identified, similarly have sour odors dominated by alcohols and esters of fatty acid derivatives (e.g., 3-methyl-1-butanol, butyl acetate), and containing distinctive compounds such as 3-methylbutanoic acid and methylpyrazines.[85]

Species heavily pollinated by flies, but also by other insects, include *Sambucus nigra* (Caprifoliaceae), with a complex floral scent dominated by 2,6,6-trimethyl-2-vinyl-5-hydroxytetrahydropyran, accompanied by lower amounts of linalool, 2-methyl-2-vinyl-tetrahydrofuran, 3,7-dimethyl-1,5,7-octatriene-3-ol, ethyl acetate, and 2-phenylethyl alcohol[86]; *Crataegus* (Rosaceae), with a fragrance consisting principally of a mixture of several fatty acid-derived alcohols, aldehydes, and ketones (e.g., 3-methyl-butanol), benzaldehyde, dimethylnonatriene, methoxylated benzenoids, accompanied by several nitrogen-containing volatiles[27]; and *Filipendula ulmaria* (Rosaceae),[87] with a floral scent of mainly methyl benzoate, followed by benzaldehyde, green-leaf volatiles, and the sesquiterpene cadalene, as well as some nitrogen-bearing benzonitriles.[88] *Silene rupestris* (Caryophyllaceae) is reportedly pollinated mainly by flies and has a scent dominated by the green-leaf volatile *cis*-hexenol, accompanied by various terpenoids.[19] To this group can be added various Apiaceae, otherwise considered to have a more generalist pollination, in which the monoterpenoid-dominated floral scents contain a variety of nitrogen-bearing compounds (e.g., nicotinaldehyde, methylanthranilate, isovaleraldoxime, 2,5-dimethylpyrazine) in large amounts (*Conopodium*, *Pastinaca* with more than 10%) to more often small amounts (*Anthriscus*, *Heracleum*, and *Aegopodium*).[26,33]

Among plants pollinated principally by syrphid flies (Syrphidae) are members of the palm genus *Asterogyne* (Arecaceae), and in the single species investigated, the floral scent was mainly monoterpenes, especially linalool and derivatives, with a low representation of benzenoid ethers.[89] In another palm, *Prestoea schultzeana*, which is visited by various flies of Ceratopogonidae, Drosophilidae, as well as Syrphidae, the floral scent is also almost exclusively terpenoids, especially linalool.[30]

8.2.3.2   Midge-Like Flies (Nonsapromyophily)

Species of *Theobroma* (Sterculiaceae) are pollinated primarily by midges, and some bees. In *Theobroma cacao*, which is pollinated by midges of Ceratopogonidae, the floral scent consists mainly of fatty acid derivatives, especially long-chain alkenes and alkanes (C_{15}–C_{17}).[90,91] In contrast, species pollinated by Cecidomyiidae midges (*Theobroma mammosum* and *Theobroma simiarum*) have floral volatiles that are more predominantly terpenoids, to which cecidomyiids seem to show greater attraction.[90,92]

8.2.3.3   Male Dacine Tephritidae

Males of many fruit fly species in the Tephritidae, subfamily Dacinae, have a special pollinator relationship with flowers that is similar to that of male euglossine bees, where the flies visit flowers to feed on floral volatile oils on the surface of petals and sepals. These volatiles and their metabolites are subsequently stored in the rectal gland and released by males, thereby increasing mating success supposedly by more successfully attracting females.[93] There are two major plant volatiles to which 90% of the species worldwide are attracted: methyl eugenol and raspberry ketone (4-(p-hydroxyphenyl)-2-butanone), with no species responding to both.[94,95] Some flower pollinating species respond to other volatiles, including zingerone[96] and *trans*-3,4-dimethoxycinnamyl alcohol.[93] Pollination by male dacine fruit flies, principally of the genus *Bactrocera*, has been reported mainly in Asia and in certain orchids (*Bulbophyllum*, *Dendrobium*) and *Fagraea berteriana* (Loganiaceae) that have floral scents heavily dominated by one of the male-attractant volatiles.[93,96–100] The flowers of *Spathiphyllum cannaefolium* (Araceae), which have a heavy sweet odor, are strongly attractive to a number of species, and this is attributed to the presence of several attractant volatiles, particularly methyl eugenol and benzyl acetate, as well as methylchavicol, p-methoxybenzyl acetate, and fatty acid derivatives.[101]

8.2.4   Insects Associated with Decaying Organic Matter

Flowers that are pollinated by insects associated with decaying organic matter have traditionally been classified under the syndrome of sapromyophily, but this term is somewhat of a misnomer because the pollinators include not only flies, but also beetles. Sapromyophilous flowers are found in various plant families, including Asclepiadaceae, Aristolochiaceae, Sterculiaceae, Rafflesiaceae, Hydnoraceae, Taccaceae, Araceae, Burmanniaceae, and Orchidaceae.[4] The flowers are characterized by colors that tend to be dull and dark (brown, purple) or greenish, often form traps that may be endowed with attractive appendages, and the pollination is typically by deceit, where flowers mimic mating and/or egg-laying habitats. They emit odors that variously resemble the smell of decaying protein, dung, urine, mushrooms, cabbage, or onions. The scents may also vary within the flower or inflorescence, as documented most extensively in species of Araceae, where volatiles from the appendix provide a typically foul-smelling odor, whereas structures inside the chamber may impart a sweet odor;[73,102] a similar situation occurs in *Aristolochia gigantea* (Aristolochiaceae).[103] This is a broad grouping of plants that are pollinated by different groups of flies and beetles, and consequently the flowers emit various types of scents.

8.2.4.1 Beetles and Flies Associated with Carrion and Excrement

Plants with flowers that mimic dead animals or excrement are pollinated by deceit: various beetles (e.g., Silphidae, Staphylinidae, Scarabaeidae, Dermestidae) and flies (e.g., Calliphoridae, Sarcophagidae, Muscidae, Anthomyiidae, Drosophilidae) are attracted by the floral smell, color, and texture, lay eggs in the flower, and thereby effect pollination.[10] Flowers with odors mimicking dead animals (carrion, decaying meat, wet hides) are typically characterized by the presence of sulfur-containing compounds (oligosulfides), together with fatty acid-derived acids, but the chemical composition of the scents vary among species. Most species investigated chemically are members of Araceae or Orchidaceae. In Araceae species with odors of decaying meat, the floral odor is often a gaseous stench that seems to be simple in composition and consists principally of sulfur-containing volatiles, as in numerous species of *Amorphophallus*.[73,104] Carrion beetles (Silphidae, Staphylinidae, Scarabaeidae) are the most often recorded insect visitors, and in a few species they have been shown to be the pollinators as well.[73] In *Amorphophallus rivieri*, past studies of floral scent chemistry reported the presence of various nitrogen compounds, including ammonia, trimethylamine, and skatole,[105] but the application of more precise methodological analyses has shown the scent to consist mainly of aliphatic hydrocarbons and dimethyl di- and trisulfide.[106] The dead-horse arum (*Helicodiceros muscivorus*), which is pollinated principally by female flies of Calliphoridae (*Calliphora* and *Lucilia*) and some Anthomyiidae,[107] emits an odor that, similar to a carcass, contains mainly dimethyl mono-, di-, and trisulfides;[108,109] these sulfur compounds were also found to be the volatiles of both the floral and carcass odors that were behaviorally active on the pollinator flies.[109] The parasitic plant *Hydnora africana* (Hydnoraceae), which is pollinated by a variety of beetles attracted to its strong fetid smell of wet decaying hides, has a more complex floral odor in which 15 volatiles seem to provide the main character of the scent. These include disulfides, as well as aliphatic acids and aldehydes, terpenoids (camphor, geranyl acetone), and benzenoids (e.g., benzaldehyde, *p*-cresol).[110]

Several orchids with unpleasant floral scents have been studied chemically, but the pollinator insects have not been determined.[111] *Cirrhopetalum fascinor* not only has a putrid scent dominated by terpenoids (especially caryophyllene), but also contains small amounts of butyric acid and short-chain aliphatic aldehydes, whereas in *Cirrhopetalum robustum*, which has a particularly strong and penetrating stench, the odor consists of a mixture of fatty acid-derived acids, monoterpenoids (e.g., nerol, limonene), benzenoids (e.g., benzaldehyde, methyl salicylate), and several amides (e.g., N,N-dimethyl formamide, N-methyl acetamide). *Cirrhopetalum gracillium* has an odor more reminiscent of wet dogs or algae, and its floral scent is composed of aliphatic aldehydes, alcohols, esters, acetic acid, geranyl acetone, and nitrogen-containing compounds (isoleucine methyl ester, indole). Another species with unknown pollinators but complex floral odors (more than 100 volatiles) is *A. gigantea* (Aristolochiaceae): on the first day of anthesis (male phase), the entrance to the floral chamber emits dimethyl disulfide and the lip, which resembles carrion, produces citral, whereas on the second day (female phase), the inside of the chamber fills with ocimenes, linalool, and acyclic sesquiterpenes.[103]

In species with floral odors resembling those of urine and excrement, the scent may contain a greater diversity of volatiles, with sulfides being rare or absent. In *Arum maculatum* (Araceae) pollinated by female owl midges (*Psychoda*, Psychodidae), up to 95 compounds have been identified.[112] The main components varied among the different regions of the inflorescence, but included the nitrogen-containing indole, various sesquiterpenes (β-caryophyllene, α-humulene, germacrene B + viridiflorine), aliphatic ketones (2-heptanone), and the benzenoid *p*-cresol. The fecal and urinous character of the floral scent is attributed to indole, *p*-cresol, and 2-heptanone, and parallel analyses of cow dung volatiles, the pollinator's normal breeding habitat, revealed a predominance of aliphatic hydrocarbons with the consistent presence of aliphatic aldehydes and *p*-cresol. All three volatiles are attractive to pollinating flies, with *p*-cresol showing the greatest activity and appearing to be a key chemical in the attraction of flies to flowers.[73] Other *Arum* species with dunglike floral odors and pollinated by owl midges, biting midges (Ceratopogonidae), other midges, dung flies (Sphaeroceridae), and beetles (especially Scarabaeidae and Staphylinidae) similarly have scents dominated by fatty acid hydrocarbons, 2-heptanone, methyl (iso)butyrate, ethanol, *p*-cresol, indole, skatole, 2-nitro-*p*-cresol, and sesquiterpenoids.[73]

The voodoo lily (*Sauromatum guttatum* [Araceae]) emits a foul odor that is described as both carrion-like and having excrement tones, and contains volatiles common to both types of odor. The inflorescence attracts a variety of insects (beetles, flies, and even bees), which might be attributed to the likewise diverse chemical profile of the odor, consisting of up to 163 volatiles,[102,113] but flies and beetles have been suggested to be the main pollinators.[113] The scent is dominated by terpenoids, both a variety of monoterpenes (e.g., limonene, β-citronellene) and sesquiterpenes (e.g., β-caryophyllene), fatty acid derivatives (e.g., ethanol, 3-hydroxy-2-butanone), and benzenoids that include toluene, *p*-cresol, and *m*-cresol, and also contains some indole as well as sulfur compounds, which in other studies were found to be prominent components.[114,115] Here again, none of the more recent studies detected any amines that were reported earlier using less precise analytical methods.[105] *Amorphophallus eichleri* also produces an odor with both dung and carrion overtones, and its scent consists of dimethyl oligosulfides together with indole, 2-heptanone, and phenylethyl alcohol.[73]

Other floral odors considered unpleasant to humans and placed under the syndrome of sapromyophily are those with fishy or cheesy tones. In the genus *Amorphophallus*, species with scents resembling those of fish typically include amines, such as trimethylamine, whereas those that smell like strong cheese include acids, such as acetic acid and isocaproic acid.[104]

8.2.4.2   Flies Associated with Decaying Vegetation

In this category are flowers pollinated by flies that are generally associated with decaying plant matter, but which visit flowers to feed on nectar and/or pollen and even to mate and lay eggs. Adult Phoridae flies are most abundant around decaying vegetation,[116] but some species visit and pollinate *Herrania cuatrecasana* (Sterculi-aceae), which has a floral scent consisting of fatty acid derivatives, sesquiterpenes, and indole.[90] Some Chloropidae breed in decaying vegetation and excrement[116] and one species (*Elachiptera formosa*) is the main pollinator of *Peltandra virginica* (Araceae), which has a brood site-based pollination system, where the flies feed on pollen, mate, oviposit, and complete larval development in the floral chamber.[117] The floral scent contains various monoterpenes and aliphatic volatiles, but is dominated by two compounds, 4,5,7-trimethyl-6,8-dioxabicyclo octane and an analog, which change in their relative representation between the female and male stages of the blooming inflorescence. Behavioral studies suggest that gravid flies searching for oviposition sites use the ratio of these two volatiles as a cue to locate inflorescences of the right stage and age.

8.2.4.3   Flies Associated with Decaying Fruits

Fruit flies in the Drosophilidae, which breed in decaying fruit, carrion, and excrement, are attracted to sapromyophilous trap flowers of *Arum* (Araceae) and *Aristolochia* (Aristolochiaceae) species,[5] and are also described as pollinators of several other species, including *Nypa* palms[118] and *Alocasia odora* (Araceae).[119] No particular fragrance profile appears to be associated with drosophilid pollination, although volatiles with fermentation-like odors have been reported in two of the three species analyzed. The floral scent of *Nypa fruticans* is principally carotenoid derivatives.[120] However, *Asimina triloba* (Annonaceae), in which *Drosophila* and other flies are potential pollinators, has a floral scent reminiscent of fermentation and composed mainly of deoxygenated fatty acid derivatives (2,3-butanediol and 3-OH-2-butanone), as well as ethanol, 3-methyl-1-butanol, and *trans*-caryophyllene.[121] Similarly *Arum palaestinum*, which differs from its dung-smelling relatives in emitting an odor of rotting fruit that attracts *Drosophila*, has a scent consisting almost entirely of ethyl acetate with lesser amounts of ethanol and acetic acid.[73]

8.2.4.4   Flies Associated with Fungi

Flies that breed in mushrooms not only are mainly fungus gnats in the Mycetophilidae, but also include members of Sciaridae, Phoridae, and Cecidomyiidae.[116,122] Some visit and pollinate flowers (e.g., in Araceae, Aristolochiaceae, Orchidaceae),[4] most of which display a pollination with brood site deception. One example is *Asarum proboscideum* (Araceae), where the spadix is transformed to a mushroom-like structure that almost fills the opening of the spathe: both sexes of fungus gnats congregate there and lay eggs, but larvae do not develop. Volatiles released from fungi, both the reproductive mushroom structures and the vegetative mycelium, vary widely among fungal species. They may include large representations of short-chain fatty acid-derived volatiles (e.g., especially 1-octen-3-ol and its derivatives, acids, ketones, and aldehydes), some benzenoids (e.g., benzyl alcohol and benzaldehyde), sulfur-containing compounds (e.g., dimethyl sulfide), selected monoterpenes, and the sesquiterpenoid geosmin.[123–125]

Only a few plant species known to be pollinated by fungivorous flies have been studied with respect to their floral scents, and they show varying similarities to scents emitted by fungi. They include *Dracula chestertonii* (Orchidaceae), which has flowers with a strong mushroom-like odor consisting primarily of 1-octen-3-ol and its derivatives, and other aliphatic short-chain alcohols, ketones, and aldehydes.[111] In members of the genus *Arisaema* (Araceae) that are pollinated mainly by Mycetophilidae and Sciaridae and have kettle trap blossoms, identified odors from five species show a predominance of various aliphatic aldehydes and alcohols (mainly C_3–C_{10}), together with alkanes, ketones and, in two species, sulfur-containing compounds; however, no terpenoids were detected and aromatic volatiles were rare.[126]

8.2.5   THYSANOPTERA

Thrips occur in vast numbers in flowers, with some being highly plant specific, and there is increasing documentation of plant species (including a range of crops) being pollinated by particular thrip species.[10,127,128] A proposed thrips pollination syndrome, thripophily, has a list of floral features that overlap with other syndromes, especially those of bee- and beetle-pollinated flowers. Accordingly, thrip flowers tend to be medium size, white to yellow, with or without nectar, produce small- to medium-size pollen grains, have floral structures providing shelter, and are sweetly scented.[127] Several floral volatiles have been found to be attractants to different thrip species, including anisaldehyde,[129] linalool oxide,[130] and methyl anthranilate,[131] and no chemical patterns are evident in the floral odors of the few thrip-pollinated species analyzed to date.

In *Xylopia aromatica* (Annonaceae), which is pollinated mainly by thrips and secondarily by small beetles, the floral scent consists predominantly of benzenoids (mainly 2-phenylethyl alcohol), accompanied by small representations of nitrogen-containing compounds, fatty acid derivatives, and isoprenoids.[53] Within the gymnosperms, some *Macrozamia* cycads are pollinated by thrips, and their cone scents were strong (nearly 1000 times greater than in congeneric species pollinated by weevils) and contained mainly monoterpenes to the exclusion of linalool; four species had scents that were dominated by β-myrcene, whereas a fifth species was codominated by *p*-cymene and α-terpinene, followed β-myrcene.[61,62]

8.2.6   HYMENOPTERA

8.2.6.1   Bees

Pollination principally by bees, referred to as melittophily, covers plants that vary immensely in floral morphology (e.g., radial to bilateral symmetry, exposed to well-hidden food rewards), color (varying most commonly from white and yellow to blue), as well as fragrance, with no trends emerging in scent chemistry to unify this broad assemblage of plant species.[4,5,10] Within the orchids alone, bee-pollinated species display a wide range of floral scents, which have been described to display tones that are rosy (citronellol, nerol/geraniol, farnesol, 6(*E*)-dihydronfarnesol, benzyl alcohol, phenylethyl alcohol, *cis*-3-hexenol, rose oxide, citronellal, α-terpineol, methyl geranate), iononic (rich in caretonoid degradation products, particularly β-ionone and its derivatives), spicy (complex of phenolic compounds, especially *p*-cresol, vinyl guaiacol, chavicol, eugenol, isoeugenol, and vanilline), aromatic (methyl benzoate, methyl salicylate, and homologues), and all combinations thereof.[111] Bees in general appear to detect a wide range of floral volatiles, and numerous studies have investigated the ability of bees, especially honeybees and bumblebees, to discriminate between individual volatiles and different combinations of volatiles.[132–135] Scented traps aimed at capturing moths and baited with phenylacetaldehyde tend to also attract large numbers of bees.[136]

In this broad category of bee pollination, a distinction can be made among the scents of flowers pollinated by food-seeking bees, by fragrance-seeking male euglossine bees, and by mate-seeking males of various species (attracted by deceit to flowers that mimic females: pollination by pseudocopulation). Given that pollination by pseudocopulation also includes other nonbee hymenopterans (as well as other insect groups), it is treated in a separate section under the Hymenoptera.

8.2.6.1.1   Food-Seeking Bees

No patterns are evident in the scent chemistry of flowers pollinated by food-seeking bees. Therefore, to bring some order to this diverse set of data, plant species are described and grouped according to the chemical classes of their dominant floral scent compounds. Further analysis is needed to tease apart these data to determine if patterns are present in relation to pollination by particular groups of bees.

Most species have floral scents in which one chemical class predominates, but many are dominated by mixtures of chemical classes. One can see a dominance of different chemical classes not only among species in the same genus, but even among subspecies and varieties, as in *Cypripedium* (Orchidaceae) pollinated by solitary bees, where separate taxa are dominated by benzenoids (1,3,5-trimethoxy benzene, benzyl alcohol, benzaldehyde, phenylacetaldehyde, methyl benzoate), fatty acid derivatives (aliphatic esters), terpenoids (lilac compounds, linalool, α-farnesene), or mixtures of terpenoids and benzenoids.[137,138] There can also be a wide variation among populations and floral (color) morphs, as shown in *Corydalis cava* (Fumariaceae) pollinated by bumblebees.[139]

Floral scents dominated by terpenoids are common among bee-pollinated plants and generally have low amounts of benzenoids and fatty acid derivatives. The clearly dominating volatiles in these scents are α- and β-pinene in *Cimicifuga* (Ranunculaceae),[140] *Polemonium* (Polemoniaceae),[141] and *Bartsia* (Scrophulariaceae)[142]; *trans*-β-ocimene in *Vicia, Lathyrus, Medicago* (Fabaceae),[143–147] *Laurus* (Lauraceae),[148] certain *Gustavia* and *Lecythis* (Lecythidaceae),[149] certain *Passiflora*,[150] and *Clarkia* (Onagraceae)[151]; linalool in Geonomeae palms (Arecaceae)[89]; β-phellandrene in *Solanum* (Solanaceae)[152]; humulene in *Cimicifuga*[140]; limonene in *Mirabilis* (Nyctaginaceae)[15] and

Primula elatior (Primulaceae),[153] which is seemingly pollinated mainly by bees[154]; caryophyllene in *Clavija* spp. (Theophrastaceae)[85]; germacrene D in *Centaurea solstitialis* (Asteraceae)[155]; *trans*-α-farnesene in *Lecythis*; α-copaene in *Grias*; and nerol in *Gustavia* (all Lecythidaceae).[149] Co-dominance of two or more terpenoids has been reported in several species. Thus, combinations of limonene and *trans*-β-caryophyllene dominate the scent in *Dactylorhiza* (Orchidaceae)[156]; *trans*-β-farnesene and *trans*-β-ocimene in *Lupinus* (Fabaceae)[35]; linalool and myrcene in *Orchis morio* (Orchidaceae)[157]; linalool and ocimene in an *Iriartea* palm (Arecaceae)[30]; linalool, citronellol, nerolidol, and limonene in *Citrus* (Rutaceae)[158]; citronellol, nerol, geraniol in *Moneses* (Pyrolaceae)[159,160]; geraniolic monoterpenes in some *Passiflora* (Passifloraceae)[150]; *trans*-β-ocimene, caryophyllene, and *trans*-nerolidol in *Disa* (Orchidaceae).[161] Finally, various terpenoids dominate the scent in *Helianthus annuus* (Asteraceae),[162] some *Clusia* species (Clusiaceae),[60] *Viola etrusca* (Violaceae),[163,164] and species in the subtribe Maxillariinae (Orchidaceae) pollinated by meliponine bees.[165] In addition, what are possibly bee-pollinated cacti (Cactaceae) have floral scents containing predominantly various terpenoids including the sesquiterpene dehydrogeosmin.[166] The presence of comparatively higher amounts of monoterpenes, especially limonene, in some species of *Silene* (Caryophyllaceae) is suggested to possibly indicate adaptation to pollination by bees.[19]

The number of species studied that have scents dominated by benzenoids or by both terpenoids and benzenoids are markedly fewer. Among the ones containing mainly benzenoids are *Cucurbita* (Cucurbitaceae), with benzyl alcohol, 1,4-dimethoxybenzene, trimethoxybenzene, as well as some green-leaf volatiles and indole[167]; *Pyrola* (Pyrolaceae), with benzaldehyde and 1,4-dimethoxybenzene[159,160]; *Petunia* (Solanaceae), with benzaldehyde[168]; *Trifolium pratense* (Fabaceae), with acetophenone, 1-phenylethanol, and methyl cinnamate[169]; one species each of *Lecythis* and *Couratari* (Lecythidaceae), with 2-phenylethanol and methyl 2-hydroxybenzoate[149]; *Antirrhinum* (Scrophulariaceae), with methyl benzoate[170]; and *Tritoniopsis* (Iridaceae), with methyl salicylate and lesser amounts of benzyl benzoate and the monoterpene limonene.[20]

Floral scents dominated by mixtures of benzenoids and terpenoids include those from several genera in the Rosaceae, namely *Rosa* species, with variously 2-phenylethanol, orcinol dimethylether, benzyl alcohol, and eugenol, together with geraniol, nerol, citronellol, and their esters and aldehydes[171-175]; *Malus*, with benzaldehyde, citral, and limonene[176]; and *Fragaria*, with benzaldehyde, benzyl alcohol, 2-phenylethanol, α-pinene, limonene, *trans*-β-ocimene, germacrene D, and α-ylangene.[177] Also in this group are members of other families, such as *Centaurea calcitrapa* (Asteraceae), with benzene, caryophyllene, and some germacrene D[178]; *Jacquinia* spp. (Theophrastaceae), with various benzenoids mixed to differing degrees with terpenoids derived from carotenoids[85]; and *Corydalis* (Fumariaceae), with limonene and linalool along with *trans*-methyl cinnamate, among others.[139]

Floral scents dominated by both terpenoids and fatty acid derivatives are represented only by kiwi fruit, *Actinidia* (Actinidiaceae) with α-farnesene, aliphatic hydrocarbons, and green-leaf volatiles,[179] although in another study the fragrances are reported to be dominated by terpenoids,[180] and *Narcissus cuatrecasasii* (Amaryllidaceae), pollinated by large Anthophoridae bees, with fatty acid-derived esters, limonene, and β-ionone.[181,182] Species with scents consisting predominantly of fatty acid derivatives include *Disa* (Orchidaceae) with short-chain (C_8–C_{10}) alcohols and acetates,[161] and some species of *Clusia* (Clusiaceae) with various aliphatic acids, hydrocarbons, and alcohols.[60] In the bee-visited populations of *Papaver rhoeas* (Papaveraceae) from northern Europe, the floral volatiles are mainly aliphatic alkanes and alkenes (especially C_9),[35] which may reflect the ancestral and odorless (to the human nose) state of populations around the Mediterranean region, where the flowers are pollinated by *Amphicoma* beetles (Glaphyridae).[183] Finally, scents with overall similar contributions of volatiles from all three chemical classes have been reported in *Passiflora ligularis*.[150]

A putatively bee-pollinated plant with unusual scent chemistry is *Robinia pseudoacacia* (Fabaceae), in which the floral fragrance consists principally of the nitrogen-containing compounds anthraniladehyde and methyl anthranilate, with lesser amounts of terpenoids and benzenoids (linalool and phenylethyl alcohol)[184]; this profile suggests pollination by Lepidoptera.

Among the plants pollinated by bees collecting oils for larval provisioning, floral scents have been analyzed in *Tritoniopsis parviflora* (Iridaceae) and *Corycium orobanchoides* (Orchidaceae). The scents in both species are similarly dominated by the benzenoid ether 3,5-dimethoxytoluene, accompanied in *T. parviflora* by benzenoid esters (benzyl benzoate and methyl salicylate).[20]

8.2.6.1.2 Fragrance-Seeking Male Euglossine Bees

Male bees in the tribe Euglossini (Apidae) are restricted to the neotropics, where they pollinate intensely fragrant flowers of species belonging primarily to the Orchidaceae, as well as to other families including Euphorbiaceae, Gesneriaceae, Bignoniaceae, Solanaceae, Araceae, and Haemodoraceae.[10] The bees have a special relationship with the flowers they pollinate, in that they collect the scent volatiles and store them in their inflated hind tibiae; the volatile compounds and their metabolites are presumed to play a role in the bee's mating. The attraction of particular bee species is based on the presence of single compounds or species-specific combinations of two or more compounds, and these floral volatiles serve as important isolation mechanisms between sympatric plant species.[185–187]

The most extensively analyzed floral scents of male euglossine-pollinated plants are those of orchids, given that bees are the pollinators of more than 600 species, including all members of the subtribes Catasetinae and Stanhopeinae. These orchids are characterized by the production of short-lived, strong fragrances that contain large amounts of relatively few compounds, mainly benzenoids and monoterpenes.[185,186,188] Most euglossine-pollinated orchid species have scents in which only 10 or fewer volatiles each represent more than 1% of the total scent. Among the common compounds are benzenoids, such as benzaldehyde, benzyl benzoate, methyl salicylate, methyl benzoate, methyl cinnamate, 2-phenylethyl acetate, 2-phenylethyl alcohol, and monoterpenes, especially *p*-cymene, limonene, camphene, α-pinene, myrcene, and 1,8-cineole.[185,186]

In nonorchid species, the fragrances are similar overall. In *Cyphomandra* (Solanaceae), where the fragrant droplets are produced on the stamens connectives, the main volatiles vary among species and include germacrene D, *trans*-β-ocimene,

1,8-cineole/limonene, ipsedienol, benzyl acetate, and benzyl alcohol,[189] whereas in *Tovomita* (Clusiaceae), with the oils produced on filaments, they are mainly terpene alcohols, such as germacrene D-ol, cubebol, and dihydrophytol.[190] In *Dalechampia* (Euphorbiaceae), the fragrance is dominated by limonene, followed by carvone oxide, benzyl acetate, and other monoterpenes[191]; in *Anthurium* and *Spathiphyllum* (Araceae) it is principally ipsdienol, α-farnesene, or eugenol[186,192]; in *Gloxinia* (Gesneriaceae) it is *trans*-carvone epoxide and 1,8-cineole[186]; and in the palm *Geonoma* (Arecaceae), the scent of different populations consistently contains mainly myrcene and derivatives, followed by (*E*)-β-farnesene and its derivatives.[193,194] Overall, nitrogen-bearing compounds occur only rarely in euglossine-pollinated flowers.[185,186,193]

8.2.6.2 Wasps

8.2.6.2.1 *Nectar-Seeking Wasps*

Studies documenting nectar-seeking wasps as primary pollinators are few, and most wasps feed on flowers with readily available nectar, which are typically plant species with generalist-type pollination, such as species of Apiaceae.[5] Flowers that rely on pollination by wasps are described as dull colored (green, reddish-brown), and the floral scent has been analyzed in only few species. The orchid *Disa sankeyi*, which is pollinated by Pompilidae wasps, has a floral scent that is described as sweet-spicy and strong to the human nose and contains at least 65 volatiles, among which (*E*)-cinnamic aldehyde (and related derivatives) and eugenol are most abundant.[195] Also included in moderate representation are monoterpenes, benzenoids (especially benzaldehyde, benzyl alcohol, and various esters), and nitrogen-containing compounds (e.g., phenylacetaldoxime). In contrast, no floral volatiles were detected in *Passiflora sexflora* (Passifloraceae), which appears to be scentless.[150]

8.2.6.2.2 *Fig Wasps*

Species of *Ficus* (Moraceae), commonly known as figs, are characterized by their unique inflorescence (syconium or fig), in which numerous tiny unisexual flowers are borne on a large receptacle shaped into a hollow vessel, with the flowers lining the inner surface and the syconium opening to the outside through a small hole, the ostiole.[5,196] They display a nursery pollination system, where the female wasps pollinate when they lay eggs in female-stage flowers, and the larvae feed on the seeds. Thus, when the syconia are receptive (i.e., at female phase), their odor attracts female wasps (emerging from male-phase syconia), which upon arrival at the fig surface are stimulated to enter and lay eggs, and in the process transfer pollen to female flowers. Adult wasps emerge during the later, male phase of the syconium, and the newly emerged female wasps, loaded with pollen, leave their natal fig and search for another receptive, female-stage syconium. Fig development is synchronous over the entire crown of a tree, requiring that wasps fly to a new tree,[197] which they locate by olfaction. There are approximately 750 species of *Ficus*, each generally with a distinct species of pollinating wasp (Chalcidoidea, Agaonidae).[198] The wasps need to have a finely tuned ability to detect and recognize volatiles of their host species, given that distances between trees can be great and that each wasp lives for only a few hours to a couple of days.[196] The volatiles released by syconia have been studied in several *Ficus* species, and their role in pollinator attraction has also been tested in a few cases.[199] In the first studies exploring syconial volatiles, only gas chromatography (GC) traces were obtained for 7 species,[200,201] but more in-depth analyses accompanied by the identification of compounds have now been conducted in close to 30 *Ficus* species.[197,199,202–205]

In the most extensive studies,[197,199,205] generally few compounds were detected in each species (1 to 47 for each species), and in each case there were usually 1 or 2 dominant compounds that were far more abundant than other volatiles. The scent composition of each species is unique: the combination of dominant volatiles differs among species, and while the identities of some floral volatiles show overlap among species, the percent representation of the shared compounds differs. The volatiles consist mainly of terpenoids (monoterpenoids and sesquiterpenoids), followed by benzenoids and aliphatic compounds. The dominant compounds, however, are mostly one or more of the following: limonene, 1,8-cineole, linalool, β-ocimene, α-copaene, β-caryophyllene, germacrene D, benzaldehyde, benzyl alcohol, alkanes; to this list can be added some compounds that were detected in only single species, such as oxygenated derivatives of linalool, γ-butyrolactone, and tentatively identified furanocoumarins.[202,203,205] Behavioral bioassays with wasps indicate that the syconial volatiles serve in both attracting the wasps and stimulating them to antennate and enter the syconium. Several compounds within the syconial scent appear to be essential for these responses, including the dominant volatiles.[199,202]

Within species, scent variation is manifested mainly between male and female trees or stages, and while the overall number and identity of compounds may differ, the scents overlap in the dominant compounds.[197,202,204] This suggests that there has been little selection for strict chemical mimicry between the sexes, which is not surprising given that male and female syconia are receptive at different times and wasps never have to choose between them.[204] The volatile profiles of individual syconia also change over time, with the number of compounds increasing as a syconium becomes receptive for pollination,[200,201] and with the relative representation of different volatiles changing after the receptive stages.[202,203] In postpollination syconia of *Ficus hispida* and *Ficus carica*, the proportional amounts of some compounds (including dibutyl phthalate, a putative insect repellent) increase while those of others, including the dominant volatiles, decrease.

8.2.6.3 Hymenopteran Pollination by Pseudocopulation

A special form of pollination is exhibited in some orchids, where the flowers attract males of certain bee or wasp species through sexual deception by mimicking the female insects with respect to shape, color, pilosity, and odor: the males attempt to copulate with the flower and in the process effect pollination (see Chapter 10). Pollination by pseudocopulation has been documented in Europe, Australia, and South Africa.[206–208] Through studies addressing the role of different floral cues in mediating this relationship, it has become clear that floral scents are of primary importance in attracting male insects and eliciting copulatory responses on the flower. The uncovering of the chemical basis of this often highly specific insect-flower relationship has been greatly aided by the development of electrophysiological methods that combine gas chromatography and electroantennographic detection (GC-EAD), which reveal which of the many scent volatiles are perceived by the antennal olfactory neurons.[209] The specificity of the floral scent vis-à-vis the insect has been found to be determined by the relative amounts of particular volatile constituents, as found in *Ophrys* species, or by the presence or absence of key compounds, as in *Chiloglottis*.[210,211]

In the European orchid genus *Ophrys*, the floral scents are typically complex mixtures of more than 100 volatiles, but only a subset of these compounds is biologically active in attracting and eliciting mating behavior in male pollinators.[212] In *Ophrys sphegodes*, the scent consists predominantly and almost exclusively of fatty acid derivatives, especially long-chain alkanes, alkenes, and esters, with only a few terpenoids in very small amounts (e.g., *trans*, *trans*-farnesol); the active compounds within this mixture, however, are only 14 alkanes and alkenes.[213] The central importance of a certain subset of aliphatic hydrocarbons in the floral scent has been found in other species, such as *Ophrys exaltata*, where linalool functions in the long-range attraction of male bees, but a subset of alkanes and alkenes is necessary to induce full mating behavior on the flower, and thus pollination.[214] In contrast, sexually deceptive orchids of the genus *Chiloglottis* and *Cryptostelis* rely on only single key volatiles, which may be among the minor components of the floral scent, to attract and elicit appropriate responses in their male wasp pollinators.[207,211]

8.2.7   LEPIDOPTERA

Flower-visiting adult Lepidoptera are mainly nectar-feeding insects, but some of the more primitive groups have chewing mouth parts and feed on pollen as well.[4] The different taxonomic groups are divided here on the basis of feeding biology, special mutualistic relationships with plants, and on whether they are diurnally or nocturnally active.

8.2.7.1   Micropterigid Moths

Members of the Micropterigidae are probably the most primitive group of Lepidoptera, and occur worldwide.[215] The adults have chewing mouthparts, are typically diurnal, and visit flowers to feed on pollen (some feed on fern spores). Plant species documented to be pollinated by Micropterigidae are also often pollinated by beetles or other insects with short mouthparts.

Most prominent are plants of the Winteraceae, particularly *Zygogynum baillonii*, *Zygogynum bicolor*, and *Zygogynum viellardii*, which are pollinated by *Sabatinca* micropterigids and have scents consisting almost entirely of mixtures of fatty acid-derived esters (e.g., ethyl acetate, 2-methyl-propyl acetate, 6-methyl-5-hepten-2-acetate) accompanied by their alcohols and a few monoterpenes.[58,59] Findings from field trials suggest that the esters are critical cues to *Sabatinca*.[59] Other plants pollinated by micropterigids are buttercups of the genus *Ranunculus* (Ranunculaceae), on which *Micropterix calthella* is a common visitor and pollinator, along with various other insects (e.g., small bees, beetles, flies), such that the flowers are usually considered to have a generalist pollination system. *Ranunculus acris* (see Figure 8.1) has a floral scent dominated by *trans*-β-ocimene, with a major representation of farnesene, and in the pollen odor of *Ranunculus* species the main volatile is the lactone protoanemonin.[29,216]

8.2.7.2 Yucca Moths

Species of *Yucca* (Agavaceae) occur in arid areas in North and Central America and rely exclusively for pollination on yucca moths of the genera *Tegeticula* and *Parategeticula* in the Prodoxidae. The plants have a nursery pollination system that involves obligate mutualism with yucca moths, where the plants rely on the adult moths for pollination and the moth larvae rely on developing seeds for food to complete their development.[217] The moths are nocturnally active and use the flowers for mating, oviposition, and larval development. When the females oviposit in the ovary, they also deposit pollen into the stigmatic cavity, thereby pollinating the flowers. The single study examining floral scent in *Yucca* shows that *Yucca filamentosa* flowers emit mainly sesquiterpenes and aliphatic hydrocarbons, with the scent being dominated by (*E*)-4,8-dimethyl-1,3,7-nonatriene, followed by a C_{11} alcohol and 1-heptadecene.[218]

8.2.7.3 Nectar-Seeking Lepidoptera

The three major groups of lepidopteran pollinators in which adult nectar feeding has evolved are the butterflies, settling moths (mainly Noctuidae), and hovering moths (Sphingidae).[10,219,220] A common feature of most flowers they visit is the production of nectar in narrow tubes or spurs (Figure 8.2), since the majority of flower-visiting Lepidoptera have a long proboscis. Flowers of many plants are visited by members of different lepidopteran groups, but information on the relative importance of each in pollination is available for only a few species. The floral scent data are therefore organized here according to the major visitor/pollinator lepidopteran groups documented for each plant species, and the boundaries between the groups are consequently blurry.

Distinction between the scents of flowers pollinated by Lepidoptera in general as compared to other insects can be seen in a survey of the daffodil genus *Narcissus* (Amaryllidaceae).[181] Species that included Lepidoptera (butterflies and moths) among their visitors had floral scents characterized by the presence of indole combined with usually high representations of benzenoid esters, and some linalool and *trans*-β-ocimene; this "Lepidoptera odor" was most typical in species that included moth (mainly sphingid) visitors.

A number of species are pollinated by a combination or different lepidopteran groups that are treated here under separate headings, namely both diurnal (mainly butterflies) and nocturnal (moths) Lepidoptera. In several species, floral scents have been documented mainly during the day. These include *Dianthus gratianopolitanus* (Caryophyllaceae),[221] with a scent containing mainly methylbenzoate, followed by methyl salicylate[222]; *Araujia sericofera* (Asclepiadaceae), with principally phenylacetaldehyde[223]; *Phlox drummondii* (Polemoniaceae), with a mixture of volatiles that

FIGURE 8.2 (See color insert following page 178.) This species of *Aquilegia* (Ranunculaceae) produces nectar in long spurs, where it is accessible mainly to lepidopteran moth pollinators, but bees can also pollinate the flowers when they forage for pollen (shown here hanging from the anthers). Photo by Dr. Heidi E.M. Dobson.

include linalool and β-caryophyllene[13]; and *Phlox paniculata*, with mainly *trans*-β-ocimene, followed by phenylacetaldehyde and 2-phenylethanol. In a few species, a temporal change in floral scent chemistry has been documented between day and night. Thus *Gymnadenia conopsea* (Orchidaceae), which is pollinated by butterflies, noctuids, and sphingids, has a diurnal floral scent consisting mainly of benzenoids (especially benzyl acetate, mixed with benzyl benzoate, eugenol, methyl eugenol, and benzyl alcohol), whereas the nocturnal scent shows changes in various volatiles, with the greatest being a decrease in benzyl alcohol and methyl eugenol, suggesting that these two volatiles might be more crucial for butterfly attraction.[224] Similarly *Gymnadenia odoratissima*, which is pollinated by a mix of butterflies and nocturnal settling moths, has a floral scent during the day that is a blend of benzenoids (especially benzyl acetate and phenylethyl acetate), but at night it displays a marked increase in phenylacetaldehyde, which was also the most attractive single compound to various nocturnal settling moth pollinators when tested in the field. The floral scent characteristics of these species are in general agreement with scent profiles associated with the different nectar-feeding Lepidoptera.

8.2.7.3.1 *Diurnal: Butterflies and Others*

Adult butterflies of many species are important pollinators, visiting flowers to feed primarily on floral nectar, and in some cases even pollen.[4] Pollination by butterflies, or psychophily, is generally associated with flowers that are upright and provide a landing platform, are solitary or clustered, have floral tubes or spurs of variable lengths containing variable quantities of nectar, have diurnal anthesis that may extend into the night, have variable colors that include often red, orange, or pink, and have odors that are described as weak, fresh, and sweet.[4,10,225] While butterflies tend to use visual cues to locate flowers, floral odor may play a role in triggering this searching behavior or in attraction at close range (see Chapter 9).[3]

Floral scents have been analyzed for approximately 30 species of plants visited and in some cases documented to be pollinated by butterflies, and some trends emerge in the floral scent chemistry. In a survey of 22 species in 13 families in Eurasia and the Americas, floral scent complexity varied widely, containing 8 to 65 volatiles,[13] but overall the scents were distinguished by the common occurrence and relatively high abundance of benzenoids, certain terpenoids (especially trans-β-ocimene and linalool), and green-leaf volatile cis-3-hexenyl acetate, the more occasional and lower representation of other fatty acid derivatives, and the rare presence of nitrogen-bearing compounds.

Some of these volatiles appear to be more important in attracting butterflies than others.[13] Among the volatiles that were common and abundant in floral scents, but emitted by both flowers and foliage, were benzaldehyde and benzyl alcohol (phenylmethanol), trans-β-ocimene, and to a lesser extent 6-methyl-5-hepten-2-one and cis-3-hexenyl acetate. However, the commonly occurring benzenoids phenylacetaldehyde and 2-phenylethanol, and the monoterpene linalool were emitted exclusively from floral tissues, suggesting that they may serve as signals to butterflies. This is supported by data on floral scent changes in aging flowers of Lantana montevidense (Verbenaceae), where linalool and its oxides were greatly reduced in emission rate, and benzenoid alcohols, aldehydes, and esters (including again phenylacetaldehyde and 2-phenylethanol) were completely lost, pointing to the likelihood that these volatiles are among the most attractive to butterfly pollinators.[22]

The general trends in scent composition found by Andersson et al.[13] seem to hold for some other floral fragrances analyzed in butterfly-visited and pollinated plants in various families,[184,226–229] including Nigritella (Orchidaceae).[111] However, numerous species deviate variously from the trends. Lantana camara (Verbenaceae), Narcissus gaditanus (Amaryllidaceae), Dianthus carthusianosum (Caryophyllaceae), and Dianthus glacialis (pollinated by diurnally active Zygaena moths) all stand out in having floral scents that are strongly dominated by terpenoids and low in benzenoids; most include rather prominently trans-β-ocimene and some cis-3-hexenyl acetate, and the main benzenoid is benzaldehyde.[13,181,222,230] In the Caryophyllaceae species of the genera Silene and Dianthus (D. deltoides, D. barbatus, and D. armeria) that are pollinated mainly by butterflies, bees, or settling moths, the scents are dominated by fatty acid derivatives (C_6 alcohols and esters), benzenoids (phenylacetaldehyde, methyl benzoate), or combinations of fatty acid derivatives (mainly cis-3-hexenyl acetate and aldehydes) and either benzenoids (benzaldehyde or mixtures) or terpenoids (e.g., limonene, β-caryophyllene, linalool, p-cymene); they often contain small amounts of nitrogen volatiles.[18,19] In Knautia arvensis (Dipsacaceae), which is strongly attractive to butterflies but also visited by bees, the floral scent is dominated by the monoterpene verbenone, which appears to be a key volatile in attracting a specialist Zygaena diurnal moth that contributes to pollination.[231] Another fragrance pattern was displayed by Buddleja davidii (Loganiaceae) and Centranthus ruber (Valerianaceae), both typical butterfly-pollinated plants that have floral scents dominated mainly or in part by the irregular terpene oxoisophorone or its oxygenated derivatives,[13] and to which temperate and tropical butterflies show high antennal sensitivity.[232,233] Accordingly, the presence of 4-oxoisophorone together with benzaldehyde, benzyl alcohol, and benzyl acetate as the main volatiles in the floral scent of *Primula farinosa* (Primulaceae) suggests pollination primarily by butterflies, although the flowers are reportedly visited by other insects as well.[153] Finally, a chemotype of *Cimicifuga simplex* (Ranunculaceae) that is pollinated by butterflies has a scent composed of equal amounts of isoeugenol and methylanthranilate, which are absent in the chemotype pollinated by bumblebees.[234]

Behavioral and to a lesser extent electrophysiological studies corroborate the important role played by the floral volatiles that are suggested here to be associated with butterfly pollination. Flower-naïve butterflies show a higher attraction both to scents of butterfly-pollinated flowers over those of other flowers[235,236] and to scents emitted by flowers over those originating from vegetative plant parts.[236] Furthermore, several species of butterflies exhibit proboscis extension responses most strongly to the benzenoids common in the floral scents of butterfly flowers, including phenylacetaldehyde, 2-phenylethanol, and benzaldehyde, and to 6-methyl-5-hepten-2-one,[226–229] and are significantly more attracted to artificial models scented with these volatiles.[226,227] In *C. simplex*, the two principal volatiles (isoeugenol and methylanthanilate), either alone or together, were shown to be critical in behaviorally attracting butterfly pollinators to the flowers.[234] In some cases, a lack of butterfly attraction to flowers pollinated by other insects may result in part from the presence of floral volatiles that are repellent to butterflies, such as γ-decalactone in *Osmanthus fragrans* (Oleaceae), which is behaviorally deterrent to *Pieris rapae*.[237]

In contrast to the behavioral studies, electrophysiological investigations suggest that butterflies can perceive a wide variety of floral volatiles, except for the highly volatile monoterpenes (e.g., pinene, carene, 1,8-cineole, sabinene, *p*-cymene, limonene).[232,233] They also indicate that antennal sensitivity is not necessarily correlated with behavioral attractiveness,[226–228] although several butterfly species show stronger responses to the volatiles typical of butterfly flowers, including phenylacetaldehyde, benzaldehyde, *trans*-β-ocimene, and oxoisophorone.[232,233]

8.2.7.3.2   Nocturnal: Moths

The nocturnally active Lepidoptera that serve as pollinators are either moths that land when they feed at flowers (i.e., settling moths), which are principally members of Noctuidae, or moths that hover (i.e., hawkmoths) of the Sphingidae family. Flowers pollinated by nocturnal moths are typified by having crepuscular or nocturnal anthesis, nectar in floral tubes or spurs, light colors, and generally pleasant and often strong scents.[4,5] While these combined floral features are generally sufficient to separate flowers pollinated by moths from those pollinated by other animals, it can be difficult to separate flowers relying mainly on hawkmoths from those relying on alighting moths.[238] Furthermore, many plant species are visited by both groups of moths, and evidence on the relative importance of each in pollination is often not available. Consequently flowers for which there are reports of visitation by both moth groups without information on their relative contributions as pollinators are discussed below separately from species reported to be pollinated by mainly one group, although these groupings are tentative and subject to change with the gathering of more complete pollination data.

The scents of flowers pollinated by moths in general are described as having a "white floral" image, whereby they contain acyclic terpene alcohols (e.g., linalool, nerolidol, farnesol) accompanied by relatively simple aromatic compounds (e.g., methyl benzoate, benzyl acetate, methyl salicylate, benzyl alcohol, 2-phenylethanol), nitrogen-bearing compounds (e.g., indole, oximes, methyl anthranilate), which can be present in large amounts, and jasmonates, tiglates, and lactones; the fragrances may also contain *trans*-ocimene as a major component or have a "rosy floral" character by emitting citronellol or geraniol.[111,239]

Among the plants reported to be visited and possibly pollinated by both groups of nocturnal moths are species of *Lonicera* (Caprifoliaceae), which have scents dominated by the terpenoid linalool and its oxides, together with other terpenoids (e.g., germacrene D, α-farnesene, α-terpineol), as well as nitrogen-containing volatiles (e.g., indole, oximes)[154,184]; *Gaura* (Onagraceae), with mainly oximes, followed by benzenoids (e.g., 2-phenylethanol, benzyl alcohol, eugenol, methoxylated compounds)[240]; and *Jasminum* (Oleaceae), with benzyl acetate accompanied by linalool, various benzenoids (e.g., eugenol, *p*-cresol, phenylethyl acetate), *cis*-jasmone, and indole.[43,241] *Platanthera chlorantha* (Orchidaceae) has two terpenoid chemotypes, both with small amounts of benzenoids (e.g., methyl benzoate); one is distinguished by a strong dominance of lilac compounds, and the other by the dominance of linalool compounds accompanied by other nongeraniolic terpenoids.[242–244]

8.2.7.3.2.1 Settling Moths
Flowers pollinated by nocturnal moths that settle on flowers while feeding fall under the syndrome of phalaenophily,[10,219] and the moth families involved include Noctuidae (of which some fly diurnally, such as *Plusia*), as well as others such as Geometridae, Tortricidae, and Pyralidae. The flowers show diurnal and nocturnal or only nocturnal anthesis, have flat platforms with floral tubes or spurs that are relatively short compared to those of hawkmoth flowers (given the shorter proboscides of the settling moths), have variable but light colors (no pure red or white), and have odors described as sweet. Floral scent plays a role in various aspects of flower visitation by settling moths, including long-distance attraction, landing, probing, and associative learning.[22,220,245]

Floral scent chemistry varies quite extensively among taxa, but certain patterns emerge from the data. In an early survey of moth-pollinated flowers in different families, the scents of species with a phalaenophilous syndrome all included both benzenoid esters and monoterpenes (e.g., limonene, ocimene, linalool, lilac compounds, but no geraniolic volatiles), each with a wide range in representation (from less than 10% to 90% of the scent); sesquiterpenes were rare and nitrogen-containing volatiles were occasional and in small amounts.[11] The general trend of scents having a predominance of either terpenoids or benzenoids seems to hold for other species subsequently studied, but the terpenoids have been found to be represented by monoterpenes as well as sesquiterpenes, and the prominent benzenoid volatiles often include aldehydes. The floral fragrances seem to display chemical compositions that are intermediate between those of flowers pollinated by butterflies and hovering moths, and show overlap with both.

In *Nicotiana* (Solanaceae), the scents of two species pollinated by small nocturnal moths are mostly terpenoids (especially irregular ones such as trimethyloxabicycloheptendione and 4-oxo-isophorone) and benzenoids (e.g., benzaldehyde, phenylmethyl acetate), with small amounts of nitrogen compounds.[16] In a survey of Nyctaginaceae, the four primarily noctuid-pollinated species had scents that were high in terpenoids, either monoterpenes (e.g., linalool derivatives, *trans*-β-ocimene) or sesquiterpenes (e.g., dimethylnonatriene, β-caryophyllene); fatty acid derivatives were mainly esters (including *cis*-hexenyl acetate); benzenoids when present were primarily benzyl alcohol, benzaldehyde, anisole, phenylacetaldehyde, and methyl salicylate; and nitrogen compounds were either present or absent.[15] Within the Caryophyllaceae, five species of *Silene* pollinated by settling moths have scents dominated by both lilac terpenoids and benzenoids (various), mainly benzenoids (especially esters), or mainly terpenoids (e.g., linalool, myrcene) along with nitrogen-containing volatiles and fatty acid-derived esters.[17] In the related genus *Dianthus, D. silvestris* contains mainly the monoterpene *trans*-ocimene accompanied by methyl benzoate[222] or a mixture of benzenoid esters,[18] and *D. monspessulanus* contains a blend of terpenoids (especially *trans*-β-ocimene, caryophyllene) together with jasmone, benzenoid esters and ethers, and some indole.[18,222] Other flowers that are known to be highly attractive to noctuids and other settling moths include *Cestrum nocturnum* (Solanaceae), with a scent dominated by benzaldehyde, benzyl acetate, and phenylacetaldehyde[246]; *Abelia grandiflora* (Caprifoliaceae), with mainly phenylacetaldehyde, benzaldehyde, 2-phenylethanol, and some benzyl alcohol[247]; and *Berberis aquifolium* (Berberidaceae), with a codominance of benzenoids (phenylacetaldehyde and some benzaldehyde) and terpenoids (α-pinene, *trans*-β-ocimene, and limonene).[248]

Behavioral studies suggest that certain volatiles are more important in attracting noctuid and other settling moths and in eliciting feeding responses. Phenylacetaldehyde by itself attracts many noctuid species, appearing to be the main attractant volatile in some flowers, and its presence in mixtures of volatiles enhances their attractiveness[136,223,246,247,249–252]; similar attractivity has been described for 2-phenylethanol.[247] The capture of *Spodoptera exigua* (Noctuidae) on sex pheromone-baited lures in the field was significantly enhanced when the lures were supplemented with various volatiles, including phenylacetaldehyde, benzaldehyde, and 3-hexenyl acetate, but the relative impact of each varied with the background crop plant.[253]

Lilac volatiles are also strong attractants. *Silene latifolia* (Caryophyllaceae) has a floral scent strongly dominated by lilac volatiles, together with various benzenoids, and in its nursery pollination by the noctuid *Hadena bicruris* (main pollinator in Europe), the lilac volatiles are the most important of the scent compounds in attracting moths and eliciting landing; however, phenylacetaldehyde is the most active in stimulating feeding (proboscis extension).[254] Interestingly, *S. latifolia* has been introduced to North America, where the scent chemistry shows greater variability than in Europe and where, possibly reflecting this variability, the flowers attract a greater diversity of moths that includes various noctuids, geometrids, as well as sphingids.[255]

Combinations of electrophysiological and behavioral studies suggest that moths detect a range of volatiles, but again point to only a few compounds as being behaviorally relevant. In electrophysiological trials, the antennae of noctuids responded to benzyl acetate, benzyl benzoate, and eugenol in the scent of *Gymnadenia conopsea* (Orchidaceae), whereas those of various settling moths visiting *G. odoratissima* responded to all benzenoids (benzaldehyde, phenylacetaldehyde, benzyl acetate, 1-phenyl-2,3-butandione, phenylethyl acetate, and eugenol); however, field traps baited with single compounds showed only phenylacetaldehyde to be behaviorally attractive.[224] In the flower generalist *Autographa gamma* (Noctuidae), the moth's antennae are sensitive to a wide diversity of floral volatiles in the fragrances of six flowers (benzenoid alcohols, esters, and aldehydes; fatty acid-derived esters; certain monoterpenoids including linalool and oxides; lilac compounds; certain sesquiterpenoids; and nitrogen-bearing compounds), but the scents of some flowers are behaviorally more attractive than others in a wind tunnel.[256] Each of the most attractive scents contains a greater diversity of benzenoids, many of which were not detected in the less attractive flowers (e.g., benzyl alcohol, 2-phenylethanol, methyl benzoate, methyl salicylate). However, in one strongly attractive species, *Platanthera bifolia* (Orchidaceae), lilac aldehydes rather than benzenoids appear to be the key volatiles mediating moth attraction and landing on the flowers,[257] thereby resembling the attraction of *H. bicruris* to *S. latifolia*.[254] In the scent of *Platanthera stricta*, the heavy representation of lilac compounds suggests that the geometrid moth pollinators might be strong selectors of scent chemistry, although the flowers are also pollinated by empidid flies and bees.[38,39]

In studies restricted to electrophysiology, *Lobesia botrana* (Tortricidae), which is strongly attracted to flowers of *Tanacetum vulgare* (Asteraceae), displayed the greatest antennal responses to the dominant floral scent volatile, namely the monoterpenoid β-thujone.[258] In *Cactoblastis cactorum* (Pyralidae), the moth antennae were sensitive to volatiles emitted from the vegetative parts of their host plant *Opuntia stricta* (Cactaceae), but also to compounds typical of floral scents, especially geraniolic monoterpenoids and benzenoids (e.g., benzylacetate, phenylacetaldehyde, 2-phenylethanol, and eugenol).[259]

8.2.7.3.2.2 Hovering Moths
The hovering moths are represented by the nectar-feeding hawkmoths, of the family Sphingidae, which constitute an important class of pollinators in warm temperate and especially tropical habitats worldwide.[220,260] Most hawkmoths fly at dusk or at night, but the hummingbird hawkmoth (*Macroglossum stellatarum*) is diurnally active and visits a wide variety of flowers favored by butterflies.[5] In hawkmoth pollination, referred to as sphingophily, flowers typically are horizontal or pendent, and have no landing platform, since hawkmoths hover while feeding. Flower anthesis is crepuscular or nocturnal, the perianth color is white or pale, the flowers secrete nectar within deep corolla tubes or spurs, and the flower odor is described as strong, heavy, and sweet.[219,220] In nocturnally feeding hawkmoths, either visual cues or floral scent are sufficient to attract naïve and experienced moths from long distances, but both cues are necessary to elicit feeding; in contrast, diurnally active species may ignore odor cues, even though they may feed from fragrant flowers.[220]

While many night-blooming hawkmoth flowers produce intense fragrances, emission rates can vary at least three orders of magnitude among species, and flowers in the field can also vary from being heavily visited to almost ignored.[22,238] There is little evidence of extreme specialization in flower choice by hawkmoths, and field studies show that hawkmoths may visit flowers adapted to other pollinator groups, including bats, hummingbirds, bees, and small moths.

Floral scent chemistry has been investigated in numerous studies of hawkmoth flowers from more than 20 families. Overall, the floral scents can be characterized by a dominance of terpenoids (especially oxygenated), benzenoid esters, nitrogen compounds, or pair-wise codominance combinations; nitrogen-bearing volatiles have been detected in most species. The "white floral" image described by Kaiser[111,239] appears to pertain probably more to hawkmoth flowers than to phalaenophilous ones, but the boundary is blurry. In general, the data suggest that when scents are dominated by terpenoids, linalool is most common, but nerolidol, *trans*-β-ocimene, β-caryophyllene, and farnesene are frequently encountered; when benzenoids dominate, they most often include methyl benzoate as a prominent compound, but others are common (e.g., salicylates, benzyl acetate); and when nitrogen-bearing compounds are abundant, they may be represented by various volatiles (e.g., oximes, indole, anthranilates). In an early chemical survey,[11] the suggestion that hawkmoth flowers differ from settling moth flowers by the presence of oxygenated sesquiterpenes (e.g., nerolidol, farnesol) has not been supported by subsequent findings. Overall, perhaps other and more consistent differences can be seen in (1) the major benzenoid compounds, with phalaenophilous floral scents having phenylacetaldehyde or benzaldehyde in addition to esters and alcohols (whereas sphingophilous flowers display more benzenoid esters); (2) nitrogen-bearing volatiles, which occur in greater frequency and abundance in sphingophilous flowers; and (3) terpenoid lilac compounds, which tend to be infrequent in sphingophilous flowers and seem to possibly reflect greater pollination by settling moths.

Within certain plant families, considerable interspecific variation exists in the floral scent chemistry of hawkmoth-pollinated species, as is well exemplified in the Orchidaceae,[111] Cactaceae,[261] and Solanaceae,[11,16,262] but in other families some patterns are apparent. In a study of three genera in Nyctaginaceae, the 14 species pollinated by hawkmoths alone or in combination with noctuids displayed scent profiles that were composed mainly of terpenoids (especially various sesquiterpenes, 4,8-dimethyl-nona-1,3,7-triene, and *trans*-β-ocimene); when benzenoids were well represented, methyl benzoate was prominent; and when fatty acid derivatives were in large amounts, they included *cis*-3-hexenyl acetate or tiglate, and *cis*-jasmone; almost all species contained nitrogen compounds, which in one species codominated with terpenoids.[15] Similarly, in the Caryophyllaceae, the genera *Dianthus* and *Saponaria* have floral scents that are mainly benzenoids (especially esters such as methyl benzoate, and some eugenol) and terpenoids (e.g., *trans*-β-ocimene, β-caryophyllene, linalool), with nitrogen compounds being common.[18,222] For the related genus *Silene*, the single species pollinated exclusively by hawkmoths has a scent consisting principally of various benzenoids, whereas flowers pollinated by both hawkmoths and settling moths all show varying amounts of benzenoids and are high in terpenoids that are either lilac compounds (possibly more typical of settling moth flowers) or other common volatiles (e.g., linalool, ocimene, α-farnesene), and some had nitrogen compounds.[17]

Some hawkmoth-pollinated species show a clear predominance of monoterpenes, especially linalool. Scents clearly dominated by linalool (45 to 80%), accompanied by other terpenoids (e.g., ocimene, nerolidol, caryophyllene), benzenoids (e.g., methyl benzoate, 2-phenylethanol), and nitrogen compounds, have been reported in *Coussarea* (Rubiaceae) and *Carica* (Caricaceae),[11] *Albizia* (Leguminosae), *Clerodendrum* (Verbenaceae), *Lonicera* (Caprifoliaceae),[263] *Crinum* (Amaryllidaceae),[264] *Tritoniopsis* (Iridaceae),[20] *Oenothera* (Onagraceae),[263,265] *Angraecum* (Orchidaceae),[111] and *Discocactus* (Cactaceae).[261] In the temperate orchid *Platanthera bifolia*, which is mainly pollinated by hawkmoths or also by settling moths,[24] analysis of separate populations revealed chemotypes that differed in the predominant types of monoterpenes, these being either linaloolic, lilac, or geraniolic; all chemotypes also contained benzenoids and other terpenoids (e.g., *trans*-β-ocimene).[243,244]

Terpenoids other than or in addition to linalool may be major volatiles in the floral scent, accompanied by various benzenoids and some nitrogen compounds. These other prominent terpenoids may be mostly *trans*-β-ocimene, as in *Brugmansia* x *candida* (Solanaceae),[262] *Plumeria* (Apocynaceae), and *Crinum* (Amaryllidaceae)[263]; both linalool and ocimene in *Hedychium* (Zingiberaceae)[11]; an overwhelming dominance of nerolidol in *Escobedia* (Scrophulariaceae)[11]; 1,8 cineole and geraniolic volatiles in *Brugmansia* (Solanaceae)[11]; geraniolic compounds in *Epidendrum* (Orchidaceae)[11]; and various terpenoids in species of Nyctaginaceae discussed above.[15]

Scents with a codominance of terpenoids and benzenoids have been found in species of *Aerangis* (Orchidaceae),[111] *Selenicereus* (Cactaceae),[261] *Nicotiana*,[16] and *Hillia* (Berberidaceae).[11] A predominance of benzenoids characterizes the floral scents of most *Angraecum* (Orchidaceae)[111] and *Nicotiana suaveolens* (Solanaceae).[16] Jasmine lactone dominates the scent of *Mystacidium venosum* (Orchidaceae), with lesser amounts of oxygenated terpenoids.[266]

A dominance of nitrogen-bearing volatiles or codominance with either benzenoids or terpenoids has been reported in several species; the benzenoids commonly include benzyl alcohol and methyl benzoate, the terpenoids linalool. In *Hylocereus* (Cactaceae), the scent of one species consists mainly of several nonoxime nitrogen compounds and another displays codominance of benzyl alcohol and indole.[261] *Delphinium leroyi* (Ranunculaceae), in a genus pollinated primarily by bees, has a scent codominated by benzyl benzoate and indole, with lesser amounts of benzyl alcohol and linalool,[267] and in *Staphanotis floribunda* (Asclepiadaceae), the fragrance is dominated by 1-nitro-2-phenylethane, accompanied by benzenoids and terpenoids (linalool, α-farnesene).[268,269] Oximes are prominent in some species, such as *Angraecum sesquipedale* (Orchidaceae), a classic species pollinated by hawkmoths, which shows a codominance of oximes and benzenoids,[111] whereas *Nicotiana longiflora*[16] and *Datura* species[11] (all Solanaceae) are dominated by oximes and *Nicotiana alata* by a codominance with terpenoids.[16]

With respect to the biological activity of scent volatiles, both wild and laboratory-reared hawkmoths show strong antennal sensitivity to a range of individual floral volatiles and complex blends.[220] In both specialist (*Sphinx perelegans*) and generalist (*Hyles lineata*) moths, the antennal responses were found to be generally similar: the greatest sensitivity was to linalool and its oxides, methyl salicylate, and benzyl acetate, and medium to large responses were to compounds such as germacrene D, nerolidol, and benzyl salicylate.[270,271] In *Manduca sexta*, female antennae were most sensitive to *trans*-β-ocimene, hexenyl acetate, phenylacetaldehyde, methyl salicylate, geranyl acetone, benzyl alcohol, followed by 3-hexenyl propionate, myrcene, linalool, and

1,8-cineole.[272] In behavioral studies, moths can track floral odors to their sources in the absence of visual cues.[220] It has been suggested that experienced moths may learn to generalize the association between nectar and complex scent blends, thereby relying mainly on a simpler scent or even single key volatiles.

8.3   VERTEBRATE POLLINATORS

8.3.1 · BIRDS

Pollination by birds, or ornithophily, is carried out in many parts of the world, both in the tropical and temperate zones, and by a wide diversity of birds in mainly 10 families.[5] However, flower visitation has been recorded in about 50 bird families, with the extent to which birds feed on flowers, and thereby play a role in pollination, varying widely.[5,273] Floral and inflorescence morphology depend in part on the type of bird pollinator,[274] which may either hover while it feeds (mainly hummingbirds in the New World) or perch (honeycreepers in the New World; sunbirds, white-eyes, sugarbirds, and honeyeaters in the Old World).[5] The flowers are generally long-tubed or spurred, vividly colored (typically red to orange, but may be yellow or green), have diurnal anthesis, produce abundant nectar,[275] and are reportedly devoid of fragrance.[4,261] With a few exceptions, birds generally have a poor sense of smell,[276] and no sense of olfaction has been reported in flower visiting other than possibly in some hummingbirds[303]; instead, birds rely primarily on vision, reinforced with learning, to locate flowers.[277,278]

Not surprisingly, most of the 24 species (18 families) of bird-pollinated flowers for which floral fragrances have been investigated show very little or no scent. All but one species are native to the New World and are pollinated by hummingbirds. The only putatively bird-pollinated species from the Old World is the mangrove *Bruguiera gymnorrhiza* (Rhizophoraceae), which contains trace amounts of five volatiles, all of which, except methyl benzoate, are thought to be emitted in response to damage when the flowers are picked.[120] In an extensive study, Knudsen et al.[14] failed to detect any floral volatiles in 9 of the 17 species examined, and the remaining 8 produced only trace amounts. Furthermore, most of these trace compounds were of terpenoid or lipoxygenase derivation and are commonly emitted by vegetative as well as floral tissues; the possible roles of these volatiles remain to be determined. Within the genus *Passiflora* (Passifloraceae), two species are reportedly scentless and one contained a single compound, butyl 2-methylpropyl-1,2-benzenedicarbox-ylate.[150,279] In contrast, two bird-pollinated species of *Nicotiana* (Solanaceae) had quite high emission rates of volatiles, especially at night; however, the putative hawkmoth attractants (linalool, nerolidol/farnesol isomers, aldoximes, benzenoid esters) that are common in other *Nicotiana* species were either absent or much reduced, suggesting that these species may be in transition states in their adaptation to bird pollination.[16] Similarly the single bird-pollinated species of *Mirabilis* (Nyctaginaceae) released some fragrance, which may be a reflection of its ancestry, since its closest relatives release scent.[15]

While species exhibiting clear adaptations to bird pollination generally emit little or no scent, the presence of floral volatiles does not inhibit bird visitation to flowers, and several cases have been reported of hummingbirds visiting and contributing pollination services to strongly fragrant flowers that are visited by other pollinators, usually mainly moths or bats.[280–283]

8.3.2 BATS

Bat pollination occurs mainly in the tropics, but also extends into temperate warm deserts and high elevations.[5] The vast majority of species are in the New World.[284] It is estimated that more than 750 plant species rely on bats for pollination. The typical floral syndrome, referred to as chiropterophily, includes flowers that have nocturnal anthesis, are whitish or drab in color, produce copious amounts of nectar, and emit strong odors that are variously described as fetid, pungent, fermented, or butter-, cabbage-, or onionlike.[12,284,285] Olfaction is probably the main sensory modality used by bats to locate flowers, although this may vary, as discussed further below. The first attempt to investigate possible chemical patterns in floral fragrances associated with bat pollination was conducted by Knudsen and Tollsten in 1995,[12] and more studies soon followed; currently approximately 32 species from 17 families have been examined: 23 species from the New World and 9 from the Old World.

No chemical compound is characteristic of all species, but sulfur-containing compounds certainly appear to be a typical trait of bat pollination, with trace to large amounts being detected in close to 75% of the species investigated.[12,120,261,286–289] Sulfur-containing volatiles are rarely found in floral scents, with the exception of sapromyophilous flowers, and their occurrence in such a diversity of plants and unrelated families appears to reflect convergent evolution in response to pollination by bats. These compounds are responsible for the characteristic pungency of the fragrances and include dimethyl disulfide (and dimethyl trisulfides and tetrasulfides, which, according to Kaiser and Tollsten,[261] form as artifacts from dimethyl disulfide during headspace trapping on different adsorbents), as well as various irregular sulfides, thiols, and isothiocyanates. The relationship between sulfur-containing volatiles and bat pollinators appears to be based on a possibly innate preference by bats for sulfur compounds, as revealed in the significantly greater attractiveness of dimethylsulfide and 2,4-dithiapentane over other individual floral scent compounds to New World *Glossophaga* bats tested in both the laboratory and field.[287]

In the Cactaceae, the family most extensively examined, five of the seven species had floral scents dominated by dimethyl disulfide, while the other two species contained no sulfur compounds.[12,261,286] In a study of nectar volatiles in a bat-pollinated parasitic plant species, *Dactylanthus taylorii* (Balanophoraceae) in New Zealand, the volatile constituents showed no presence of sulfur compounds.[290] Interestingly, the frequency of sulfur compounds and their prominence in floral scents appears to be higher in plants from the New World than from the Old World,[288,289] and this may be related to the biology of the bat pollinators.

In the New World, pollination is carried out by nectariferous microchiropteran bats of the family Phyllostomidae, which have a mixed diet that includes insects and which use ultrasonic echolocation for orientation, whereas in the Old World (including South Pacific islands), pollination is performed by megachiropterans of the family Pteropodidae, which have diets that range from mainly fruits to exclusively

COLOR FIGURE 8.1 Example of generalist pollination: *R. acris* (Ranunculaceae) has bowl-shaped flowers with pollen and nectar readily available to various insects; the flowers are visited and pollinated by flies, beetles, micropterigid moths, and solitary bees. Shown here is the bee *Chelostoma florisomne* (Megachilidae), which collects pollen only on *Ranunculus* species. Photo by Dr. Heidi E.M. Dobson.

COLOR FIGURE 8.2 This species of *Aquilegia* (Ranunculaceae) produces nectar in long spurs, where it is accessible mainly to lepidopteran moth pollinators, but bees can also pollinate the flowers when they forage for pollen (shown here hanging from the anthers). Photo by Dr. Heidi E.M. Dobson.

COLOR FIGURE 9.1 A butterfly visiting *K. arvensis* (Dipsacaceae). Photo by Dr. Heidi E.M. Dobson.

COLOR FIGURE 10.1 Males of *C. ciliata* pseudocopulating with a flower of *Ophrys speculum* (head pollination). Photo by H.F. Paulus.

COLOR FIGURE 10.2 *Chiloglottis trapeziformis* attracts males of its pollinator species, the thynnine wasp *Neozeleboria cryptoides*. Photo by F.P. Schiestl.

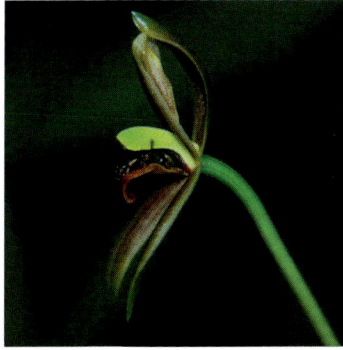

COLOR FIGURE 10.3 Pseudocopulating drones of *N. testaceicornis* on a flower of *M. ringens*. Photo by R.B. Singer.

COLOR FIGURE 10.9 In the sympatrically occurring *O. sphegodes* and *O. fusca*, the location of attachment of the pollinia to the male provides a mechanical component to the isolating mechanism via odor. *A. nigroaenea* males are the pollinators of both species. However, while the pollinia are deposited on the head of the male bee in *O. sphegodes* (top), males visiting *O. fusca* back into the flower and receive pollinia on the tip of the abdomen (bottom).

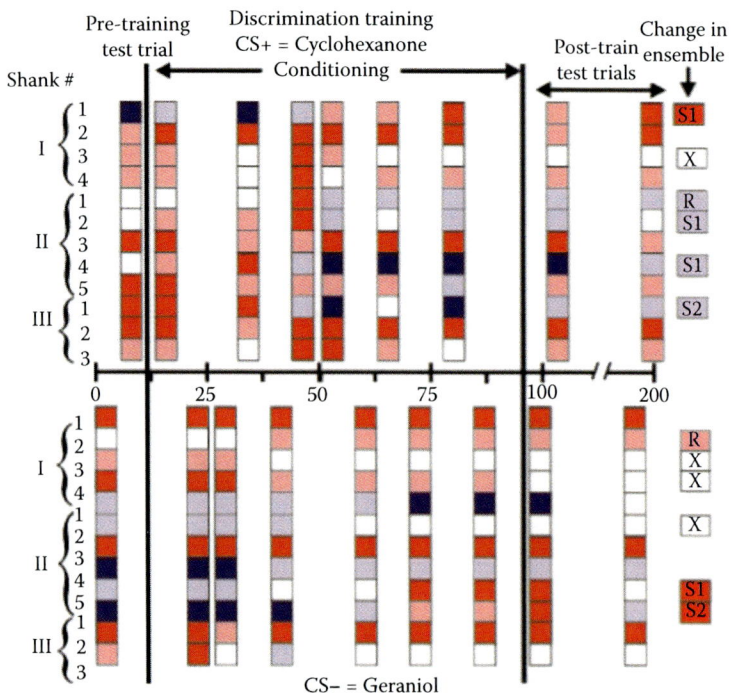

COLOR FIGURE 12.10 (b) The trial-by-trial mean changes in neural responses to odor stimulation during differential conditioning (CS+ and CS– odors as described in the text) of the moth, *M. sexta* (from Daly et al.[112]). Unit response data by trial are from a single animal over the course of differential conditioning. Twelve different units (putatively neurons) were identified and recorded on each trial with each odor. Offset of columns indicates that the CS+ and CS odorants were tested on separate trials. Pre- and post-training trials were conducted without reinforcement. Boxes are color coded at each trial to indicate the statistical change in a unit's response to odor stimulation on those trials: red, excitatory; blue, inhibitory; clear, no change in background firing rate. Overall changes in unit responses over the course of the experiment are summarized on the right: (red "R") an excitatory unit was recruited; (blue "R") an inhibitory unit was recruited; (clear X), an initially responsive unit was dropped; (blue "S1") switched response polarity to one odor and from an excitatory to an inhibitory response; (clear "S2") switched response polarity to both odors (color in figure denotes response at the first post-test). (From Daly, K.C., Christensen, T.A., Lei, H., Smith, B.H., and Hildebrand, J.G., *Proc. Nat. Acad. Sci. USA* 101, 10476–10481, 2004, with permission.)

nectar and which do not use ultrasonic calls.[291] Petersson et al.[292] propose that selection for sulfur compounds may have been especially strong in the New World, and neotropics, given that glossophagine bats, by virtue of their ability to echolocate, often forage in more cluttered vegetation where they rely more heavily on olfaction to locate flowers, since they have no or limited color vision. In contrast, Old World bats feed more commonly in open areas, such as above the canopy, where flowers are readily located visually and where olfactory cues would presumably play a lesser role (contradicting the prediction by Faegri and van der Pijl[4]). Scent chemistry seems to support this view: sulfur compounds reported by Petersson et al. in five of the eight West African species examined were prominent in only one species and were in either very low or trace amounts in the other species; in addition, two species had very weak scents. As still further corroboration, *Ceiba pentandra* (Bombacaceae), which is native to both South America and Africa, has a floral scent that is dominated by dimethylsulfide in specimens from Costa Rica,[287] but contains no sulfur volatiles in West African samples.[289]

As in floral scents in general, the number of compounds detected in each floral scent varies widely among species, and the nonsulfur volatiles belong to chemical classes commonly reported in fragrances. Among the terpenoids, the most frequently detected volatiles include limonene, pinene, myrcene, and sabinene, with sesquiterpenes being only occasional and usually in low representation. Fatty acid derivatives vary widely in occurrence, from undetected to predominant scent constituents, with short-chain ketones generally most common. Benzenoids are often in very small amounts and do not exceed 30% of the volatiles; their identities vary widely across species, with perhaps benzaldehyde being most common. Nitrogen-bearing compounds are reported in a few species, and in *Parkia bicolor* (Leguminosae) they are the dominant volatiles.[289] In species with little or no detectable amounts of sulfur compounds, the scent is dominated variously by terpenoids, benzenoids, and fatty acid derivatives.[12,279,286,288,289]

8.3.3 BIRDS, BATS, AND MOTHS

Switches in pollinators, as well as mixed pollination systems involving bats have been documented for several species, and the pollinators are usually either bats and bird or bats and moths; correspondingly, the floral scents may show various intermediate features.

In switches from bird to bat pollination, the floral scent is often weak, as reviewed by Knudsen and Tollsten.[12] One example is *Heliconia solomonensis* (Heliconiaceae),[293] which belongs to a genus that is mainly hummingbird pollinated, and has green flowers that open at night and produce copious nectar, but are odorless to the human nose. On the plant family level, Faegri and van der Pijl[4] suggest that in the Bignoniaceae, the change has probably been from birds to bats, but all three bat-pollinated species analyzed to date displayed sulfur compounds in their floral scents.[12,286,289] In contrast, in the Bombacaceae and the genus *Musa* (Musaceae), the switch is proposed to have been in the opposite direction.[4]

Floral traits of bat-pollinated plants share much in common with moth-pollinated species, and switches between moth and bat pollination, most evident in the Cactaceae, appear to be relatively easy given that nocturnal anthesis, a multitude of anthers, and often considerably large amounts of nectar are present in flowers pollinated by either animal group.[4] One species visited by both groups and displaying a mixed pollination syndrome is the caesalpinioid *Browneopsis disepala* (Leguminosae), in which the floral scent is dominated by fatty acid derivatives (especially 2-heptanol and 2-heptanone), followed by monoterpenes (e.g., α-, β-pinene, *trans*-β-ocimene), sesquiterpenes (e.g., *trans*-α-bergamotene), and small amounts of benzenoids.[294] The scent lacks sulfur compounds and resembles some bat flowers that have scents with fatty acid-derived ketones as the major constituents.[286,289]

It would be interesting to analyze the floral scents of other species that are pollinated to different extents by both bats and other animals (e.g., birds, nonflying mammals, insects), such as species of Agavaceae,[295,296] Cactaceae,[281,297] and others[280,298] to determine the degree to which chemical adaptation in floral fragrance has been selected.

8.4 CONCLUSION

The current information on floral scent chemistry reviewed here suggests that there are general trends in the patterns of scent volatiles vis-à-vis the types (guilds, functional groups) of animals that serve as pollinators (Table 8.1). These trends do not apply to all plants in each pollination category, nor do they form clear-cut boundaries with trends found in plants pollinated by other animal types. Nevertheless, it is becoming apparent that pollinator animals are acting as selective agents in determining the kinds of volatile compounds emitted by flowers, with some groups of animals exerting more selective pressure than others. Clearly the chemistry of floral scents has a large impact on which animals visit the flowers, whether the flowers be generalist in attracting a variety of insects or specialist in attracting only a select group of animals.[7,299] In this respect, the concept of general floral syndromes can be applied to floral scent, and floral scent chemistry can be used to tentatively predict pollinators, but only, however, when it is considered in concert with other floral traits, including morphology, rewards, and phenology, which together, as a suite, define a pollination syndrome.[300] Floral scent, similar to other floral traits, is not a reliable indication of pollination systems when considered alone.[301] Indeed, it must always be borne in mind that no trends in floral scent chemistry are clear-cut; variability occurs in all the different syndromes addressed here, with some being more clearly defined than others, and the boundaries between syndromes can be quite blurry.

A multivariate analysis of the phenological states of various floral traits used to characterize pollination syndromes revealed four distinct groups of syndromes and led to the conclusion that differentiating clearly between syndromes on the basis of floral traits is very difficult.[302] Some of the overlaps found in the study are similar to those apparent here in the floral scent chemistry, such as those among the butterflies and moth groups, but the degree of differentiation may also depend heavily on the breadth of the syndromes. In the scent data, dividing both the flies and beetles into more groups based on their biology revealed more fine-tuned differences that would be lost had the plants pollinated by these insects been pooled together.

TABLE 8.1
Characteristic Chemical Profiles of Floral Scents in Species Pollinated by Particular Animal Groups, Based on a Review of the Literature (The profiles are tentative and do not encompass all species, but are provided to show general suggestive trends in scent chemistry.)

Animal Pollinator Group	Typical Floral Volatiles
Invertebrates	
A. Generalist (diverse insects)	Fatty acid derivatives, terpenoids, and benzenoids all present, with one often dominant
B. Coleoptera	
Tropical dynastine Scarabaeidae	Methoxylated benzenoids common
Other tropical beetles	Esters (fatty acid derived, benzenoid) and terpenoids common
Beetles of temperate regions	Variable; *N*-compounds frequent
C. Diptera	
Typical myophily	Fatty acid-derived acids, alcohols, and *N*-compounds common
Midge-like flies (nonsapromyophily)	Variable
Male dacine Tephritidae	Methyl eugenol or 4-(*p*-hydroxyphenyl)-2-butanone
D. Insects associated with decaying organic matter	
Beetles and flies on carrion or excrements	*S*- or *N*-compounds, fatty acid-derived acids, alcohols, ketones; *p*-cresol (excrement)
Flies on decaying vegetation	Variable (few data)
Flies on decaying fruits	Variable, with fatty acid-derived alcohols frequent (few data)
Flies on fungi	Fatty acid-derived alcohols, aldehydes, ketones; *S*-compounds occasional
E. Thysanoptera	Variable (few data)
F. Hymenoptera	
Food-seeking bees	Variable, with terpenoids commonly abundant
Fragrance-seeking male euglossine bees	Usually few volatiles, mainly benzenoids and monoterpenes
Nectar-seeking wasps	Variable (few data)
Fig wasps	Usually few volatiles, with one or two strongly dominant, especially terpenoids
G. Lepidoptera	
Micropterigid moths	Fatty acid-derived esters frequent
Yucca moths	Fatty acid-derived hydrocarbons, alcohols, and sesquiterpenes (few data)

TABLE 8.1 (*Continued*)
Characteristic Chemical Profiles of Floral Scents in Species Pollinated by Particular Animal Groups, Based on a Review of the Literature (The profiles are tentative and do not encompass all species, but are provided to show general suggestive trends in scent chemistry.)

Animal Pollinator Group		Typical Floral Volatiles
Diurnal butterflies		Benzenoids (phenylacetaldehyde, 2-phenyl ethanol, benzaldehyde, benzyl alcohol), terpenoids (linalool, *trans*-β-ocimene, *cis*-3-hexenyl acetate, oxoisophorone), N-compounds occasional, in small amounts
Nocturnal settling moths		Benzenoids (benzaldehyde, phenylacetaldehyde, esters), terpenoids (linalool, β-ocimene, lilac compounds); sometimes fatty acid-derived esters, N-compounds
Nocturnal hovering moths		Abundant benzenoids (esters, especially methyl benzoate), terpenoids (especially linalool), and N-compounds
Vertebrates		
A. Birds		Weak or no scent; some scent in cases of mixed pollination
B. Bats	New World Microchiroptera	S-compounds common and prevalent
	Old World Megachiroptera	S-compounds not prevalent; odors may be weak

Overall, most of the volatiles in each floral scent are commonly occurring components of floral fragrances in general and have been reported in other species,[21] but it is the combinations of particular volatiles, together with the occurrence of rarely encountered volatiles that gives a species its unique signature, and in turn unites groups of species with similar types of pollinators. While each species has a species-specific blend of volatiles, comprised of both widespread and less common compounds, findings suggest that plants with similar pollinator guilds converge on certain core constituents that are required attractants.[2] Thus, the scents of hawkmoth-pollinated flowers tend to be strong, sweet, and share certain compound classes (benzenoid esters, oxygenated terpenoids, nitrogenous compounds), but at the same time, interspecific variation is high within both the core volatiles and the background volatiles common to flowers pollinated by other animals or diverse insects.

The floral volatile data summarized here are being expanded on rapidly, and with this will emerge new syndrome classes as well as changes in the currently presented ones. It is hoped that this review will stimulate investigations into poorly understood pollination systems as well as more extensively documented ones in order to expand as well as solidify our understanding of how the chemistry of floral scents is linked to the biology of the principal pollinator animals. Integration of chemical analyses with investigations into both a plant's pollination ecology and the behavior of pollinator animals in response to floral fragrances is crucial in determining the role of floral volatiles in pollination and the selective forces operating in the evolution of particular floral scent chemical compositions; and increased emphasis needs to be placed on this synthesis of experimental approaches.

ACKNOWLEDGMENTS

I thank Whitman College for financial support during the writing of this chapter on my sabbatical leave, Jiulina and Lisi for their patience, and Rob Raguso and Andreas Erhardt for their input and sharing of unpublished data. I dedicate this to all whose enthusiastic interest and scientific endeavors have done so much to advance the field of floral scent biology in pollination.

REFERENCES

1. Dobson, H.E.M., Floral volatiles in insect biology, in *Insect-Plant Interactions*, vol. 5, Bernays, E.A., Ed., CRC Press, Boca Raton, FL, 1994, p. 47.
2. Raguso, R.A., Floral scent, olfaction, and scent-driven foraging behavior, in *Cognitive Ecology of Pollination*, Chittka, L. and Thomson, J.D., Eds., Cambridge University Press, 2001, p. 83.
3. Weiss, M.R., Vision and learning in some neglected pollinators: beetles, flies, moths, and butterflies, in *Cognitive Ecology of Pollination*, Chittka, L. and Thomson, J.D., Eds., Cambridge University Press, 2001, p. 171
4. Faegri, K. and van der Pijl, L., *The Principles of Pollination Ecology*, 3rd ed., Pergamon, Oxford, 1979.
5. Proctor, M., Yeo, P., and Lack, A., *The Natural History of Pollination*, Timber Press, Portland, OR, 1996.
6. Kandori, I., Diverse visitors with various pollinator importance and temporal change in the important pollinators of *Geranium thunbergii* (Geraniaceae), *Ecol. Res.* 17, 283, 2002.
7. Waser, N.M., Chittka, L., Price, M.V., Williams, N.M., and Ollerton, J., Generalization in pollination systems, and why it matters, *Ecology* 77, 1043, 1996.
8. Armbruster, W.S., Fenster, C.B., and Dudash, M.R., Pollination "principles" revisited: specialization, pollination syndromes, and the evolution of flowers, *Norske Vidensk. – Akad. I. Mat. Naturvid. Kl. Skrif., Ny Ser.* 39, 179, 2000.
9. Fenster, C.B., Armbruster, W.S., Wilson, P., Dudash, M.R., and Thomson, J.D., Pollination syndromes and floral specialization, *Annu. Rev. Ecol. Syst.* 35, 375, 2004.
10. Endress, P.K., *Diversity and Evolutionary Biology of Tropical Flowers*, Cambridge University Press, 1994.
11. Knudsen, J.T. and Tollsten, L., Trends in floral scent chemistry in pollination syndromes: floral scent composition in moth-pollinated taxa, *Bot. J. Linn. Soc.* 113, 263, 1993.
12. Knudsen, J.T. and Tollsten, L., Floral scent in bat-pollinated plants: a case of convergent evolution, *Bot. J. Linn. Soc.* 119, 45, 1995.

13. Andersson, S., Nilsson, L.A., Groth, I., and Bergström, G., Floral scents in butterfly-pollinated plants: possible convergence in chemical composition, *Bot. J. Linn. Soc.* 140, 129, 2002.

14. Knudsen, J.T., Tollsten, L., Groth, I., Bergström, G., and Raguso, R.A., Trends in floral scent chemistry in pollination syndromes: floral scent composition in hummingbird-pollinated taxa, *Bot. J. Linn. Soc.* 146, 191, 2004.

15. Levin, R.A., Raguso, R.A., and McDade, L.A., Fragrance chemistry and pollinator affinities in Nyctaginaceae, *Phytochemistry* 58, 429, 2001.

16. Raguso, R.A., Levin, R.A., Foose, S.E., Holmberg, M.W., and McDade, L.A., Fragrance chemistry, nocturnal rhythms and pollination "syndromes" in *Nicotiana*, *Phytochemistry* 63, 265, 2003.

17. Jürgens, A., Witt, T., and Gottsberger, G., Flower scent composition in night-flowering *Silene* species (Caryophyllaceae), *Biochem. Syst. Ecol.* 30, 383, 2002.

18. Jürgens, A., Witt, T., and Gottsberger, G., Flower scent composition in *Dianthus* and *Saponaria* species (Caryophyllaceae) and its relevance for pollination biology and taxonomy, *Biochem. Syst. Ecol.* 31, 345, 2003.

19. Jürgens, A., Flower scent composition in diurnal *Silene* species (Caryophyllaceae): phylogenetic constraints or adaptation to flower visitors?, *Biochem. Syst. Ecol.* 32, 841, 2004.

20. Manning, J.C. and Goldblatt, P., Radiation of pollination systems in the Cape genus *Tritoniopsis* (Iridaceae: Crocoideae) and the development of bimodal pollination strategies, *Int. J. Plant Sci.* 166, 459, 2005.

21. Knudsen, J.T., Tollsten, L., and Bergström, G., Floral scents: a checklist of volatile compounds isolated by head-space techniques, *Phytochemistry* 33, 253, 1993.

22. Raguso, R.A., Why do flowers smell? The chemical ecology of fragrance-driven pollination, in *Advances in Insect Chemical Ecology*, Cardé, R.T. and Millar, J.G., Eds., Cambridge University Press, 2004, p. 151.

23. Pelletier, L., Brown, A., Otrysko, B., and McNeil, J.N., Entomophily of the cloudberry (*Rubus chamaemorus*), *Entomol. Exp. Appl.* 101, 219, 2001.

24. Zych, M., Pollination biology of *Heracleum sphondylium* L. (Apiaceae): the advantages of being white and compact, *Acta Soc. Bot. Poloniae* 71, 163, 2002.

25. Tollsten, L., Knudsen, J.T., and Bergström, G., Floral scent in generalistic *Angelica* (Apiaceae): an adaptive character?, *Biochem. Syst. Ecol.* 22, 161, 1994.

26. Tollsten, L., and Øvstedal, D.O., Differentiation in floral scent chemistry among populations of *Conopodium majus* (Apiaceae), *Nord. J. Bot.* 14, 361, 1994.

27. Robertson, G.W., Griffiths, D.W., Woodford, J.A.T., and Birch, A.N.E., A comparison of the flower volatiles from hawthorn and four raspberry cultivars, *Phytochemistry* 33, 1047, 1993.

28. Robertson, G.W., Griffiths D.W., Woodford, J.A.T., and Birch, A.N.E., Changes in the chemical composition of volatiles released by the flowers and fruits of the red raspberry (*Rubus idaeus*) cultivar Glen Prosen, *Phytochemistry* 38, 1175, 1995.

29. Bergström, G., Dobson, H.E.M., and Groth, I., Spatial fragrance patterns within the flowers of *Ranunculus acris* (Ranunculaceae), *Plant Syst. Evol.* 195, 221, 1995.

30. Knudsen, J.T., Tollsten, L., and Ervik, F., Flower scent and pollination in selected neotropical palms, *Plant Biol.* 3, 642, 2001.

31. Nilsson, L.A., The pollination ecology of *Listera ovata* (Orchidaceae), *Nord. J. Bot.* 1, 461, 1981.

32. Custódio, L., Nogueira, J.M.F., and Romano, A., Sex and developmental stage of carob flowers affects composition of volatiles, *J. Hort. Sci. Technol.* 79, 689, 2004.

33. Borg-Karlson, A.-K., Valterová, I., and Nilsson, L.A., Volatile compounds from flowers of six species in the family Apiaceae: bouquets for different pollinators, *Phytochemistry* 35, 111, 1994.

34. Tollsten, L. and Knudsen, J.T., Floral scent in dioecious *Salix* (Salicaceae): a cue determining the pollination system?, *Plant Syst. Evol.* 182, 229, 1992.

35. Dobson, H.E.M., Groth, I., and Bergström, G., Pollen advertisement: chemical contrasts between whole-flower and pollen odors, *Am. J. Bot.* 83, 877, 1996.

36. Bernhardt, P., Sage, T., Weston, P., Azuma, H., Lam, M., Thien, L.B., and Bruhl, J., The pollination of *Trimenia moorei* (Trimeniaceae): floral volatiles, insect/wind pollen vectors and stigmatic self-incompatibility in a basal angiosperm, *Ann. Bot.* 92, 445, 2003.

37. Borg-Karlson, A.-K., Unelius, C.R., Valterová, I., and Nilsson, L.A., Floral fragrance chemistry in the early flowering shrub *Daphne mezereum*, *Phytochemistry* 41, 1477, 1996.

38. Patt, J.M., Rhoades, D.F., and Corkill, J.A., Analysis of the floral fragrance of *Platanthera stricta*, *Phytochemistry* 27, 91, 1988.

39. Patt, J.M., Merchant, M.W., Williams, D.R.E., and Meeuse, B.J.D., Pollination biology of *Platanthera stricta* (Orchidaceae) in Olympic National Park, Washington, *Am. J. Bot.* 76, 1097, 1989.

40. Raguso, R.A. and Roy, B.A., "Floral" scent production by *Puccinia* rust fungi that mimic flowers, *Mol. Ecol.* 7, 1127, 1998.

41. Bernhardt, P., Convergent evolution and adaptive radiation of beetle-pollinated angiosperms, *Plant Syst. Evol.* 222, 293, 2000.

42. Silberbauer-Gottsberger, I., Gottsberger, G., and Webber, A.C., Morphological and functional flower characteristics of New and Old World Annonaceae with respect to their mode of pollination, *Taxon* 52, 701, 2003.

43. Mookherjee, B.D., Trenkle, R.W., and Wilson, R.A., The chemistry of flowers, fruits and spices: live vs. dead a new dimension in fragrance research, *Pure Appl. Chem.* 62, 1357, 1990.

44. Ervik, F. and Knudsen, J.T., Water lilies and scarabs: faithful partners for 100 million years?, *Biol. J. Linn. Soc.* 80, 539, 2003.

45. Giberneau, M., Barabé, D., Labat, D., Cerdan, P., and Dejean, A., Reproductive biology of *Montrichardia arborescens* (Araceae) in French Guiana, *J. Trop. Ecol.* 19, 1, 2003.

46. Ervik, F., Tollsten, L., and Knudsen, J.T., Floral scent chemistry and pollination ecology in phytelephantoid palms (Arecaceae), *Plant Syst. Evol.* 217, 279, 1999.

47. Kite, G., Reynolds, T., and Prance, G.T., Potential pollinator-attracting chemicals from *Victoria* (Nymphaeaceae), *Biochem. Syst. Ecol.* 19, 535, 1991.

48. Schultz, K., Kaiser, R., and Knudsen, J.T., Cyclanthone and derivatives, new natural products in the flower scent of *Cyclanthus bipartitus* Poit., *Flavour Fragr. J.* 14, 185, 1999.

49. Dieringer, G. and Espinosa, J.E.S., Reproductive ecology of *Magnolia schiedeana* (Magnoliaceae), a threatened cloud forest tree species in Veracruz, Mexico, *Bull. Torrey Bot. Club* 121, 154, 1994.

50. Azuma, H., Toyota, M., Asakawa, Y., Yamaoka, R., Garcia-Franco, J.G., Dieringer, G., Thien, L.B., and Kawano, S., Chemical divergence in floral scents in *Magnolia* and allied genera (Magnoliaceae), *Plant Species Biol.* 12, 69, 1997.

51. Dieringer, G., Cabrera, L., Lara, M., Loya, L., and Reyes-Castillo, P., Beetle pollination and floral thermogenicity in *Magnolia tamaulipana* (Magnoliaceae), *Int. J. Plant Sci.* 160, 64, 1999.

52. Larsson, M.C., Stensmyr, M.C., Bice, S., and Hansson, B.S., Attractiveness of fruit and flower odorants detected by olfactory receptor neurons in the fruit chafer *Pachnoda marginata*, *J. Chem. Ecol.* 29, 1253, 2003.

53. Jürgens, A., Weber, A.C., and Gottsberger, G., Floral scent compounds of Amazonian Annonaceae species pollinated by small beetles and thrips, *Phytochemistry* 55, 551, 2000.

54. Thien, L.B., Heimermann, W.H., and Holman, R.T., Floral odors and quantitative taxonomy of *Magnolia* and *Liriodendron*, *Taxon* 24, 557, 1975.

55. Azuma, H., Thien, L.B., and Kawano, S., Molecular phylogeny of *Magnolia* (Magnoliaceae) inferred from cpDNA sequences and evolutionary divergence of the floral scents. *J. Plant Res.* 112, 291, 1999.

56. Azuma, H., Toyota, M., and Asakawa, Y., Intraspecific variation of floral scent chemistry in *Magnolia kobus* DC. (Magnoliaceae), *J. Plant Res.* 114, 411, 2001.

57. Bergström, G., Groth, I., Pellmyr, O., Endress, P.K., Thien, L.B., Hubener, A., and Francke, W., Chemical basis of a highly specific mutualism: chiral esters attract pollinating beetles in Eupomatiaceae, *Phytochemistry* 30, 3221, 1991.

58. Thien, L.B., Bernhardt, P., Gibbs, G.W., Pellmyr, O.M., Bergstrom, G., Groth, I., and McPherson, G., The pollination of *Zygogynum* (Winteraceae) by a moth, *Sabatinca* (Micropterigidae): an ancient association?, *Science* 227, 540, 1985.

59. Pellmyr, O., Thien, L.B., Bergström, G., and Groth, I., Pollination of New Caledonian Winteraceae: opportunistic shifts or parallel radiation with their pollinators?, *Plant Syst. Evol.* 173, 143, 1990.

60. Nogueira, P.C. de L., Bittrich, V., Shepard, G.J., Lopes, A.V., and Marsaioli, A.J., The ecological and taxonomic importance of flower volatiles of *Clusia* species (Guttiferae), *Phytochemistry* 56, 443, 2001.

61. Terry, I., Moore, C.J., Forster, P.I., Walter, G.H., Machin P.J., and Donaldson, J.S., Pollination ecology of the genus *Macrozamia*: cone volatiles and pollinator specificity, in *Proceedings of the 6th International Conference on Cycad Biology*, Pattaya, Thailand, 2002, Lindstrom, A., Ed., Nong Nooch Tropical Botanical Garden, Thailand, 2004, p. 155.

62. Terry, I., Moore, C.J., Walter, G.H., Forster, P.I., Roemer, R.B., Donaldson, J.S., and Machin, P.J., Association of cone thermogenesis and volatiles with pollinator specificity in *Macrozamia* cycads, *Plant Syst. Evol.* 243, 233, 2004.

63. Pellmyr, O., Tang, W., Groth, I., Bergstrom, G., and Thien, L., Cycad cone and angiosperm floral volatiles: inferences for the evolution of insect pollination, *Biochem. Syst. Evol.* 19, 623, 1991.

64. Lajis, N.H., Hussein, M.Y., and Toia, R.F., Extraction and identification of the main compound present in *Elaeis guineensis* flower volatiles, *Pertanika* 8, 105, 1985.

65. Dufaÿ, M., Anstett, M., and Hossaert-McKey, M.C., When leaves act like flowers: how dwarf palms attract their pollinators, *Ecol. Lett.* 6, 28, 2003.

66. Dufaÿ, M., Hossaert-McKey, M.C., and Anstett, M., Temporal and sexual variation of leaf-produced pollinator-attracting odours in the dwarf palm, *Oecologia* 139, 392, 2004.

67. Caissard, J.-C., Meekijjironenroj, A., Baudino, S., and Anstett, M.-C., Localization of production and emission of pollinators on whole leaves of *Chamaerops humilis* (Arecaeae), *Am. J. Bot.* 91, 1190, 2004.

68. Mawdsley, J.R., The importance of species of Dasytinae (Coleoptera: Melyridae) as pollinators in western North America. *Coleopt. Bull.* 57, 154, 2003.

69. Pellmyr, O., Groth, I., and Bergström, G., Comparative analysis of the floral odors of *Actaea spicata* and *A. erythrocarpa* (Ranunculaceae), *Nova Acta Reg. Soc. Sci. Ups. Series V:C* 3, 157, 1984.

70. Pellmyr, O., The pollination biology of *Actaea pachypoda* and *A. rubra* (including *A. erythrocarpa*) in northern Michigan and Finland, *Bull. Torrey Bot. Club* 112, 265, 1985.

71. Pellmyr, O., Bergström, G., and Groth, I., Floral fragrances in *Actaea*, using differential chromatograms to discern between floral and vegetative volatiles, *Phytochemistry* 26, 1603, 1987.

72. Dahl, Å.E., Wassgren, A.-B., and Bergström, G., Floral scents in *Hypecoum* sect. *Hypecoum* (Papaveraceae): chemical composition and relevance to taxonomy and mating system, *Biochem. Syst. Ecol.* 18, 157, 1990.

73. Kite, G.C., Hetterscheid, W.L.A., Lewis, M.J., Boyce, P.C., Ollerton, J., Cocklin, E., Diaz, A., and Simmonds, M.S.J., Inflorescence odours and pollinators of *Arum* and *Amorphophallus* (Araceae), in *Reproductive Biology in Systematics, Conservation and Economic Botany*, Owens, S.J. and Rudall, P.J., Eds., Royal Botanic Gardens, Kew, 1998, p. 295.

74. Lewis, G.P., Knudsen, J.T., Klitgaard, B.B., and Pennington, R.T., The floral scent of *Cyathostegia mathewsii* (Leguminosae, Papilionoideae) and preliminary observations on reproductive biology, *Biochem. Syst. Ecol.* 31, 951, 2003.

75. Ruther, J., Male-biased response of garden chafer, *Phyllopertha horticola* L., to leaf alcohol and attraction of both sexes to floral plant volatiles, *Chemoecology* 14, 187, 2004.

76. Larson, B.M.H., Kevan, P.G., and Inouye, D.W., Flies and flowers: taxonomic diversity of anthophiles and pollinators, *Can. Entomol.* 133, 439, 2001.

77. Kearns, C.A., Anthophilous fly distribution across an elevation gradient, *Am. Midl. Nat.* 127, 172, 1992.

78. Eberling, H. and Olesen, J.M., The structure of a high latitude plant-flower visitor system: the dominance of flies, *Ecography* 22, 314, 1999.

79. Goldblatt, P. and Manning, J.C., The long-proboscid fly pollination system in southern Africa, *Ann. Mo. Bot. Gard.* 87, 146, 2000.

80. Szucsich, N.U. and Krenn, H.W., Flies and concealed nectar sources: morphological innovations in the proboscis of Bombyliidae (Diptera), *Acta Zool.* 83, 183, 2002.

81. Kugler, H., Über die optische Wirkung von Fliegenblumen auf Fliegen, *Dtsch. Bot. Gesell. Ber.* 69, 387, 1956.

82. Liebermann, A., Korrelation zwischen den antennalen Geruchsorganen und der Biologie der Musciden, *Z. Morph. Ökol. Tiere* 5, 1, 1925.

83. Erhardt, A., Pollination of the edelweiss, *Leontopodium alpinum*, *Bot. J. Linn. Soc.* 111, 229, 1993.

84. Nimitkeatkai, H., Doi, M., Sugihara, Y., Inamoto, K., Ueda, Y., and Imanishi, H., Characteristics of unpleasant odor emitted by *Gypsophila* inflorescences, *J. Jpn. Soc. Hort. Sci.* 74, 139, 2005.

85. Knudsen, J.T. and Ståhl, B., Floral odours in the Theophrastaceae, *Biochem. Syst. Ecol.* 22, 259, 1994.

86. Velíšek, J., Kubelka, V., Pudil, F., Svobodová, Z., and Davídek, J., Volatile constituents of elder (*Sambucus nigra* L.) I. Flowers and leaves, *Lebensm. Wiss. u. Technol.*, 14, 309, 1981.

87. Dreyer, M.K. and Kirsch, T., Der Insektenkomplex des Mädesüss (*Filipendula ulmaria* L.), *Zool. Anz.* 218, 49, 1987.

88. Brunke, E.J., Hammerschmidt, F.J., and Schmaus, G., Flower scent of some traditional medicinal plants, in *Bioactive Volatile Compounds from Plants*, Teranishi, R., Buttery, R.G., and Sugisawa, H., Eds., American Chemical Society Symposium Series 525, American Chemical Society, Washington, DC, 1993, p. 282.

89. Knudsen, J.T., Floral scent chemistry in Geonomoid palms (Palmae: Geonomeae) and its importance in maintaining reproductive isolation, *Mem. N. Y. Bot. Gard.* 83, 141, 1999.

90. Erickson, B.J., Young, A.M., Strand, M.A., and Erickson, E.H., Jr., Pollination biology of *Theobroma* and *Herrania* (Sterculiaceae), *Insect Sci. Appl.* 8, 301, 1987.

91. Young, A.M., and Severson, D.W., Comparative analysis of steam distilled floral oils of cacao cultivars (*Theobroma cacao* L., Sterculiaceae) and attraction of flying insects: implications for a *Theobroma* pollination syndrome, *J. Chem. Ecol.* 20, 2687, 1994.

92. Young, A.M., Pollination biology of *Theobroma* and *Herrania* (Sterculiaceae). IV. Major volatile constituents of steam-distilled floral oils as field attractants to cacao-associated midges (Diptera: Cecidomyiidae and Ceratopogonidae) in Costa Rica, *Turrialba* 39, 454, 1989.

93. Nishida, R., Shelly, T.E., and Kaneshiro, K.Y., Acquisition of female-attracting fragrance by males of oriental fruit fly from a Hawaiian lei flower, *Fagraea berteriana*, *J. Chem. Ecol.* 23, 2275, 1997.

94. Metcalf, R.L., Chemical ecology of Dacinae fruit flies (Diptera: Tephritidae), *Ann. Entomol. Soc. Am.* 83, 1017, 1990.

95. Clarke, A.R., Balagawi, S., Clifford, B., Drew, R.A.I., Leblanc, L., Mararuai, A., McGuire, D., Putulan, D., Sar, S.A., and Tenakanai, D., Evidence of orchid visitation by *Bactrocera* species (Diptera: Tephritidae) in Papua New Guinea, *J. Trop. Ecol.* 18, 441, 2002.

96. Tan, K.-H. and Nishida, R., Mutual reproductive benefits between a wild orchid, *Bulbophyllum patens*, and *Batrocera* fruit flies via a floral synomone, *J. Chem. Ecol.* 26, 533, 2000.

97. Nishida, R., Iwahashi, O., and Tan, K.-H., Accumulation of *Dendrobium superbum* (Orchidaceae) fragrance in the rectal glands by males of the melon fly, *Dacus cucurbitae*, *J. Chem. Ecol.* 19, 713, 1993.

98. Tan, K.-H., Nishida, R., and Toong, Y.C., Floral synomone of a wild orchid, *Bulbophyllum cheiri*, lures *Bactrocera* fruit flies for pollination, *J. Chem. Ecol.* 28, 1161, 2002.

99. Nishida, R., Tan, K.-H., Wee, S.L., Hee, A.K.W., and Toong, Y.C., Phenylpropanoids in the fragrance of the fruit fly orchid, *Bulbophyllum cheiri*, and their relationship to the pollinator, *Bactrocera papayae*, *Biochem. Syst. Ecol.* 32, 245, 2004.

100. Keng-Hong, T. and Nishida, R., Synomone or kairomone? – *Bulbophyllum apertum* flower releases raspberry ketone to attract *Bactrocera* fruit flies, *J. Chem. Ecol.* 31, 497, 2005.

101. Lewis, J.A., Moore, C.J., Fletcher, M.T., Drew, R.A.I., and Kitching, W., Volatile compounds from the flowers of *Spathiphyllum cannaefolium*, *Phytochemistry* 27, 2755, 1988.

102. Hadacek, F. and Weber, M., Club-shaped organs as additional osmophores within the *Sauromatum* inflorescence: odour analysis, ultrastructural changes and pollination aspects, *Plant Biol.* 4, 367, 2002.

103. Raguso, R.A., Floral organ and stage-specific odors of *Aristolochia gigantea*: the potential for scent-driven division of labor in trap-pollination, unpublished, 2005.

104. Kite, G.C. and Hetterscheid, W.L.A., Inflorescence odours of *Amorphophallus* and *Pseudodracontium* (Araceae), *Phytochemistry* 46, 71, 1997.

105. Smith, B.N. and Meeuse, B.J.D., Production of volatile amines and skatole at anthesis in some arum lily species, *Plant Physiol.* 41, 343, 1966.

106. Stránský, K. and Valterová, I., Release of volatiles during the flowering period of *Hydrosme rivieri* (Araceae), *Phytochemistry* 52, 1387, 1999.

107. Seymour, R.S., Giberneau, M., and Ito, K., Thermogenesis and respiration of inflorescences of the dead horse arum *Helicodiceros muscivorus*, a pseudothermoregulatory aroid associated with fly pollination, *Funct. Ecol.* 17, 886, 2003.

108. Kite, G.C., Inflorescence odour of the foul-smelling aroid *Helicodiceros muscivorus*, *Kew Bull.* 55, 237, 2000.

109. Stensmyr, M.C., Urru, I., Collu, I., Celander, M., Hansson, B.S., and Angioy, A.M., Rotting smell of dead-horse arum florets, *Nature* 420, 625, 2002.

110. Burger, B.V., Munro, J.M., and Visser, Z.H., Determination of plant volatiles 1: analysis of the insect-attracting allomone of the parasitic plant *Hydnora africana* using Grob-Habich actibated charcoal traps, *J. High Resolut. Chromatogr. Commun.* 11, 496, 1988.

111. Kaiser, R., *The Scent of Orchids*, Hoffmann-La Roche, Basel, 1993.

112. Kite, G.C., The floral odour of *Arum maculatum*, *Biochem. Syst. Ecol.* 23, 343, 1995.

113. Skubatz, H., Kunkel, D.D., Howald, W.N., Trenkle, R., and Mookherjee, B., The *Sauromatum guttatum* appendix as an osmophore: excretory pathways, composition of volatiles and attractiveness to insects, *New Phytol.* 134, 631, 1996.

114. Chen, J. and Meeuse, B.J.D., Production of free indole by some aroids, *Acta Bot. Neerl.* 20, 627, 1971.

115. Borg-Karlson, A.-K., Englund, F.O., and Unelius, C.R., Dimethyl oligosulphides, major volatiles released from *Sauromatum guttatum* and *Phallus impudicus*, *Phytochemistry* 35, 321, 1994.

116. Borror, D.J., Triplehorn, C.A., and Johnson, N.F., *An Introduction to the Study of Insects*, 6th ed., Saunders, Fort Worth, TX, 1989.

117. Patt, J.M., French, J.C., Schal, C., Lech, J., and Hartman, T.G., The pollination biology of Tuckahoe, *Peltandra virginica* (Araceae), *Am. J. Bot.* 82, 1230, 1995.

118. Essig, F.B., Pollination in some New Guinea palms, *Principes* 17, 75, 1973.

119. Miyake, T. and Yafuso, M., Floral scents affect reproductive success in fly-pollinated *Alocasia odora* (Araceae), *Am. J. Bot.* 90, 370, 2003.

120. Azuma, H., Toyota, M., Asakawa, Y., Takaso, T., and Tobe, H., Floral scent chemistry of mangrove plants, *J. Plant Res.* 115, 47, 2002.

121. Goodrich, K.R., Zjhra, M.L., and Raguso, R.A., Sex and scentibility: protogyny and yeasty odor in flowers of pawpas (*Asimina triloba*: Annonaceae). *Int. J. Plant Sci.*, in press.

122. Gullan, P.J. and Cranston, P.S., *The Insects: An Outline of Entomology*, Chapman & Hall, London, 1994.

123. Talou, T., Delmas, M., and Gaset, A., Analysis of headspace volatiles from entire black truffle (*Tuber melanosporum*), *J. Sci. Food Agric.* 48, 57, 1989.

124. Mau, J.-L., Beelman, R.B., and Ziegler, G.R., Aroma and flavor components of cultivated mushrooms, in *Spices, Herbs and Edible Fungi*, Charalambous, G., Ed., Elsevier, Amsterdam, 1994, p. 657.

125. Rapior, S., Breheret, S., Talou, T., Pelissier, Y., Milhau, M., and Bessiere, J.-M., Volatile components of fresh *Agrocybe aegerita* and *Tricholoma sulfureum*. *Cryptogam. Mycol.* 19, 15, 1998.

126. Vogel, S. and Martens, J., A survey of the function of the lethal kettle traps of *Arisaema* (Araceae), with records of pollinating fungus gnats from Nepal, *Bot. J. Linn. Soc.* 133, 61, 2000.

127. Kirk, W.D.J., Feeding, in *Thrips as Crop Pests*, Lewis, T., Ed., Cab International, Wallington, U.K., 1997, p. 119.

128. Mound, L.A., Thysanoptera: diversity and interactions, *Annu. Rev. Entomol.* 50, 247, 2005.

129. Kirk, W.D.J., Effect of some floral scents on host finding by thrips (Insecta: Thysanoptera), *J. Chem. Ecol.* 11, 35, 1985.

130. Hooper, A.M., Bennison, J.A., Luszniak, M.C., Pickett, J.A., Pow, E.M., and Wadhams, L. J., *Verbena* x *hybrida* flower volatiles attractive to western flower thrips, *Frankliniella occidentalis*, *Pest. Sci.* 55, 633, 1999.

131. Imai, T., Maekawa, M., and Murai, T., Attractiveness of methyl anthranilate and its related compounds to the flower thrips, *Thrips hawaiisensis* (Morgan), *T. coloratus* Schmutz, *T. flavus* Schrank, and *Megalurothrips distalis* (Karny) (Thysanoptera: Thripidae), *Appl. Entomol. Zool.* 36, 475, 2001.

132. Galizia, C.G., Küttner, A., Joerges, J., and Menzel, R., Odour representation in honeybee olfactory glomeruli shows slow temporal dynamics: an optical recording study using a voltage-sensitive dye, *J. Insect Physiol.* 46, 877, 2000.

133. Wright, G.A., Skinner, B.D., and Smith, B.H., Ability of honeybee, *Apis mellifera*, to detect and discriminate odors of varieties of canol (*Brassica rapa* and *Brassica napus*) and snapdragon flowers (*Antirrhinum majus*), *J. Chem. Ecol.* 28, 721, 2002.

134. Paldi, N., Zilber, S., and Shafir, S., Associative olfactory learning of honeybees to differential rewards in multiple contexts—effect of odor component and mixture similarity, *J. Chem. Ecol.* 29, 2515, 2003.

135. Laloi, D. and Pham-Delègue, M.H., Bumble bees show asymmetrical discrimination between two odors in a classical conditioning procedure, *J. Insect Behav.* 17, 385, 2004.

136. Meagher, R.L., Trapping noctuid moths with synthetic floral volatile lures, *Entomol. Exp. Appl.* 103, 219, 2002.

137. Bergström, G., Birgersson, G., Groth, I., and Nilsson, L.A., Floral fragrance disparity between three taxa of lady's slipper *Cypripedium calceolus* (Orchidaceae), *Phytochemistry* 31, 2315, 1992.

138. Barkman, T.J., Beaman, J.H., and Gage, D.A., Floral fragrance variation in *Cypripedium*: implications for evolutionary and ecology studies, *Phytochemistry* 44, 875, 1997.

139. Olesen, J.M. and Knudsen, J.T., Scent profiles of flower colour morphs of *Corydalis cava* (Fumariaceae) in relation to foraging behaviour of bumblebee queens (*Bombus terrestris*), *Biochem. Syst. Ecol.* 22, 231, 1994.

140. Groth, I., Bergström, G., and Pellmyr, O., Floral fragrances in *Cimicifuga*: chemical polymorphism and incipient speciation in *Cimicifuga simplex*, *Biochem. Syst. Ecol.* 15, 441, 1987.

141. Irwin, R.E. and Dorsett, B., Volatile production by buds and corollas of two sympatric, confamiliar plants, *Ipomopsis aggregata* and *Polemonium foliosissimum*, *J. Chem. Ecol.* 28, 565, 2002.

142. Bergström, G. and Bergström, J., Floral scents of *Bartsia alpina* (Scrophulariaceae): chemical composition and variation between individual plants, *Nord. J. Bot.* 9, 363, 1989.

143. Buttery, R.G., Kamm, J.A., and Ling, L.C., Volatile components of alfalfa flowers and pods, *J. Agric. Food Chem.* 30, 739, 1982.

144. Sutton, C.J., Keegans, S.J., Kirk, W.D.J., and Morgan, E.D., Floral volatiles of *Vicia faba*, *Phytochemistry* 31, 3427, 1992.

145. Griffiths, D.W., Robertson, G.W., Shepherd, T., and Ramsay, G., Epicuticular waxes and volatiles from faba bean (*Vicia faba*) flowers, *Phytochemistry* 52, 607, 1999.
146. Porter, A.E.A., Griffiths, D.W., Robertson, G.W., and Sexton, R., Floral volatiles of the sweet pea *Lathyrus odoratus*, *Phytochemistry* 51, 211, 1999.
147. Blackmer, J.L., Rodriguez-Saona, C., Byers, J.A., Shope, K.L., and Smith, J., Behavioral response of *Lygus hesperus* to conspecifics and headspace volatiles of alfalfa in a Y-tube olfactometer, *J. Chem. Ecol.* 30, 1547, 2004.
148. Flamini, G., Cioni, P.L., and Morelli, I., Differences in the fragrances of pollen and different floral parts of male and female flowers of *Laurus nobilis*, *J. Agric. Food Chem.* 50, 4647, 2002.
149. Knudsen, J.T. and Mori, S.A., Floral scents and pollination in neotropical Lecythidaceae, *Biotropica* 28, 42, 1996.
150. Lindberg, A.B., Knudsen, J.T., and Olesen, J.M., Independence of floral morphology and scent chemistry as trait groups in a set of *Passiflora* species, in Totland, Ø., Armbruster, W.S., Fenster, C., Molau, U., Nilsson, L.A., Olesen, J.M., Ollerton, J., Philipp, M., and Ågren, J., Eds., *The Scandinavian Association for Pollination Ecology honours Knut Fægri*, The Norwegian Academy of Science and Letters, Oslo, 2000, p. 91.
151. Raguso, R.A. and Pichersky, E., Floral volatiles from *Clarkia breweri* and *C. concinna* (Onagraceae): recent evolution of floral scent and moth pollination, *Plant Syst. Evol.* 194, 55, 1995.
152. Dobson, H.E.M., Floral odor changes associated with male sterility in tomato plants (*Lycopersicon esculentum*, Solanaceae), unpublished data, 2005.
153. Gaskett, A.C., Conti, E., and Schiestl, F.P., Floral odor variation in two heterostylous species of *Primula*, *J. Chem. Ecol.* 31, 1223, 2005.
154. Knuth, P., *Handbuch der Blütenbiologie*, von Wilhelm Engelmann, Leipzig, 1898.
155. Buttery, R.G., Maddox, D.M., Light, D.M., and Ling, L.C., Volatile components of yellow starthistle, *J. Agric. Food Chem.* 34, 786, 1986.
156. Nilsson, L.A., The pollination ecology of *Dactylorhiza sambucina* (Orchidaceae), *Bot. Notiser* 133, 367, 1980.
157. Nilsson, L.A., Anthecology of *Orchis morio* (Orchidaceae) at its outpost in the north, *Nova Acta Reg. Soc. Sci. Ups. Series V:C* 3, 167, 1984.
158. Altenburger, R. and Matile, P., Further observations on rhythmic emission of fragrance in flowers, *Planta* 180, 194, 1990.
159. Knudsen, J.T. and Tollsten, L., Floral scent and intrafloral scent differentiation in *Moneses* and *Pyrola* (Pyrolaceae), *Plant Syst. Evol.* 177, 81, 1991.
160. Knudsen, J.T. and Olesen, J.M., Buzz-pollination and patterns in sexual traits in north European Pyrolaceae, *Am. J. Bot.* 80, 900, 1993.
161. Johnson, S.D., Steiner, K.E., and Kaiser, R., Deceptive pollination in two subspecies of *Disa spathulata* (Orchidaceae) differing in morphology and floral fragrance, *Plant Syst. Evol.* 255, 87, 2005.
162. Etievant, P.X., Azar, M., Pham-Delegue, M.H., and Masson, C.J., Isolation and identification of volatile constituents of sunflowers (*Helianthus annuus* L.), *J. Agric. Food Chem.* 32, 503, 1984.
163. Flamini, G., Cioni, P.L., and Morelli, I., Analysis of the essential oil of the aerial parts of *Viola etrusca* from Monte Labbro (south Tuscany, Italy) and *in vivo* analysis of flower volatiles using SPME, *Flavour Fragr. J.* 17, 147, 2002.
164. Selvi, F., Foggi, B., and Di Fazio, L., Patterns of phenotypic variation in *Viola etrusca* Erben (Violaceae), *Candollea* 50, 309, 1995.

165. Flach, A., Dondon, R.C., Singer, R.B., Koehler, S., Amaral, M.C.E., and Marsaioli, A., The chemistry of pollination in selected Brazilian Maxillariinae orchids: floral rewards and fragrance, *J. Chem. Ecol.* 30, 1045, 2004.

166. Schlumpberger, B.O., Jux, A., Kunert, M., Boland, W., and Wittmann, D., Musty-earthy scent in cactus flowers: characteristics of floral scent production in dehydro-geosmin-producing cacti, *Int. J. Plant Sci.* 165, 1007, 2004.

167. Andersen, J.F., Composition of the floral odor of *Cucurbita maxima* Duchesne (Cucurbitaceae), *J. Agric. Food Chem.* 35, 60, 1987.

168. Stuurman, J., Hoballah, M.-E., Broger, L., Moore, J., Basten, C., and Kuhlemeier, C., Dissection of floral pollination syndromes in *Petunia, Genetics* 168, 1585, 2004.

169. Buttery, R.G., Kamm, J.A., and Ling, L.C., Volatile components of red clover leaves, flowers, and seed pods: possible insect attractants, *J. Agric. Food Chem.* 32, 254, 1984.

170. Dudareva, N., Murfitt, L.M., Mann, C.J., Gorenstein, N., Kolosova, N., Kish, C.M., Bonham, C., and Wood, K., Developmental regulation of methyl benzoate biosynthesis and emission in snapdragon flowers, *Plant Cell* 12, 949, 2000.

171. Dobson, H.E.M., Bergström, G., and Groth, I., Differences in fragrance chemistry between flower parts of *Rosa rugosa* Thunb. (Rosaceae), *Isr. J. Bot.* 39, 143, 1990.

172. Flament, I., Debonneville, C., and Furrer, A., Volatile constituents of roses, in *Bioactive Volatile Compounds from Plants*, Teranishi, R., Buttery, R.G., and Sugisawa, H., Eds., American Chemical Society Symposium Series 525, American Chemical Society, Washington, DC, 1993, p. 269.

173. Antonelli, A., Fabbri, C., Giorgioni, M.E., and Bazzocchi, R., Characterization of 24 old garden roses from their volatile compositions, *J. Agric. Food Chem.* 45, 4435, 1997.

174. Helsper, J.P.F.G., Davies, J.A., Bouwmeester, H.J., Krol, A.F., and van Kampen, M.H., Circadian rhythmicity in emission of volatile compounds by flowers of *Rosa hybrida* L. cv. Honesty, *Planta* 207, 88, 1998.

175. Shalit, M., Shafir, S., Larkov, O., Bar, E., Kaslassi, D., Adam, Z., Zamir, D., Vainstein, A., Weiss, D., Ravid, U., and Lewinsohn, E., Volatile compounds emitted by rose cultivars: fragrance perception by man and honeybees, *Isr. J. Plant Sci.* 52, 245, 2004.

176. Buchbauer, G., Jirovetz, L., Wasicky, M., and Nikforov, A., Headspace and essential oil analysis of apple flowers, *J. Agric. Food Chem.* 41, 116, 1993.

177. Ashman, T.-L., Bradburn, M., Cole, D.H., Blaney, B.H., and Raguso, R.A., The scent of a male: the role of floral volatiles in pollination of a gender dimorphic plant, *Ecology* 86, 2099, 2005.

178. Binder, R.G., Turner, C.E., and Flath, R.A., Volatile components of purple starthistle, *J. Agric. Food Chem.* 38, 1053, 1990.

179. Tatsuka, K., Suekane, S., Sakai, Y., and Sumitani, H., Volatile constituents of kiwi fruit flowers: simultaneous distillation and extraction versus headspace sampling, *J. Agric. Food Chem.* 38, 2176, 1990.

180. Samadi-Maybodi, A., Shariat, M.R., Zarei, M., Rezai, M.B., Headspace analysis of the male and female flowers of kiwifruit grown in Iran, *J. Essent. Oil. Res.* 14, 414, 2002.

181. Dobson, H.E.M., Arroyo, J., Bergström, G., and Groth, I., Interspecific variation in floral fragrances within the genus *Narcissus* (Amaryllidaceae), *Biochem. Syst. Ecol.* 25, 685, 1997.

182. Dobson, H.E.M., Arroyo, J., and Kephart, S.R., Pollination biology of *Narcissus assoanus* and *N. cuatrecasasii* in southern Spain, unpublished data, 2001.

183. Dafni, A., Bernhardt, P., Shmida, A., Ivri, Y., Greenbaum, S., O'Toole, C.H., and Losito, L., Red bowl-shaped flowers: convergence for beetle pollination in the Mediterranean region, *Isr. J. Bot.* 39, 81, 1990.

184. Joulain, D., Study of the fragrance given off by certain springtime flowers, in *Progress in Essential Oil Research*, Brunke, E.-J., Ed., Walter de Gruyter, Berlin, 1986, p. 57.

185. Williams, N.H. and Whitten, W.M., Orchid floral fragrances and male euglossine bees: methods and advances in the last sesquidecade, *Biol. Bull.* 164, 355, 1983.

186. Gerlach, G. and Schill, R., Composition of orchid scents attracting euglossine bees, *Bot. Acta* 104, 379, 1991.

187. Schiestl, F.P. and Roubik, D.W., Odor compound detection in male euglossine bees, *J. Chem. Ecol.* 29, 253, 2003.

188. Hills, H.G., Fragrance cycling in *Stanhopea pulla* (Orchidaceae, Stanhopeinae) and identificiation of trans-limonene oxide as a major fragrance component, *Lindleyana* 4, 61, 1989.

189. Sazima, M, Vogel, S., Cocucci, A.A., and Hausner, G., The perfume flowers of *Cyphomandra* (Solanaceae): pollination by euglossine bees, bellows mechanism, osmophores, and volatiles, *Plant Syst. Evol.* 187, 51, 1993.

190. Nogueira, P.C. de L., Marsaioli, A.J., Emaral, M. do C.E., and Bittrich, V., The fragrant floral oils of *Tovomita* species, *Phytochemistry* 49, 1009, 1998.

191. Armbruster, W.S., Keller, S., Matsuki, M., and Clausen, T.P., Pollination of *Dalechampia magnoliifolia* (Euphorbiaceae) by male euglossine bees, *Am. J. Bot.* 76, 1279, 1989.

192. Whitten, M.W., Hills, H.G., and Williams, N.H., Occurrence of ipsdienol in floral fragrances, *Phytochemistry* 27, 2759, 1988.

193. Knudsen, J.T., Andersson, S., and Bergman, P., Floral scent attraction in *Geonoma macrostachys*, an understorey palm of the Amazonian rain forest, *Oikos* 85, 409, 1999.

194. Knudsen, J.T., Variation in floral scent composition within and between populations of *Geonoma macrostachys* (Arecaceae) in the western amazon, *Am. J. Bot.* 89, 1772, 2002.

195. Johnson, S.D., Specialized pollination by spider-hunting wasps in the African orchid *Disa sankeyi*, *Plant Syst. Evol.* 251, 153, 2005.

196. Anstett, M.C., Hossaert-McKey, M., and Kjellberg, F., Figs and fig pollinators: evolutionary conflicts in a coevolved mutualism, *Trends Ecol. Evol.* 12, 94, 1997.

197. Grison, L., Edwards, A.A., and Hossaert-McKey, M., Interspecies variation in floral fragrances emitted by tropical *Ficus* species, *Phytochemistry* 52, 1293, 1999.

198. Weiblen, G.D., How to be a fig wasp, *Annu. Rev. Entomol.* 47, 299, 2002.

199. Grison-Pigé, L., Bessière, J.-M., and Hossaert-McKey, M., Specific attraction of fig-pollinating wasps: role of volatile compounds released by tropical figs, *J. Chem. Ecol.* 28, 283, 2002.

200. Barker, N.P., Evidence of a volatile attractant in *Ficus ingens* (Moraceae), *Bothalia* 15, 607, 1985.

201. Ware, A.B., Kaye, P.T., Compton, S.G., and van Noort, S., Fig volatiles: their role in attracting pollinators and maintaining pollinator specificity, *Plant Syst. Evol.* 186, 147, 1993.

202. Giberneau, M., Buser, H.R., Frey, J.E., and Hossaert-McKey, M., Volatile compounds from extracts of figs of *Ficus carica*, *Phytochemistry* 46, 241, 1997.

203. Song, Q.S., Yang, D.R., Zhang, G.M., and Yang, C.R., Volatiles from *Ficus hispida* and their attractiveness to fig wasps, *J. Chem. Ecol.* 27, 1929, 2001.

204. Grison-Pigé, L., Bessière, J.-M., Turlings, C.J., Kjellberg, F., Roy, J., and Hossaert-McKey, M., Limited intersex mimicry of floral odour in *Ficus carica*, *Funct. Ecol.* 15, 551, 2001.

205. Grison-Pigé, L., Hossaert-McKey, M., Greeff, J.M., and Bessière, J.-M., Fig volatile compounds: a first comparative study, *Phytochemistry* 61, 61, 2002.
206. Paulus, H.F. and Gack, C., Pollinators as prepollinating isolation factors: evolution and speciation in *Ophrys* (Orchidaceae), *Isr. J. Bot.* 39, 43, 1990.
207. Schiestl, F.P., Peakall, R., and Mant, J.G., Chemical communication in the sexually deceptive orchid genus *Cryptostelis*, *Bot. J. Linn. Soc.* 144, 199, 2004.
208. Steiner, K.E., Whitehead, V.B., and Johnson, S.D., Floral pollinator divergence in two sexually deceptive South African orchids, *Am. J. Bot.* 81, 185, 1994.
209. Schiestl, F.P. and Marion-Poll, F., Detection of physiologically active flower volatiles using gas chromatography coupled with electroantennography, in *Molecular Methods of Plant Analysis*, vol. 21, *Analysis of Taste and Aroma*, Jackson, J.F., Linskens, H.F., and Inman, R., Eds., Springer, Berlin, 2002, p. 173.
210. Schiestl, F.P. and Ayasse, M., Do changes in floral odor cause speciation in sexually deceptive orchids?, *Plant Syst. Evol.* 234, 111, 2002.
211. Mant, J.G., Schiestl, F.P., Peakall, R., and Weston, P.H., A phylogenetic study of pollinator conservatism among sexually deceptive orchids, *Evolution* 56, 888, 2002.
212. Ayasse, M., Schiestl, F.P., Paulus, H.F., Löfstedt, C., Hansson, B., Ibarra, F., and Francke, W., Evolution of reproductive strategies in the sexually deceptive orchid *Ophrys sphecodes*: how does flower-specific variation of odor signals influence reproductive success?, *Evolution* 54, 1995, 2000.
213. Schiestl, F.P., Ayasse, M., Paulus, H.F., Löfstedt, C., Hansson, B.S., Ibarra, F., and Francke, W., Orchid pollination by sexual swindle, *Nature* 399, 421, 1999.
214. Mant, J., Brandli, C., Vereecken, N.J., Schulz, C.M., Francke, W., Schiestl, F.P., Cuticular hydrocarbons as sex pheromone of the bee *Colletes cunicularius* and the key to its mimicry by the sexually deceptive orchid, *Ophrys exaltata*, *J. Chem. Ecol.*, 31, 1765, 2005.
215. Scoble, M.J., *The Lepidoptera: Form, Function and Diversity*, Oxford University Press, Oxford, 1992, 192.
216. Jürgens, A. and Dötterl, S., Chemical composition of anther volatiles in Ranunculaceae: genera-specific profiles in *Anemone*, *Aquilegia*, *Caltha*, *Pulsatilla*, *Ranunculus*, and *Trollius* species, *Am. J. Bot.* 91, 1969, 2004.
217. Pellmyr, O., Yuccas, yucca moths, and coevolution: a review, *Ann. Mo. Bot. Gard.* 90, 35, 2003.
218. Svensson, G.P., Hickman, M.O., Jr., Bartram, S., Boland, W., Pellmyr, O., and Raguso, R.A., Chemistry and geographic variation of floral scent in *Yucca filamentosa* (Agavaceae). *Am. J. Bot.* 92, 1624, 2005.
219. Erhardt, A., Flower preferences, nectar preferences and pollination effect of Lepidoptera, *Adv. Ecol.* 1, 239, 1991.
220. Raguso, R.A. and Willis, M.A., Hawkmoth pollination in Arizona's Sonoran Desert: behavioral responses to floral traits, in *Butterflies: Ecology and Evolution Taking Flight*, Boggs, C.L., Watt, W.B., and Ehrlich, P.R., Eds., 2003, p. 43.
221. Erhardt, A., Pollination of *Dianthus gratianopolitanus* (Caryophyllaceae), *Plant Syst. Evol.* 170, 125, 1990.
222. Erhardt, A., Unpublished data, 1993.
223. Cantelo, W.W. and Jacobson, M., Phenylacetaldehyde attracts moths to bladder flower and to blacklight traps, *Environ. Entomol.* 8, 444, 1978.
224. Huber, F.K., Kaiser, R., Sauter, W., and Schiestl, F.P., Floral scent emission and specific pollinator attraction in two species of *Gymnadenia* (Orchidaceae), *Oecologia* 142, 564, 2005.

225. Corbet, S.A., Butterfly nectaring flowers: butterfly morphology and flower form, *Entomol. Exp. Appl.* 96, 289, 2000.

226. Honda, K., Ômura, H., and Hayashi, N., Identification of floral volatiles from *Ligustrum japonicum* that stimulate flower-visiting by cabbage butterfly, *Pieris rapae*, *J. Chem. Ecol.* 24, 2167, 1998.

227. Ômura, H., Honda, K., and Hayashi, N., Chemical and chromatic bases for preferential visiting by the cabbage butterfly, *Pieris rapae*, to rape flowers, *J. Chem. Ecol.* 25, 1895, 1999.

228. Ômura, H., Honda, K., Nakagawa, A., and Hayashi, N., The role of floral scent of the cherry tree, *Prunus yedoensis*, in the foraging behavior of *Luehdorfia japonica* (Lepidoptera: Papilionidae), *Appl. Entomol. Zool.* 34, 309, 1999.

229. Ômura, H. and Honda, K., Priority of color over scent during flower visitation by adult *Vanessa indica* butterflies, *Oecologia* 142, 588, 2005.

230. Erhardt, A. and Jäggi, B., From pollination by Lepidoptera to selfing: the case of *Dianthus glacialis* (Caryophyllaceae), *Plant Syst. Evol.* 195, 67, 1995.

231. Naumann, C.M., Ockenfels, P., Schmitz, J., Schmidt, F., and Francke, W., Reactions of *Zygaena* moths to volatile compounds of *Knautia arvensis* (Lepidoptera: Zygaenidae), *Entomol. Gener.* 15, 255, 1991.

232. Andersson, S., Antennal responses to floral scent in the butterflies *Inachis io*, *Aglais urticae* (Nymphalidae), and *Gonepteryx rhamni* (Pieridae), *Chemoecology* 13, 13, 2003.

233. Andersson, S. and Dobson, H.E.M., Antennal responses to floral scents in the butterfly *Heliconius melpomene*, *J. Chem. Ecol.* 29, 2319, 2003.

234. Pellmyr, O., Three pollination morphs in *Cimicifuga simplex*; incipient speciation due to inferiority in competition, *Oecologia* 68, 304, 1986.

235. Andersson, S., Foraging responses in the butterflies *Inachis io*, *Aglais urticae* (Nymphalidae), and *Gonepteryx rhamni* (Pieridae) to floral scents, *Chemoecology* 13, 1, 2003.

236. Andersson, S. and Dobson, H.E.M., Behavioral foraging responses by the butterfly *Heliconius melpomene* to *Lantana camara* floral scent, *J. Chem. Ecol.* 29, 2303, 2003.

237. Ômura, H., Honda, K., and Hayashi, N., Floral scent of *Osmanthus fragrans* discourages foraging behavior of cabbage butterfly, *Pieris rapae*, *J. Chem Ecol.* 26, 655, 2000.

238. Haber, W.A. and Frankie, G.W., A tropical hawkmoth community: Costa Rican dry forest Sphingidae, *Biotropica* 21, 155, 1989.

239. Kaiser, R.A.J., On the scent of orchids, in *Bioactive Volatile Compounds from Plants*, Teranishi, R., Buttery, R.G., and Sagisawa, H., Eds., American Chemical Society Symposium Series 525, American Chemical Society, Washington, DC, 1993, p. 240.

240. Shaver, T.N., Lindgren, P.D., and Marshall, H.F., Nighttime variation in volatile content of flowers of the night blooming plant *Gaura drummondii*, *J. Chem. Ecol.* 23, 2673, 1997.

241. Christensen, L.P., Jakobsen, H.B., Kristiansen, K., and Møller, J., Volatiles emitted from flowers of γ-radiated and nonradiated *Jasminum polyanthum* Franch. in situ, *J. Agric. Food Chem.* 45, 2199, 1997.

242. Nilsson, L.A., Processes of isolation and introgressive interplay between *Platanthera bifolia* (L.) Rich and *P. chlorantha* (Custer) Reichb. (Orchidaceae), *Bot. J. Linn. Soc.* 87, 325, 1983.

243. Nilsson, L.A., Characteristics and distribution of intermediates between *Platanthera bifolia* and *P. chlorantha* (Orchidaceae) in the nordic countries, *Nord. J. Bot.* 5, 407, 1985.

244. Tollsten, L. and Bergström, G., Fragrance chemotypes of *Platanthera* (Orchidaceae): the result of adaptation to pollinating moths?, *Nord. J. Bot.* 13, 607, 1993.

245. Cunningham, J.P., Moore, C.J., Zalucki, M.P., and West, S.A., Learning, odour preference and flower foraging in moths, *J. Exp. Biol.* 207, 87, 2004.

246. Heath, R.R., Landolt, P.J., Dueben, B., and Lenczewski, B., Identification of floral compounds of night-blooming Jessamine attractive to cabbage looper moths, *Environ. Entomol.* 21, 854, 1992.

247. Haynes, K.F., Zhao, J.Z., and Latif, A., Identification of floral compounds from *Abelia grandiflora* that stimulate upwind flight in cabbage looper moths, *J. Chem. Ecol.* 17, 637, 1991.

248. Landolt, P.J. and Smithhisler, C.L., Characterization of the floral odor of Oregongrape: possible feeding attractants for moths, *Northwest Sci.* 77, 81, 2003.

249. Creighton, C.S., McFadden, T.L., and Cuthbert, E.R., Supplementary data on phenylacetaldehyde: an attractant for Lepidoptera, *J. Econ. Entomol.* 66, 114, 1973.

250. Landolt, P.J., Adams, T., Reed, H.C., and Zack, R.S., Trapping alfalfa looper moths (Lepidoptera: Noctuidae) with single and double component floral chemical lures, *Environ. Entomol.* 30, 667, 2001.

251. Meagher, R.L., Trapping fall armyworm (Lepidoptera: Noctuidae) adults in traps baited with pheromone and a synthetic floral volatile compound, *Fla. Entomol.* 84, 288, 2001.

252. Meagher, R.L., Collection of soybean looper and other noctuids in phenylacetaldehyde-baited field traps, *Fla. Entomol.* 84, 154, 2001.

253. Deng, J.-Y., Wei, H.-Y., Huang, Y.-P., and Du, J.-W., Enhancement of attraction to sex pheromones of *Spodoptera exigua* by volatile compounds produced by host plants, *J. Chem. Ecol.* 30, 2037, 2004.

254. Dötterl, S., Jürgens, A., Seifert, K., Laube, T., Weißbecker, B., and Schütz, S., Nursery pollination by a moth in *Silene latifolia*: the role of odours in eliciting antennal and behavioural responses, *New Phytol* 2005, doi:10.1111/j.1469-8137.2005.01509.

255. Dötterl, S., Wolfe, L.M., and Jürgens, A., Qualitative and quantitative analyses of flower scent in *Silene latifolia*, *Phytochemistry* 66, 195, 2005.

256. Plepys, D., Ibarra, F., Francke, W., and Lofstedt, C., Odour-mediated nectar foraging in the silver Y moth, *Autographa gamma* (Lepidoptera: Noctuidae): behavioural and electrophysiological responses to floral volatiles, *Oikos* 99, 75, 2002.

257. Plepys, D., Ibarra, F., and Löfstedt, C., Volatiles from flowers of Platanthera bifolia (Orchidaceae) attractive to the silver Y moth, *Autographa gamma* (Lepidoptera: Noctuidae), *Oikos* 99, 69, 2002.

258. Gabel, B., Thiery, D., Suchy, V., Marion-Poll, F., Hradsky, P., and Farkas, P., Floral volatiles of *Tanacetum vulgare* L. attractive to *Lobesia botrana* Den. et Schiff. females, *J. Chem. Ecol.* 18, 693, 1992.

259. Pophof, B., Stange, G., and Abrell, L., Volatile organic compounds as signals in a plant-herbivore system: electrophysiological responses in olfactory sensilla of the moth *Cactoblastis cactorum*, *Chem. Senses* 30, 51, 2005.

260. Grant, V., The systematics and geographical distribution of hawkmoth flowers in the temperate North America flora, *Bot. Gaz.* 144, 439, 1983.

261. Kaiser, R. and Tollsten, L., An introduction to the scent of cacti, *Flavour Fragr. J.* 10, 153, 1995.

262. Kite, G.C. and Leon, C., Volatile compounds emitted from flowers and leaves of *Brugmansia* x *candida* (Solanaceae), *Phytochemistry* 40, 1093, 1995.

263. Miyake, T., Yamaoka, R., and Yahara, T., Floral scents of hawkmoth-pollinated flowers in Japan, *J. Plant Res.* 111, 199, 1998.

264. Manning, J.C. and Snijman, D., Hawkmoth-pollination in *Crinum variabile* (Amaryllidaceae) and the biogeography of sphingophily in southern African Amaryllidaceae, *S. Afr. J. Bot.* 68, 212, 2002.

265. Kawano, S., Odaki, M., Yamaoka, R., Oda-Tanabe, M., Takeuchi, M., and Kawano, N., Pollination biology of *Oenothera* (Onagraceae): the interplay between floral UV-absorbancy patterns and floral volatiles as signals to nocturnal insects, *Plant Species Biol.* 10, 31, 1995.

266. Luyt, R. and Johnson, S.D., Hawkmoth pollination of the African epiphytic orchid *Mystacidium venosum*, with special reference to flower and pollen longevity, *Plant Syst. Evol.* 228, 49, 2001.

267. Johnson, S.D., Hawkmoth pollination and hybridization in *Delphinium leroyi* (Ranunculaceae) on the Nyika Plateau, Malawi, *Nord. J. Bot.* 21, 599, 2001.

268. Matile, P. and Altenburger, R., Rhythms of fragrance emission in flowers, *Planta* 174, 242, 1988.

269. Pott, M.B., Pichersky, E., and Piechulla, B., Evening specific oscillations of scent emission, SAMT enzyme activity, and SAMT mRNA in flowers of *Staphanotis floribunda*, *J. Plant Physiol.* 159, 925, 2002.

270. Raguso, R.A., Light, D.M., and Pichersky, E., Electroantennogram responses of *Hyles lineata* (Sphingidae: Lepidoptera) to volatile compounds from *Clarkia breweri* (Onagraceae) and other moth-pollinated flowers, *J. Chem. Ecol.* 22, 1735, 1996.

271. Raguso, R.A. and Light, D.M., Electroantennogram responses of male *Sphinx perelegans* hawkmoths to floral and "green-leaf volatiles," *Entomol. Exp. Appl.* 86, 287, 1998.

272. Fraser, A.M., Mechaber, W.L., and Hildebrand, J.G., Electroantennographic and behavioral responses of the sphinx moth *Manduca sexta* to host plant headspace volatiles, *J. Chem. Ecol.* 29, 1813, 2003.

273. Schwilch, R., Mantovani, R., Spina, F., and Jenni, L., Nectar consumption of warblers after long-distance flights during spring migration, *Ibis* 143, 24, 2001.

274. Bruneau, A., Evolution and homology of bird pollination syndromes in *Erythrina* (Leguminosae), *Am. J. Bot.* 84, 54, 1997.

275. Martinez del Rio, C., Baker, H.G., and Baker, I., Ecological and evolutionary implications of digestive processes: bird preferences and the sugar constituents of floral nectar and fruit pulp, *Experientia* 48, 544, 1992.

276. Evans, H.E. and Heiser, J.B., What's inside: anatomy and physiology, in *Handbook of Bird Biology*, 2nd ed., Podulka, S., Rohybaugh, R.W., and Bonney, R., Eds., Princeton University Press, Princeton, NJ, 2004.

277. Hurly, T.A. and Healy, S.D., Memory for flowers in rufous hummingbirds: location or local visual cues?, *Anim. Behav.* 51, 1149, 1996.

278. Meléndez-Ackerman, E.J. and Campbell, D.R., Adaptive significance of flower color and inter-trait correlations in an *Ipomopsis* hybrid zone, *Evolution* 52, 1293, 1998.

279. Varassin, I.G., Trigo, J.R., and Sazima, M., The role of nectar production, flower pigments and odour in the pollination of four species of *Passiflora* (Passifloraceae) in south-eastern Brazil, *Bot. J. Linn. Soc.* 136, 139, 2001.

280. Sazima, M., Sazima, I., and Buzato, S., Nectar by day and night: *Siphocampylus sulfureus* (Lobeliaceae) pollinated by hummingbirds and bats, *Plant Syst. Evol.* 191, 237, 1994.

281. Sahley, C.T., Bat and hummingbird pollination of an autotetraploid columnar cactus, *Weberbauerocereus weberbaueri* (Cactaceae), *Am. J. Bot.* 83, 1329, 1996.

282. Aigner, P.A. and Scott, P.E., Use and pollination of a hawkmoth plant, *Nicotiana attenuata*, by migrant hummingbirds, *Southwest. Nat.* 47, 1, 2002.

283. Wolff, D., Braun, M., and Liede, S., Nocturnal versus diurnal pollination success in *Isertia laevis* (Rubiaceae): a sphingophilous plant visited by hummingbirds, *Plant Biol.* 5, 71, 2003.

284. Winter, Y. and von Helversen, O., Bats as pollinators: foraging energetics and floral adaptations, in *Cognitive Ecology and Pollination*, Chittka, L. and Thomson, J.D., Eds., Cambridge University Press, 2001.

285. Wyatt, R., Pollinator-plant interactions in the evolution of breeding systems, in *Pollination Biology*, Real, L., Ed., Academic Press, Orlando, FL, 1983, p. 51.

286. Bestmann, H.J., Winkler, L., and von Helversen, O., Headspace analysis of flower scent constituents of bat-pollinated plants, *Phytochemistry* 46, 1169, 1997.

287. von Helversen, O., Winkler, L., and Bestmann, H.J., Suphur-containing "perfumes" attract flower-visiting bats, *J. Comp. Physiol. A* 186, 143, 2000.

288. Pettersson, S. and Knudsen, J.T., Floral scent and nectar production in *Parkia biglobosa* Jacq. (Leguminosae: Mimosoideae), *Bot. J. Linn. Soc.* 135, 97, 2001.

289. Pettersson, S.. Ervik, F., and Knudsen, J.T., Floral scent of bat-pollinated species:West Africa vs. the New World, *Biol. J. Linn. Soc.* 82, 168, 2004.

290. Ecroyd, C.E., Franich, R.A., Kroese, H.W., and Steward, D., Volatile constituents of *Dactylanthus taylorii* flower nectar in relation to flower pollination and browsing by animals, *Phytochemistry* 40, 1387, 1995.

291. Fleming, T.H., Plant-visiting bats, *Am. Sci.* 81, 460, 1993.

292. Winter, Y., López, J., and von Helversen, O., Ultraviolet vision in a bat, *Nature* 425, 612, 2003.

293. Kress, W.J., Bat pollination of an Old World *Heliconia*, *Biotropica* 17, 302, 1985.

294. Knudsen, J.T. and Klitgaard, B.B., Floral scent and pollination in *Browneopsis disepala* (Leguminosae: Caesalpinioideae) in western Ecuador, *Brittonia* 50, 174, 1998.

295. Arizaga, S., Ezcurra, E., Peters, E., de Arellano, F.R., and Vega, E., Pollination ecology of *Agave macroacantha* (Agavaceae) in a Mexican tropical desert. I. Floral biology and pollination mechanisms, *Am. J. Bot.* 87, 1004, 2000.

296. Slauson, L.A., Pollination biology of two chiropterophilous agaves in Arizona, *Am. J. Bot.* 87, 825, 2000.

297. Fleming, T.H., Sahley, C.T., Holland, J.N., Nason, J.D., and Hamrick, J.L., Sonoran desert columnar cacti and the evolution of generalized pollination systems, *Ecol. Monogr.* 71, 511, 2001.

298. Tschapka, M. and von Helversen, O., Pollinators of syntopic *Marcgravia* species in Costa Rican lowland rain forest: bats and opossums, *Plant Biol.* 1, 382, 1999.

299. Johnson, S.D. and Steiner, K.E., Specialized pollination systems in southern Africa, *S. Afr. J. Sci.* 99, 345, 2003.

300. Hargreaves, A., Johnson, S.D., and Nol, E., Do floral syndromes predict specialization in plant pollination systems? An experimental test in an "ornithophilous" African *Protea*, *Oecologia* 140, 295, 2004.

301. Ramírez, N., Floral specialization and pollination: a quantitative analysis and comparison of the Leppik and the Faegri and van der Pijl classification systems, *Taxon* 52, 687, 2003.

302. Ollerton, J. and Watts, S., Phenotypic space and floral typology: towards an objective assessment of pollination syndromes. *Norske Vidensk.-Akad. I. Mat. Naturvid. Kl. Skrif., Ny Ser.* 39, 149, 2000.

303. Goldsmith, K.M., and Goldsmith, T.H., Sense of smell in the black-chinned hummingbird, *Condor* 84, 237, 1982.

9 Floral Scent and Butterfly Pollinators

Susanna Andersson

CONTENTS

9.1 INTRODUCTION

The previous chapter provided an overview of the role of floral scent in interactions with the gamut of animal visitors. In this chapter and the next, two specific types of insect pollinators and their interactions with scented flowers are examined in detail. This chapter focuses on butterflies. Adult butterflies visit flowering plants in search of food, in most cases nectar and in a few cases pollen. From the insect perspective, one may ask: Which plants are attractive to nectar-seeking butterflies and how do butterflies find them? Do butterflies use floral scent in their search for food, and which are these floral scents? From the plant perspective, one may ask: Which plants are butterflies actually pollinating at foraging visits? What are the floral scent compositions of butterfly-pollinated plants? Why are plants emitting these floral scent compounds? These issues are discussed below.

9.2   CHEMICAL SIGNALS IN BUTTERFLY LIFE

In butterfly adult life, chemical signals play a major role. Most olfactory receptors are found at the antennae, while the butterfly's own production of volatiles emits from glands close to the genitalia or on the wings in proximity to scent scales. In courtship, males locate females, first at long distance, by visual cues, especially size and color, whereas volatile and tactile chemical signals become important at close range. These chemicals are mostly male pheromones emitted from wing scales with glandular bases or from glands near the genitalia, the androconial organs.[1-6] Male display is often unique for each species and serves to maintain reproductive isolation between closely related species, permitting females to assess the quality of a potential mate.[7] Gilbert[8] and Forsberg and Wiklund[9] identified repellent pheromones from females, emitted after mating, that are thought to originate from males. During oviposition, females use both gustatory and olfactory stimulants in host-plant identification.[10-19] In defense, chemicals are central in all butterfly stages, with both volatile emissions and liquid secretions against predators and parasites.[5,20] Considering all the chemicals entangling butterflies in different life situations, it is reasonable to believe they are also sensitive to floral volatile compounds from food plants.

9.3   BUTTERFLY NECTAR PLANTS

Most butterfly species need to feed in the adult stage. They feed on nectar, fruits, tree sap, and mud to fulfill their nutritional demands of nitrogen, carbohydrates, fats, minerals, and vitamins.[21] A special case is the tropical butterfly genus *Heliconius*, which has the ability to feed on pollen by dissolving nitrogen-containing amino acids and the minerals therein.[22,23] In northern Europe, nectar is the main food of adult butterflies.[7] Butterflies have preferences regarding nectar sources, and choose specific plant species to forage from. Several of these nectar plants are also being pollinated by butterflies.

In a study of day-flying Lepidoptera nectar sources in central Illinois,[24] it was found that the diversity of Lepidoptera was highest on Asclepiadaceae and Verbenaceae, and that day-flying lepidopterans were particularly abundant on *Aster pilosus* Willdenow (Asteraceae) and *Cephalanthus occidentalis* L. (Rubiaceae). The most polyphagous butterfly species—*Colias philodice* Godart (Pieridae), *Danaus plexippus* (L.) (Danaidae), and *Artogeia rapae* (L.) [= *Pieris rapae* (L.); Pieridae]—visited flowers of 50 or more plant species. However, most lepidopteran species visited a limited range of nectar plants.

Nectar-foraging butterflies frequently visit Asteraceae species, such as *Centaurea scabiosa* L., *Cirsium arvense* (L) Scop., and *Inula salicina* L. *Eupatorium cannabinum* L. is yet another favorite butterfly nectar plant, which displays obvious

FIGURE 9.1 (See color insert following page 178.) A butterfly visiting *K. arvensis* (Dipsacaceae). Photo by Dr. Heidi E.M. Dobson.

floral specializations toward butterfly pollination, such as tube-shaped flowers.[25–27] The Caryophyllaceae (e.g., *Agrostemma githago* L., *Dianthus armeria* L., *Dianthus deltoides* L., and *Silene flos-cuculi* (L) Greuter & Burdet) display butterfly-pollination adaptive traits.[26–29] *Silene acaulis* L. (Caryophyllaceae) seems to be an important nectar source for butterflies in the Swedish high alpine mountains, as judged from the butterflies' choice of *S. acaulis* over other accessible plants such as *Rhododendron lapponicum*.[30] Butterflies also show a preference for *Knautia arvensis* (L) Coulter (Dipsacaceae) (Figure 9.1)[25,31,32] and *Scabiosa columbaria* L. (Dipsacaceae).[33] Other plants often used as nectar sources by butterflies are *Origanum vulgare* L. (Lamiaceae),[34] *Buddleja davidii* Franchet (Loganiaceae),[26,35] *Primula farinosa* L. (Primulaceae),[27] and *Centranthus ruber* (L.) DC (Valerianaceae), all showing adaptations in morphological traits to butterflies.[26]

The European orchid *Gymnadenia conopsea* (L.) R. BR. depends on different Lepidoptera species for pollination.[26] In Sweden, two subspecies occur, the early flowering, smaller ssp. *conopsea*, and the late flowering, coarser ssp. *densiflora* (Wahlenb.) A. Dietr.[36] Interestingly, it seems as if ssp. *conopsea* is used as a nectar plant and is pollinated by moths while butterflies forage from and pollinate ssp. *densiflora*.[27]

Butterflies and moths forage from and pollinate *Phlox drummondii* Hook. (Polemoniaceae) and *Phlox paniculata* L. cv., species native to the North American continent.[33] *C. occidentalis* L. (Rubiaceae), also native to North America, is most often visited by butterflies, bumblebees, and bees.[24]

In South America, *Warszewiczia coccinea* (Vahl) Kl. (Rubiaceae) is an attractive nectar plant to butterflies,[37] and butterflies and hummingbirds are probable pollinators.[38] *Heliconius* spp. and several other tropical butterfly species forage from *Lantana camara* L. (Verbenaceae) and butterflies mainly pollinate the plant.[22,23,35,37,39,40]

Some plants attract foraging butterflies but are food deceptive, and these plants thus do not provide any food reward. *Daphne mezereum* L. (Thymelaceaceae) is one such example.[27] *Anacamptis pyramidalis* (L) Rich. (Orchidaceae) and *Centaurium erythraea* Raf. (Gentianaceae) are two other food frauds that display floral traits that could be adaptations to butterfly pollination.[26,41,42]

Butterflies are able to assess the nutritional quality of their food, such as amino acid and nectar sugar content.[43–48] The capacity to efficiently judge and evaluate a potential food source leads to reduced food searching time, thereby permitting more time for direct reproductive activities. Butterflies do choose certain plant species as nectar sources.

9.4 BUTTERFLY-POLLINATED PLANTS

Butterflies can be efficient pollinators of plants they visit for nectar or pollen feeding. Adult butterflies often fly longer distances when searching for food and may therefore transfer pollen between different populations of a plant species, thus achieving long-distance gene flow; this is in contrast to colony-dependent insects, such as bumblebees, which typically fly fairly short distances, resulting in only localized pollen dispersal.[42,49–51]

One plant that is typically butterfly pollinated is the woody shrub *L. camara* (Verbenaceae), native to tropical and subtropical America and naturalized in other areas.[39,52] Flowers of *L. camara* change from yellow to red when pollinated (because of the appearance of anthocyanin), thereby directing pollinators, mainly butterflies, to unpollinated and rewarding yellow flowers.[53] *L. camara* is a favorite nectar plant for several tropical butterfly species (e.g., *Heliconius* butterflies).[22,23,37]

At arctic and alpine sites, bumblebees and flies are important pollinators.[54–58] However, butterflies can also act as pollinators. At a site in the northern Swedish Lapland, a potential plant–butterfly pollination relationship could be the case between *Silene acaulis* and butterflies; 23 out of 34 butterfly flower visits were observed on *S. acaulis*, while for bumblebees this ratio was only 7 out of 1089 visits.[59] Indeed, *S. acaulis* displays features that are characteristic of butterfly-adapted plants:[28,35,60] sweet and heavy floral scent, reddish-pink flower color, nectar concealed in narrow corolla tubes, continuous yellowish nectar guide at the base of the petals, erect flower with flat rims as a landing platform, and stamens and styles that protrude out of the flowers. An examination of butterflies for the presence of pollen on the ventral body surface has shown that they regularly carry *S. acaulis* pollen.[30] However, bumblebees and Dipteran species have also been reported as pollinators of *S. acaulis*.[56,57,61–63]

9.5 SPECIES SPECIFICITY OF FLORAL SCENTS
AND FLOWER CONSTANCY

Every plant species emits a unique floral scent composition,[64] although variations are present.[65] Such species-specific fragrances could be important for an insect's food searching ability to locate rewarding and avoid nonrewarding plants. A plant's characteristic floral scent composition, when recognized by an insect, might also be of importance for flower constancy, and may be explained by an insect's memory constraints or learning ability.[66–69] The butterfly *Thymelicus flavus* (Hesperiidae) revisits certain nectar plants as long as sufficient nectar is provided, and it is possible that the butterfly locate the plants in using the floral volatiles.[70]

9.6 CONVERGENT EVOLUTION OF FLORAL CHARACTERS

Butterflies may benefit from gauging the rewards of a potential food source because it reduces the amount of time spent searching, giving them more time for reproductive activities. Butterflies are mobile, and as pollinators put selective pressure on the plant to adapt to the butterflies' physiological requirements. The adaptive pressure to specialize is assumed to be stronger in the sessile plant than in the mobile pollinator.[71,72] The imbalance in the interaction grows more profound since, for plants, pollination is a step closer in space and time to reproduction, compared to butterflies, where foraging is one prerequisite for gaining energy for reproduction, but otherwise is distant in space and time.

A wide range of olfactory and visual floral signals is believed to have evolved as adaptations to promote efficient pollination by animals.[35,66,71,73–75] In the synchronized diversification of angiosperms and animal pollinators, plant floral traits probably evolved from being a defense against herbivory to being an attractant to the most efficient pollinators.[28,35,73,75] Reproductive gains for plants and pollinators may promote specialization into coadapted reproductive features[77] or so-called pollination syndromes, when resulting from convergent evolution of suites of floral traits among unrelated plant species to fit related pollinators.[35] Floral characteristics that typify butterfly-adapted plants include brightly colored flowers, diurnal anthesis, and a narrow corolla tube.[35,60,81,82]

9.7 FROM GENERALIZATION TO SPECIALIZATION OF INTERACTIONS

The degree of morphological adaptation in flowers to butterfly pollination differs among plant species, since interactions range from broadly to narrowly generalized.[29,32,83,84] Indeed, pollination interactions in general range from being specialized, with a very tight connection to one functional type of pollinator, as with several species of the Orchidaceae, to generalized cases where many various species act as pollinators.[72,84–86] There seems to be a relationship between the degree of specialization and regional area, with specialized relationships being more common in the Southern Hemisphere and generalized relationships being more common in the Northern Hemisphere.[84,87] In a specialist butterfly pollination relationship, plant and butterfly adaptations should be more easily identified compared to the generalist cases.

9.8 FLORAL SCENT CHARACTERISTICS OF BUTTERFLY-POLLINATED PLANTS

Compounds from butterfly pollinated plants of exclusive floral origin, in relatively great abundance, may be potential food signals for butterflies. If such compounds are butterfly foraging elicitors, they would be potential candidates for being plant adaptations matching the olfactory requirements of butterflies. Certain benzenoid compounds, such as phenylacetaldehyde, and monoterpenoid alcohols, for example, linalool-related compounds, are emitted in relatively large amounts from several butterfly-pollinated plants of different families and regions of the world.[80] The predominance of benzenoids in European temperate plants can be contrasted against linalool and its derivatives in American species, constituting two discernible groups. Convergent evolution within, but with differences between, continents might be a possibility. However, although floral scents, representatives of ecological signaling molecules, may have evolved partly from adaptations to the requirements of interacting species, their expression is still embedded in a phylogenetic structure.[88]

9.8.1 BUTTERFLY AND MOTH POLLINATION

Similarities in the fragrance compositions of butterfly- and moth-pollinated plants because of the taxonomic similarities of the pollinators is possible. Plant adaptations to the two different pollinator groups may follow a line of least resistance.[71] While similarities in floral traits exist among butterfly- and moth-pollinated plants, differences in floral stimuli have been found in several genera, including *Phlox* (Polemoniaceae),[33] *Disa* (Orchidaceae),[89] *Dianthus* (Caryophyllaceae),[62,90,91] and *Platantera* (Orchidaceae).[92] Some common features in floral scent compositions of butterfly- and moth-pollinated plants are that benzenoid- and linalool-related compounds are commonly emitted, while fatty acid derivatives are not.[78–80] Nitrogenous compounds are emitted from the butterfly-pollinated *C. scabiosa*, *P. paniculata*, *W. coccinea*, and *C. ruber*[80] and from several moth-pollinated species[78,93] convergence resulting in similarities of scent profiles for butterfly- and moth-pollinated.

The European orchid *G. conopsea* is reported to be dependent on different Lepidoptera species for pollination.[26] The two subspecies occurring in Sweden seem to depend on different pollinator faunas; pollinia of ssp. *conopsea* have been found mostly on moths, while pollinia of ssp. *densiflora* have been found mostly on butterflies.[27] The flower volatiles of the suspected butterfly-pollinated ssp. *densiflora* are dominated by, in number and amount, benzenoid compounds, clearly different from the fatty acid derivative-dominated floral scent of ssp. *conopsea*.[94,95] High benzenoid emissions are also characteristic of other temperate butterfly-pollinated orchids, such as *A. pyramidalis*[80] and *Gymnadenia* (Nigritella) *nigra* (L.) Reichb. f.[94] The ecological significance of certain benzenoid compounds, such as phenylacetaldehyde, as butterfly foraging attractants has been demonstrated in the butterflies *P. rapae*[96] and *Luehdorfia japonica* (Papilionidae).[97] Certain benzenoid compounds seem to be one characteristic of the floral scents of plants attractive to foraging butterflies; plants that often are also pollinated by butterflies. From ssp. *conopsea* the nitrogen-containing floral scent compound indole was emitted in large amounts, which agrees with the findings of indole in floral scents from other moth-pollinated plants.[78] Indole elicited relatively low proboscis extension activity in the butterfly *P. rapae* L. (Pieridae) compared to benzenoid compounds.[96,98] Thus the presence of indole in the floral scent could be one diverging trait toward moth pollination and away from butterfly pollination. Adaptive divergence in floral traits has probably been of importance in pollinator-mediated selection; for example, as demonstrated in *Mimulus* (Scrophulariaceae) (monkeyflowers), with one hummingbird and one bee pollinator guild differing in their preferences of floral color and nectar amount.[99]

Ethological isolation by pollinators mediated through floral scent is thought to have occurred in *Cimicifuga simplex* Wormsk. (Ranunculaceae), with populations that have evolved into either butterfly or bumblebee chemotypes,[100] and in the similar and partly sympatric orchids *Platanthera bifolia* and *P. chlorantha* that have become adapted to different moth faunas.[92] Moreover, the floral scent chemistry variation in sympatric taxa of geonomoid palms is concluded to be of importance in maintaining reproductive isolation by different pollinators.[101] Since both moths and butterflies are sensitive to floral scents, the differences in the floral scent compositions of the two subspecies of *G. conopsea* could be an important factor in a possible process of ethological isolation and ongoing speciation.

9.8.2 FOOD-DECEPTIVE SPECIES

Food-deceptive plants mask their food deficiency by displaying floral traits that normally signal rewards to pollinating insects. The floral scent compositions of the three food-frauds *C. erythraea*, *A. pyramidalis*, and *D. mezereum*[26,102] were analyzed.[80] Interestingly, compounds in common for these species have been shown to be attractive to foraging insects. For example, *C. erythraea* and *A. pyramidalis* emit 2-phenylethanol, a compound that elicits foraging behavior in *P. rapae*[96] and in moths.[103] *Daphne mezereum* and *A. pyramidalis* both emit linalool, which is attractive to foraging bees.[102,104] *A. pyramidalis* depends to a great degree on butterflies for pollination, and its flower blend is distinctive in that it is composed of only a few compounds (phenylacetaldehyde, 2-phenylethanol, linalool, verbenone, α-pinene, and oxoisophorone) and varies greatly between individuals. This may indicate that no single specific compound, but rather several different ones have key functions in butterfly attraction. The large scent variation found in *A. pyramidalis* is also found in another food-deceptive orchid pollinated by moths, *Epidendrum ciliare*.[105] In food-deceptive pollination systems, large variations in scent composition are generally thought to be a strategy to prevent pollinators from learning to avoid the non-rewarding flowers.

9.9 BUTTERFLY BEHAVIORAL AND PHYSIOLOGICAL RESPONSES TO FLORAL SCENTS

Behavioral foraging reactions to floral scents emphasize the ecological function of floral scents. That floral scent is a valuable pollinator attractant is supported by its emission peak correlations with pollinator specificity and activity in, for example, wasps and bees,[106–114] moths,[65,78,105,115–117] bats,[118] and butterflies.[96] Moreover, pollen odor attraction in bee flower selection has been demonstrated to specify from which flower parts pollinator-attractive compounds are emitted.[119–122]

9.9.1 OLFACTORY AND VISUAL FLOWER SIGNALS

What is the relative use of floral scent and color in association with rewards for pollinators? The honeybee (*Apis mellifera* [Hymenoptera]) is capable of using one or more sensory traits to enhance its foraging efficiency in different environmental situations.[123] In naïve individuals of the moth *Manduca sexta*, visual or olfactory flower cues presented singly elicit attraction, but both stimuli are required for the complete behavioral chain from attraction to feeding.[124] Lewis and Lipani[125] proposed that while floral scents override colors as flower rewarding associations in bees, the opposite might be true for butterflies.

Butterflies have vision for a broad spectra of wavelengths from ultraviolet (UV) to red (300 to 700 nm)[126] and are highly sensitive to floral colors. Indeed, butterflies demonstrate innate preferences to certain colors, and the ability of butterflies to associate colors with food reward is strong.[37,69,125,127–135] Red and yellow are most easily associated with food for the butterfly *Papilio xuthus*, while other colors take longer to learn, most likely because of innate preferences for red and yellow.[136] Weiss[134] suggests that a combination of innate and learned behavior is used to locate and forage on flowers, as butterflies learn to associate reward more quickly with an innately preferred color.

Because of the excellent vision of butterflies and their use of floral colors in identifying rewarding plants, one might assume that butterflies respond to chemical flower signals to a lesser extent. However, floral volatiles are, in addition to visual stimuli, important foraging guides for butterflies.[26,96,127,137]

The relative reliance on color and scent varies among butterfly species. Ilse[127] found, for example, that *Inachis io L.* (*Nymphalidae*) locates food based on both visual and olfactory cues, while *Aglais urticae L.* (*Nymphalidae*) only responds to visual stimuli and *Gonepteryx rhamni L.* (*Pieridae*) reacts mostly on vision and hardly ever on olfaction. Moreover, depending on whether the adult butterfly is flower inexperienced (naïve) or experienced, the reliance on different floral cues may differ. Newly emerged *Heliconius melpomene* L. (Lepidoptera: Nymphalidae: Heliconiinae) seem to rely mainly on olfactory floral cues from one preferred nectar plants *L. camara* to select flowers, while, upon gaining foraging experience, the butterflies learn to associate a food source more with visual stimuli of which the yellow color seems to be higher ranked than the *L. camara* complete floral scent blend.[138] However, when butterflies were conditioned to green scented models, they also learned to associate green with food, but they did so to a lesser extent than with yellow. When both choice models were green in the tests, the scent of flowers was preferred over that of the vegetative parts. When offered a choice between green floral-scented and yellow vegetative-scented models, butterflies did not show any preference between the two models. The reasons for this could be either diminished association of food with yellow to the advantage of green or a stronger association of food with floral scent than with green.

The three common North European butterflies, *I. io*, *A. urticae*, and *G. rhamni* demonstrated, when flower naïve, behavioral foraging preferences, in terms of probing following visitation and probing time, to floral scents of the butterfly nectar plant *C. arvense* compared to the nonbutterfly plants *Achillea millefolium* L. (Asteraceae) and *Philadelphus coronarius* L. (Hydrangiaceae), regardless of whether the scents were offered in mauve or green models.[139] Thus, even though green color by itself is not associated with flower food rewards in butterflies,[127] foraging preferences were shown to green models when these also emitted floral scents of nectar plants. However, when mixing scents from nectar plants with either the assumed flower color (mauve) or the assumed nonflower color (green), species differed in their responses. For *I. io* and *A. urticae*, a stronger association was shown between floral scent and sugar water reward compared to mauve color and reward, while for *G. rhamni*, mauve color was more strongly associated with reward than scent.

9.9.2 FLORAL VERSUS VEGETATIVE PLANT SCENT BLENDS

For several butterfly nectar plants, some volatiles are in common for both flowers and vegetative parts.[80] Interestingly, flower naïve individuals of *I. io*, *A. urticae*, and *G. rhamni* showed stronger foraging responses to volatiles from flowers over those from vegetative parts of the butterfly nectar plants *C. scabiosa*, *C. arvense*, *K. arvensis*, *B. davidii*, and *O. vulgare*.[139] Thus without any experience, the butterflies discerned and preferred scent compositions of flowers over those of vegetative parts. *H. eliconius melpomene* could distinguish between the floral and vegetative blends of *L. camara*,[138] reflected in the fact that more individuals made deep probes on the floral scented models than on the vegetative scented ones, which also emphasizes the crucial role and clearly stimulating effect of floral scent in the recognition of food-rewarding sources at a close distance. Thus the specific floral scent composition could be distinguished from the vegetative scent composition. Floral scent compounds such as the monoterpene *trans*-β-ocimene and the sesquiterpenes α-and β-caryophyllene from *L. camara*, even though they are emitted in relatively large amounts, might be of less importance as food signals because they are also emitted from vegetative parts. Such a distinction between compounds was found regarding butterfly antennal responses using gas chromatography-electroantennographic detection (GC-EAD) for *I. io*, *A. urticae*, *G. rhamni*, and *H. melpomene*.[140,141] Indeed, abundant pure floral scent compounds from *C. arvense* and *B. davidii*, such as phenylacetaldehyde, linalool, linalool-related compounds, oxoisophorone, and dihydrooxoisophorone, elicited large antennal responses, in contrast to the often small antennal responses to high abundant compounds that are not exclusively of floral origin from *L. camara*, such as α-humulene and β-caryophyllene.

9.9.3 GENDER DIMORPHISM

No differences in the behavioral responses such as probing to floral scents were found between females and males of autumn generation *G. rhamni* and *I. io* individuals[139] with the lack of gender differences in terms of the plants selected for foraging found in an autumn population of *G. rhamni*.[32] That females and males in these adult hibernating species show similarities in foraging preferences for plants and in the use of floral scent might be explained by the autumn being the time when both females and males need to maximize their food intake to prepare for hibernation. In spring, when mating occurs, gender differences in foraging are more likely to be present because females and males differ in their primary tasks. Males are more active in seeking mates,[142] while the females are often busy searching for acceptable host plants. In addition, the eggs of the female need to mature during mating time, as the eggs are not mature on emergence,[143] and food finding may be essential for this.

In *H. melpomene*, antennal responses to floral scent compounds were constantly stronger in females,[141] which may be understood in the context of the high sensitivity of females to chemicals in their surroundings: male sex pheromones must be judged[5] as well as host plant compounds.[14,15] In a food searching context, for *Heliconius* butterflies, because females feed more often on pollen than do males, which may increase a female's nutritional investment in eggs,[22,23] females might have a more refined ability to chemically locate pollen sources. The greater sensitivity of butterfly females to nectar amino acids compared to males[47,48,144] suggests that female butter- flies in general may also be more sensitive to floral chemicals at the gustatory as well as the olfactory levels.

9.9.4   BUTTERFLY LEARNING AND SWITCHING BETWEEN PLANTS

In a natural habitat with a diversity of available plant species, the ability of butterflies to be flexible in their foraging preferences is clearly of adaptive value and allows efficient tracking of the most rewarding flowers. When there is a shortage of the preferred plants, butterflies can shift to other rewarding plants and forage on a wide range of different species, thus acting as generalists.[29,125] Learning abilities may be more important for behaviors, such as foraging, that in each performance are less directly related to fitness, in comparison with behaviors where each performance has a greater outcome on fitness, such as oviposition. McNeely and Singer[145] found that the butterfly *Euphydryas editha* (Nymphalidae) learned to prefer and exploit particular nectar species, but did not learn to find or prefer different oviposition plants. However, *Battus philenor* (Papilionidae) learned to distinguish colors in both foraging and ovipositing behavioral contexts.[146] Flexible foraging preferences have been observed in *Pieris napi* regarding their use of flower color.[69] For the butterfly *Heodes virgaurea* L. (Lycaenidae), Douwes[31] discovered profound differences in visual characteristics of the flowers visited, which suggested that this butterfly could handle very complex search images.

Butterflies may also be able to learn to handle complexity in floral scents. The butterflies *I. io* and *G. rhamni* learned to switch between the floral scents of *C. arvense*, dominated by benzenoid compounds and monoterpenoids, and those of *B. davidii*, dominated by irregular terpenoids,[139] implying an ability to handle very complex scent blends. The butterflies learned to reverse their preferences to the most recently rewarding floral scent blend.

9.9.5   PHYSIOLOGICAL RESPONSES AND POSSIBLE CONVERGENT EVOLUTION

There might be a characteristic olfactory sensitivity in butterflies as part of floral scent convergent evolution. A strikingly low and even absent antennal response to the most volatile monoterpene alkenes where found for *H. melpomene*, native to tropical South America, as well as for the temperate butterflies *I. io*, *A. urticae*, and *G. rhamni*, perhaps indicating a general feature in butterflies.

However, differences in antennae response between tropical and temperate butterfly species are expected, considering their different habitats and exploitation of nectar plants. The tropical *H. melpomene* showed strong antennal responses to linalool-related compounds, such as the furanoid linalool oxide II,[141] while this compound elicited no or little response in the temperate butterflies *I. io*, *A. urticae*, and *G. rhamni*.[140] The temperate butterflies instead show greater antennal responses to phenylacetaldehyde and 2-phenylethanol. Interestingly, this differential antennal receptivity is also reflected in the floral scent compositions of the tropical and temperate plants, where phenylacetaldehyde and 2-phenylethanol are released in relatively large amounts from the temperate *B. davidii* and *C. arvense*, while not at all from the tropical *L. camara* and *W. coccinea*, which might be adaptations of the floral scent composition toward local pollinators.

However, temperate butterfly species do have differences in their physiological response to flower volatiles. Overall the antennal responses were more similar in the Nymphalid butterflies *I. io* and *A. urticae* compared to the response in the pierid *G. rhamini*. Although all three butterflies use nectar plants, such as *C. arvense*, *B. davidii*, *K. arvensis*, and *O. vulgare*, in common, their larval diets differ, *Urtica dioica* L. (Urticaceae) for *I. io* and *A. urtica* and *Frangula alnus* Mill (Rhamnaceae) for *G. rhamni*, which may influence each species' sensitivity to chemicals as adults.

9.10 FUNCTIONAL BASIS OF FLORAL SCENT FOR BUTTERFLY POLLINATION

Floral scent compositions generally show high complexity, often including more than 40 different compounds.[80] Usually some but not all of the compounds in a plant's floral scent contribute to the food signal for insect pollinators.[75,122]

9.10.1 SINGLE COMPOUNDS VERSUS BLENDS

In studying the learning capacity of the moth *Spodoptera littoralis*, Hartlieb et al.[147] found that males were able to associate single pheromone components with food, but that the association was impaired when the complete pheromone blend was presented. They argued that single-compound neurons might mediate a general arousal, while the neural integration of all compounds in a pheromone blend must be there to generate the behavior. Regarding behavioral responses of butterflies to floral scents, synergistic effects of compounds have been shown; for example, in *P. rapae*, foraging responses to *Ligustrum japonicum* Thunb. (Oleaceae) floral scent, a blend of phenylacetaldehyde, 2-phenylethanol, 6-methyl-5-hepten-2-one, benzaldehyde, and methyl phenylacetate, was more attractive than when these compounds were offered separately.[96] Similarly Pellmyr[100] found synergistic effects of isoeugenol and methyl anthranilate from one morph of *C. simplex* in the attraction of three nymphalid butterfly species. The individual compounds had a limited effect. However, contradicting findings have been observed in bees,[102] where the food signaling potential for linalool, the main constituent in the blend of *D. mezereum*, was as high as for the whole blend. Thus the compound of greatest abundance in the floral scent blend was also a major constituent in the food signal for bees. Phenylacetaldehyde elicits strong antennal responses in *A. urticae*, *I. io*, and *G. rahmni*.[140] The compound elicits foraging behavior in *P. rapae*[96,98] and *L. japonica* Leech (Papilionidae),[97] as well as in several moth species.[103,148–151] Furthermore, Honda[2] suggested that phenylacetaldehyde could be a male sex pheromone since the compound is found on the wings of the papilionid butterfly *Atrophaneura alcinous alcinous* Klug (Papilionidae) in larger amounts on males than on females.

9.10.2 COMPOUNDS OF COMMON BIOSYNTHETIC ORIGIN

The functional unit of the floral scent that attracts butterflies may lie in the biosynthetic origin of compounds, rendering single compounds relatively unimportant. Species-specific floral scents are recognized by several butterflies.[138,139] This discriminative ability may be connected to the presence of benzenoid compounds or irregular terpenoids, characterizing the butterfly-favored plants, in contrast to the sesquiterpenes and fatty acid derivatives common in the nonfavored plants such as *P. coronarius* and *A. millefolium*.[80,138,139]

Inachis io, *A. urticae*, and *H. melpomene* show strong antennal responses to the three characteristic irregular terpenoid ketones oxoisophoroneoxide, oxoisophorone, and dihydrooxoisophorone from *B. davidii* floral scents,[140,141] although these compounds are emitted in different relative amounts; oxoisophorone at about 46 times that of dihydrooxoisophorone and almost 5 times that of oxoisophoroneoxide. The fact that no differential preference was made between the compounds points to the possibility that compounds with the same biosynthetic background may be the functional unit of attraction. In *B. davidii*, these compounds together constitute a considerable percentage of all compounds, thus they make an obvious signal together.

9.11 CONCLUDING REMARKS

Floral scent is one among other floral characters important in butterfly pollination. In butterflies, olfaction, together with the other senses, are involved to varying degrees, depending on the species that are interacting, as well as their condition and motivation.

To fully understand the floral scent function in butterfly pollination, one needs to study plant reproductive success resulting from butterfly visits as a response to floral scents and other stimuli of altering expressions, as well as how butterfly use of floral scents influences butterfly fitness.

ACKNOWLEDGMENTS

Thanks to Maria and Jonas Sandstedt for valuable comments on an earlier version of the manuscript, and to Heidi Dobson for the photo.

REFERENCES

1. Bergström, G. and Lundgren, L, Androconial secretions of three species of butterflies of the genus *Pieris* (Lep., Pieridae). *Zoon Suppl* 1, 67, 1973.
2. Honda, K., Odor of a papilionid butterfly: odoriferous substances emitted by *Atrophaneura alcinous alcinous* (Lepidoptera: Papilionidae), *J. Chem. Ecol.* 6, 867, 1980.
3. Schulz, S. and Francke, W., Volatile compounds from androconial organs of Danaine and Ithomiine butterflies, *Z. Naturforsch.* 43c, 99, 1988.
4. Boppré, M., Chemically mediated interactions between butterflies, in *The Biology of Butterflies*, Vane-Wright, R.I. and Ackery, P.R., Eds., Princeton University Press, Princeton, NJ, 1989, p. 259.
5. Schulz, S., Boppré, M., and Vane-Wright, R.I., Specific mixtures of secretions from male scent organs of African milkweed butterflies (Danainae), *Philos. Trans. R. Soc. Lond. B* 342, 161, 1993.
6. Nishida, R., Schulz, S., Kim, C.S., Fukami, H., Kuwahara, Y., Honda, K., and Hayashi, N., Male sex pheromone of a giant danaine butterfly, *Idea leuconoe*, *J. Chem. Ecol.* 22, 949, 1996.
7. Shreeve, T.G., Adult behaviour, in *The Ecology of Butterflies in Britain*, Dennis, R.L.H., Ed., Oxford Science Publications, Oxford, 1992, p. 22.
8. Gilbert, L.E., Postmating female odor in Heliconius butterflies: a male-contributed antiaphrodisiac?, *Science* 193, 419, 1976.
9. Forsberg, J., and Wiklund, C., Mating in the afternoon: time-saving in courtship and remating by females of a polyandrous butterfly *Pieris napi* L., *Behav. Ecol. Sociobiol.* 25, 349, 1989.
10. Den Otter, C.J., Behan, M., and Maes, F.W., Single cell responses in female *Pieris brassicae* (Lepidoptera: Pieridae) to plant volatiles and conspecific egg odours, *J. Insect Physiol.* 26, 465, 1980.
11. Spencer, K.C., Chemical mediation of coevolution in the Passiflora-Heliconius interaction, in *Chemical Mediation of Coevolution*, Spencer, K.C., Ed., Academic Press, San Diego, 1988, p. 167.
12. Feeny, P., Städler, E., Åhman, I., and Carter, M., Effects of plant odor on oviposition by the black swallowtail butterfly, *Papilio polyxenes* (Lepidoptera: Papilionidae), *J. Insect Behav.* 2, 803, 1989.
13. Topazzini, A., Mazza, M., and Pelosi, P., Electroantennogram responses of five lepidoptera species to 26 general odourants, *J. Insect Physiol.* 36, 619, 1990.
14. van Loon, J.J.A., Frentz, W.H., and van Eeuwijk, F.A., Electroantennogram responses to plant volatiles in two species of Pieris butterflies, *Entomol. Exp. Appl.* 62, 253, 1992.
15. Baur, R., Feeny, P., and Staedler, E., Oviposition stimulants for the black swallowtail butterfly: identification of electrophysiologically active compounds in carrot volatiles, *J. Chem. Ecol.* 19, 919, 1993.
16. Huang, X., Renwick, J.A.A., and Sachdev-Gupta, K., Oviposition stimulants in *Barbarea vulgaris* for *Pieris rapae* and *P. napi oleracea*: isolation, identification and differential activity, *J. Chem. Ecol.* 20, 423, 1994.
17. Chew, F.S. and Renwick, J.A.A., Host plant choice in Pieris butterflies, in *Chemical Ecology of Insects*, vol. 2, Cardé, R.T. and Bell, W.J., Eds., Chapman & Hall, New York, 1995, p. 214.

18. Baur, R. and Feeny, P., Comparative electrophysiological analysis of plant odor perception in females of three Papilio species, *Chemoecology* 5/6, 26, 1995.

19. Honda, K., Hayashi, N., Abe, F., and Yamauchi, T., Pyrrolizidine alkaloids mediate host-plant recognition by ovipositing females of an Old World danaid butterfly, *Idea leuconoe*, *J. Chem. Ecol.* 23, 1703, 1997.

20. Brower, L.P., Chemical defence in butterflies, in *The Biology of Butterflies*, Vane-Wright, R.I. and Ackery, P.R., Eds., Princeton University Press, Princeton, NJ, 1989, p. 109.

21. McFarlane, J.E., Nutrition and digestive organs, in *Fundamentals of Insect Physiology*, Blum, M.S., Ed., John Wiley & Sons, New York, 1985, p. 59.

22. Gilbert, L.E., Pollen feeding and reproductive biology of Heliconius butterflies, *Proc. Natl. Acad. Sci. USA* 69, 1403, 1972.

23. Boggs, C.L., Smiley, J.T., and Gilbert, L.E., Patterns of pollen exploitation by Heliconius butterflies, *Oecologia* 48, 284, 1981.

24. Tooker, J.F., Reagel, P.F., and Hanks, L.M., Nectar sources of day-flying Lepidoptera of central Illinois, *Ann. Entomol. Soc. Am.* 95, 84, 2002.

25. Lack, A.J., The ecology of flowers of chalk grassland and their insect pollinators, *J. Ecol.* 70, 773, 1982.

26. Proctor, M., Yeo, P., and Lack, A., *The Natural History of Pollination*, Harper Collins, London, 1996.

27. Nilsson, L.A., personal communication, 2000.

28. Müller, H., *Befruchtung der Blumen durch Insekten und die gegensitigen anpassungen beider*, Verlag von Wilhelm Engelmann, Leipzig, 1873.

29. Jennersten, O., Flower visitation and pollination efficiency of some north European butterflies, *Oecologia* 63, 80, 1984.

30. Andersson, S. and Bergman, P., unpublished data, 2001.

31. Douwes, P., Adult feeding in the scarce copper, *Heodes virgaureae* L. (Lep., Lycaenidae), *Entomol. Tidskr.* 99, 1, 1978.

32. Jennersten, O., Nectar source plant selection and distribution pattern in an autumn population of *Gonepteryx rhamni* (Lep. Pieridae), *Entomol. Tidskr.* 101, 109, 1980.

33. Grant, V. and Grant, K.A., *Flower Pollination of in the Phlox Family*, Columbia University Press, New York, 1965.

34. Proctor, M. and Yeo, P., *The Pollination of Flowers*, William Collins Sons, Glasgow, 1975.

35. Faegri, K. and van der Pijl, L, *The Principles of Pollination Ecology*, Pergamon Press, Oxford, 1979.

36. Delforge, P., *Europas Orkideer*, G.E.C. Gads Forlag, Köpenhamn, 1995.

37. Crane, J., Imaginal behaviour of a Trinidad butterfly, *Heliconius erato* hydara Hewitson, with special reference to the social use of color, *Zoologica* 40, 167, 1955.

38. Bawa, K.S. and Beach, J.H., Self-incompatibility systems in the Rubiaceae of a tropical lowland wet forest, *Am. J. Bot.* 70, 1281, 1983.

39. Barrows, E.M., Nectar robbing and pollination of *Lantana camara* (Verbenaceae), *Biotropica* 8, 132, 1976.

40. Schemske, D.W., Pollinator specificity in *Lantana camara* and *L. trifolia* (Verbenaceae), *Biotropica* 8, 260, 1976.

41. Lind, H. and Lindeborg, M., Lepidopterans as presumptive pollinators of *Anacamptis pyramidalis*, *Entomol. Tidskr.* 110, 156, 1989.

42. Lind, H., Occurrence, population trends and fruit setting in *Anacamptis pyramidalis* on Öland, Sweden, *Svensk Bot. Tidskr.* 86, 329, 1992.

43. Baker, H.G. and Baker, I., Floral nectar sugar constituents in relation to pollinator type, in *Handbook of Experimental Pollination Biology*, Jones, C.E. and Little, R.J., Eds., Van Nostrand Reinhold, New York, 1983, p. 117.

44. Erhardt, A. and Baker, I., Pollen amino acids—an additional diet for a nectar feeding butterfly?, *Plant Syst. Evol.* 169, 111, 1990.

45. Erhardt, A., Nectar sugar and amino acid preferences of *Battus philenor* (Lepidoptera, Papilionidae), *Ecol. Entomol.* 16, 425, 1991.

46. Erhardt, A., Preference and non-preferences for nectar constituents in *Ornithoptera priamus poseidon* (Lepidoptera, papilionidae), *Oecologia* 90, 581, 1992.

47. Erhardt, A. and Rusterholz, H.P., Do peacock butterflies (*Inachis io* L.) detect and prefer nectar amino acids and other nitrogenous compounds?, *Oecologia* 117, 536, 1998.

48. Rusterholz, H.P. and Erhardt, A., Can nectar properties explain sex-specific flower preferences in the Adonis blue butterfly *Lysandra bellargus*?, *Ecol. Entomol.* 25, 81, 2000.

49. Schmitt, J., Pollinator foraging behaviour and gene dispersal in *Senecio* (Compositae), *Evolution* 34, 934, 1980.

50. Courtney, S.P., Hill, C.J., and Westerman, A., Pollen carried for long periods by butterflies, *Oikos* 38, 260, 1982.

51. Lind, H., Lepidoptera: important long-distance pollinators for plants in fragmented habitats, *Svensk Bot. Tidskr.* 88, 185, 1994.

52. Denton, G.R.W., Muniappan, R., and Marutani, M., The distribution and biological control of *Lantana camara* in Micronesia, *Micronesica* 3(suppl), 71, 1991.

53. Weiss, M.R., Floral colour change: a widespread functional convergence, *Am. J. Bot.* 82, 167, 1995.

54. Kevan, P.G., Insect pollination of high arctic flowers, *J. Ecol.* 60, 831, 1972.

55. Kevan, P.G., Flower, insect, and pollination ecology in the Canadian high arctic, *Polar Rec.* 16, 667, 1973.

56. Philipp, M., Bocher, J., Mattsson, O., and Woodell, S.R.J., A quantitative approach to the sexual reproductive biology and population structure in some arctic flowering plants: *Dryas integrifolia*, *Silene acaulis* and *Ranunculus nivalis*, *BioScience* 34, 1, 1990.

57. Totland, O., Pollination in alpine Norway: flowering phenology, insect visitors, and visitation rates in two plant communities, *Can. J. Bot.* 71, 1072, 1993.

58. Kearns, C.A. and Inouye, D.W., Fly pollination of *Linum lewisii* (Linaceae), *Am. J. Bot.* 81, 1091, 1994.

59. Bergman, P., Molau, U., and Holmgren, B., Micrometeorological impacts on insect activity and plant reproductive success in an alpine environment, Swedish Lapland, *Arctic Alpine Res.* 28, 196, 1996.

60. Cruden, R.W. and Hermann-Parker, S.M., Butterfly pollination of *Caesalpinia pulcherrima*, with observations on a psychophilous syndrome, *J. Ecol.* 67, 155, 1979.

61. Shykoff, J.A., Sex polymorphism in *Silene acaulis* (Caryophyllaceae) and the possible role of sexual selection in maintaining females, *Am. J. Bot.* 79, 138, 1992.

62. Erhardt, A. and Jäggi, B., From pollination by Lepidoptera to selfing: the case of *Dianthus glacialis* (Caryophyllaceae), *Plant Syst. Evol.* 195, 67, 1995.

63. Marr, D.L., Impact of a pollinator-transmitted disease on reproduction in healthy *Silene acaulis*, *Ecology* 78, 1471, 1997.

64. Knudsen, J.T., Tollsten, L., and Bergström, G., Floral scents: a checklist of volatile compounds isolated by head-space techniques, *Phytochemistry* 33, 253, 1993.

65. Tollsten, L. and Bergström, G., Variation and post-pollination changes in floral odours released by *Platanthera bifolia* (Orchidaceae), *Nord. J. Bot.* 9, 359, 1989.

66. Grant, V., Pollination systems as isolating mechanisms in angiosperms, *Evolution* 3, 82, 1949.

67. Lewis, A.C., Memory constraints and flower choice in *Pieris rapae*, *Science* 232, 863, 1986.

68. Waser, N.M., Flower constancy: definition, cause, and measurement, *Am. Nat.* 127, 593, 1986.

69. Goulson, D. and Cory, J.S., Flower constancy and learning in foraging preferences of the green-veined white butterfly *Pieris napi*, *Ecol. Entomol.* 18, 315, 1993.

70. Goulson, D., Ollerton, J., and Sluman, C., Foraging strategies in the small skipper butterfly, *Thymelicus flavus*: when to switch?, *Anim. Behav.* 53, 1009, 1997.

71. Stebbins, G.L., Adaptive radiation in reproductive characteristics in angiosperms, I: pollination mechanisms, *Annu. Rev. Ecol. Syst.* 1, 307, 1970.

72. Nilsson, L.A., Orchid pollination biology, *Trends Ecol. Evol.* 7, 255, 1992.

73. van der Pijl, L., Ecological aspects of flower evolution, II. Zoophilous flower classes, *Evolution* 15, 44, 1960.

74. Crepet, W.L., The role of insect pollination in the evolution of the angiosperms, in *Pollination Biology*, Real, L., Ed., Academic Press, Orlando, FL, 1983, p. 31.

75. Pellmyr, O. and Thien, L.B., Insect reproduction and floral fragrances: keys to the evolution of the angiosperms?, *Taxon* 35, 76, 1986.

76. Firn, R.D. and Jones, C.G., Natural products: a simple model to explain chemical diversity, *Nat. Prod. Rep.* 20, 382, 2003.

77. Feinsinger, P., Coevolution and pollination, in *Coevolution*, Futuyma, D.J. and Slatkin, M., Eds., Sinauer Associates, Sunderland, MA, 1983, p. 282.

78. Knudsen, J.T. and Tollsten, L., Trends in floral scent chemistry in pollination syndromes: floral scent composition in moth pollinated taxa, *Bot. J. Linn. Soc.* 113, 263, 1993.

79. Raguso, R.A. and Pichersky, E., Floral volatiles from *Clarkia breweri* and *C. concinna* (Onagraceae): recent evolution of floral scent and moth pollination, *Plant Syst. Evol.* 194, 55, 1995.

80. Andersson, S., Nilsson, L.A., Groth, I., and Bergström, G., Floral scents in butterfly-pollinated plants: possible convergence in chemical composition, *Bot. J. Linn. Soc.* 140, 129, 2002.

81. Jennersten, O., Pollination in *Dianthus deltoids* (Caryophyllaceae): effects of habitat fragmentation on visitation and seed set, *Conserv. Biol.* 2, 359, 1988.

82. Cocucci, A.A., Galetto, L., and Sersic, A., The floral syndrome of *Caesalpinia gilliesii* (Fabaceae-Caesalpinioideae), *Darwiniana (San Isidro)* 31, 111, 1992.

83. Johnson, S.D. and Bond, W.J., Red flowers and butterfly pollination in the fynbos of South Africa, in *Plant-Animal Interactions in Mediterranean-Type Ecosystems*, Arianoutsou, M. and Groves, R., Eds., Kluwer Academic, Dordrecht, 1994, p. 137.

84. Johnson, S.D. and Steiner, K.E., Generalization versus specialization in plant pollination systems, *Trends Ecol. Evol.* 15, 140, 2000.

85. Nilsson, L.A., Jonsson, L., Ralison, L., and Randrianjohany, E., Angraecoid orchids and hawkmoths in central Madagascar: specialized pollination systems and generalist foragers, *Biotropica* 19, 310, 1987.

86. Waser, N.M., Chittka, L., Price, M.V., Williams, N.M., and Ollerton, J., Generalization in pollination systems, and why it matters, *Ecology* 77, 1043, 1996.

87. Johnson, S.D. and Steiner, K.E., Specialized pollination systems in southern Africa, *S. Afr. J. Sci.* 99, 345, 2003.

88. Wink, M., Evolution of secondary metabolites from an ecological and molecular phylogenetic perspective, *Phytochemistry* 64, 3, 2003.
89. Johnson, S.D., Linder, H.P., and Steiner, K.E., Phylogeny and radiation of pollination systems in *Disa* (Orchidaceae), *Am. J. Bot.* 85, 402, 1998.
90. Erhardt, A., Pollination of *Dianthus gratianopolitanus* (Caryophyllaceae), *Plant Syst. Evol.* 170, 125, 1990.
91. Erhardt, A., Pollination of *Dianthus superbus* L., *Flora* 185, 99, 1991.
92. Nilsson, L.A., Processes of isolation and introgressive interplay between *Platanthera bifolia* (L.) Rich. and *P. chlorantha* (Custer) Reichb. (Orchidaceae), *Bot. J. Linn. Soc.* 87, 325, 1983.
93. Joulain, D., Study of the fragrance given off by certain springtime flowers, in *Progress in Essential Oil Research*, Brunkel, E.-J., Ed., Walter de Gruyter, Berlin, 1986, p. 57.
94. Kaiser, R., *The Scent of Orchids: Olfactory and Chemical Investigations*, Elsevier, Amsterdam, 1993.
95. Andersson, S., unpublished data, 2000.
96. Honda, K., Ômura, H., and Hayashi, N., Identification of floral volatiles from *Ligustrum japonicum* that stimulate flower-visiting by cabbage butterfly, *Pieris rapae*, *J. Chem. Ecol.* 24, 2167, 1998.
97. Ômura, H., Honda, K., Nakagawa, A., and Hayashi, N., The role of floral scent of the cherry tree, *Prunus yedoensis*, in the foraging behavior of *Luehdorfia japonica* (Lepidoptera: Papilionidae), *Appl. Entomol. Zool. (Jpn.)* 34, 309, 1999.
98. Ômura, H., Honda, K., and Hayashi, N., Chemical and chromatic bases for preferential visiting by the cabbage butterfly, *Pieris rapae*, to rape flowers, *J. Chem. Ecol.* 25, 1895, 1999.
99. Schemske, D.W. and Bradshaw, H.D., Jr., Pollinator preference and the evolution of floral traits in monkeyflowers (Mimulus), *Proc. Natl. Acad. Sci. USA* 96, 11910, 1999.
100. Pellmyr, O., Three pollination morphs in *Cimicifuga simplex*; incipient speciation due to inferiority in competition, *Oecologia* 68, 304, 1986.
101. Knudsen, J.T., Floral scent chemistry in geonomoid palms (Palmae: Geonomeae) and its importance in maintaining reproductive isolation, in *Evolution, Variation, and Classification of Palms Memoirs of the New York Botanical Garden*, Menderson, A. and Borchsenius, F., Eds., New York Botanical Garden Press, New York, 1999, p. 141.
102. Borg-Karlson, A.-K., Unelius, C.R., Valterova, I., and Nilsson, L.A., Floral fragrance chemistry in the early flowering shrub *Daphne mezereum* (Thymelaeaceae), *Phytochemistry* 41, 1477, 1996.
103. Heath, R.R., Landolt, P.J., Dueben, B., and Lenczewski, B., Identification of floral compounds of night-blooming jessamine attractive to cabbage looper moths, *Environ. Entomol.* 21, 854, 1992.
104. Pham-Delegue, M.H., Behavioural discrimination of oilseed rape volatiles by the honeybee *Apis mellifera* L., *Chem. Senses* 18, 483, 1993.
105. Moya, S. and Ackerman, J.D., Variation in the floral fragrance of *Epidendrum-ciliare* (Orchidaceae), *Nord. J. Bot.* 13, 41, 1993.
106. Kullenberg, B. and Bergström, G., The pollination of Ophrys orchids, in *Chemistry in Botanical Classification, Proceedings of the 25th Nobel Symposium*, Bendz, G. and Santesson, J., Eds., Academic Press, New York, 1974, p. 253.
107. Williams, N.H. and Whitten, W.M., Orchid floral fragrances and male euglossine bees: methods and advances in the last sesquidecade, *Biol. Bull.* 164, 355, 1983.
108. Kullenberg, B., Borg-Karlson, A.-K., and Kullenberg, A.L., Field studies on the behaviour of the *Eucera nigrilabris* male in the odour flow from flower labellum extract of *Ophrys tenthredinifera*, *Nova Acta Reg. Soc. Sci. Ups. Ser. V:C* 3, 79, 1984.

109. Borg-Karlson, A.-K., Chemical and ethological studies of pollination in the genus Ophrys (Orchidaceae), *Phytochemistry* 29, 1359, 1990.
110. Gerlach, G. and Schill, R., Composition of orchid scents attracting euglossine bees, *Bot. Acta* 104, 379, 1991.
111. Sazima, M., Vogel, S., Cocucci, A.A., and Hausner, G., The perfume flowers of Cyphomandra (Solanaceae): pollination by euglossine bees, bellows mechanism, osmophores, and volatiles, *Plant Syst. Evol.* 187, 51, 1993.
112. Schiestl, F.P., Ayasse, M., Paulus, H.F., Erdmann, D., and Francke, W., Variation of floral scent emission and postpollination changes in individual flowers of *Ophrys sphegodes* subsp. *sphegodes*, *J. Chem. Ecol.* 23, 2881, 1997.
113. Schiestl, F.P. and Ayasse, M., Post-mating odor in females of the solitary bee, *Andrena nigroaenea* (Apoidea, Andrenidae), inhibits male mating behavior, *Behav. Ecol. Sociobiol.* 48, 303, 2000.
114. Knudsen, J.T., Andersson, S., and Bergman, P., Floral scent attraction in *Geonoma macrostachys*, an understorey palm of the Amazonian rain forest, *Oikos* 85, 409, 1999.
115. Brantjes, N.B.M., Sphingophilous flowers: function of their scent, in *Pollination and Dispersal*, Brantjes, N.B.M., and Linskens, H.F., Eds., 1973, p. 27.
116. Brantjes, N.B.M., Sensory responses to flower in night-flying moths, in *The Pollination of Flowers by Insects*, Richards, A.J., Ed., Academic Press, London, 1978, p. 13.
117. Loughrin, J.H., Hamilton-Kemp, T.R., Andersen, R.A., and Hildebrand, D.F., Circadian rhythm of volatile emission from flowers of *Nicotiana sylvestris* and *N. suaveolens*, *Physiol. Plant.* 83, 492, 1991.
118. Knudsen, J.T. and Tollsten, L., Floral scent in bat-pollinated plants: a case of convergent evolution, *Bot. J. Linn. Soc.* 119, 45, 1995.
119. Dobson, H.E.M., Role of flower and pollen aromas in host plant recognition by solitary bees, *Oecologia* 72, 618, 1987.
120. Dobson, H.E.M., Groth, I., and Bergström, G., Pollen advertisement: chemical contrasts between whole-flower and pollen odors, *Am. J. Bot.* 83, 877, 1996.
121. Dobson, H.E.M., Danielson, E.M., and van Wesep, I.D., Pollen odor chemicals as modulators of bumble bee foraging on *Rosa rugosa* Thunb. (Rosaceae), *Plant Species Biol.* 14, 153, 1999.
122. Dobson, H.E.M. and Bergström, G., The ecology and evolution of pollen odors, *Plant Syst. Evol.* 222, 63, 2000.
123. Giurfa, M., Núñez, J.A., and Backhaus, W., Odour and colour information in the foraging choice behaviour of the honeybee, *J. Comp. Physiol. A* 175, 773, 1994.
124. Raguso, R.A. and Willis, M.A., Synergy between visual and olfactory cues in nectar feeding by naïve hawkmoths, *Manduca sexta*, *Anim. Behav.* 63, 1, 2002.
125. Lewis, A.C. and Lipani, G.A., Learning and flower use in butterflies: hypothesis from honey bees, in *Insect-Plant Interactions*, Bernays, E.A., Ed., CRC Press, Boca Raton, FL, 1990, p. 95.
126. Frazier, J.L., Nervous system: sensory system, in *Fundamentals of Insect Physiology*, Blum, M.S. Ed., John Wiley & Sons, New York, 1985, p. 287.
127. Ilse, D., Über den Farbensinn der Tagfalter, *J. Comp. Physiol.* 8, 658, 1928.
128. Ilse, D. and Vaidya, V.G., Spontaneous feeding response to colours in *Papilio demoleus* L., *Proc. Indian Acad. Sci. B* 43, 23, 1956.
129. Swihart, C.A. and Swihart, S.L., Colour selection and learned feeding preferences in the butterfly *Heliconius charitonius* Linn., *Anim. Behav.* 18, 60, 1970.
130. Swihart, C.A., Colour discrimination by the butterfly *Heliconius charitonius* Linn., *Anim. Behav.* 19, 156, 1971.

131. Swihart, S.L., The neural basis of color vision in the butterfly, *Heliconius erato*, *J. Insect Physiol.* 18, 1015, 1972.

132. Weiss, M.R., Floral colour changes as cues for pollinators, *Nature* 354, 227, 1991.

133. Weiss, M.R., Associative colour learning in a nymphalid butterfly, *Ecol. Entomol.* 20, 298, 1995.

134. Weiss, M.R., Innate colour preferences and flexible colour learning in the pipevine swallowtail, *Anim. Behav.* 53, 1043, 1997.

135. Kandori, I. and Ohsaki, N., The learning abilities of the white cabbage butterfly, *Pieris rapae*, foraging for flowers, *Res. Popul. Ecol.* 38, 111, 1996.

136. Kinoshita, M., Shimada, N., and Arikawa, K., Colour vision of the foraging swallowtail butterfly *Papilio xuthus*, *J. Exp. Biol.* 202, 95, 1999.

137. Groth, I., Bergström, G., and Pellmyr, O., Floral fragrances in *Cimicifuga*: chemical polymorphism and incipient speciation in *Cimicifuga simplex*, *Biochem. Syst. Ecol.* 15, 441, 1987.

138. Andersson, S. and Dobson, H.E.M., Behavioural foraging responses by the butterfly *Heliconius melpomene* to *Lantana camara* floral scent, *J. Chem. Ecol.* 29, 2303, 2003.

139. Andersson, S., Foraging responses in the butterflies *Inachis io*, *Aglais urticae* (Nymphalidae), and *Gonepteryx rhamni* (Pieridae) to floral scents, *Chemoecology* 13, 1, 2003.

140. Andersson, S., Antennal responses to floral scents in the butterflies *Inachis io*, *Aglais urticae* (Nymphalidae), and *Gonepteryx rhamni* (Pieridae), *Chemoecology* 13, 13, 2003.

141. Andersson, S. and Dobson, H.E.M., Antennal responses to floral scents in the butterfly *Heliconius melpomene*, *J. Chem. Ecol.* 29, 2319, 2003.

142. Wickman, P.-O., Butterfly leks, *Entomol. Tidskr.* 117, 73, 1996.

143. Porter, K., Eggs and egg-laying, in *The Ecology of Butterflies in Britain*, Dennis, R.L.H., Ed., Oxford University Press, Oxford, 1992, p. 46.

144. Mevi-Schutz, J. and Erhardt, A., Can *Inachis io* detect nectar amino acids at low concentrations?, *Physiol. Entomol.* 27, 256, 2002.

145. McNeely, C. and Singer, M.C., Contrasting the roles of learning in butterflies foraging for nectar and oviposition sites, *Anim. Behav.* 61, 847, 2001.

146. Weiss, M.R. and Papaj, D.R., Colour learning in two behavioural contexts: how much can a butterfly keep in mind?, *Anim. Behav.* 65, 425, 2003.

147. Hartlieb, E., Anderson, P., and Hansson, B.S., Appetitive learning of odours with different behavioural meaning in moths, *Physiol. Behav.* 67, 671, 1999.

148. Creighton, C.S., McFadden, T.L., and Cuthbert, E.R., Supplementary data on phenylacetaldehyde: an attractant for Lepidoptera, *J. Econ. Entomol.* 66, 114, 1973.

149. Cantelo, W.W. and Jacobson, M., Phenylacetaldehyde attracts moths to bladder flower and to blacklight traps, *Environ. Entomol.* 8, 444, 1979.

150. Haynes, K.F., Zhao, J.Z., and Latif, A., Identification of floral compounds from *Abelia grandiflora* that stimulate upwind flight in cabbage looper moths, *J. Chem. Ecol.* 17, 637, 1991.

151. Plepys, D., Odour-mediated nectar foraging in the silver Y moth, *Autographa gamma*, PhD dissertation, Lund University, Sweden, 2001.

10 Floral Scent and Pollinator Attraction in Sexually Deceptive Orchids

Manfred Ayasse

CONTENTS

10.1   INTRODUCTION

With about 20,000 to 30,000 species, the orchids comprise one of the largest angiosperm families.[1] An outstanding feature of orchid flowers is their variety in form and function and their diversity in pollination systems, a phenomenon that was noticed by Charles Darwin.[2] Most remarkable are those species that employ deceit rather than food rewards to attract pollinators.[3,4] Among the approximately 10,000 deceptive species,[5] food deceptive orchids are the most numerous. Certain flowers of the food deceptive species mimic the floral structure of food providing species.[1] Most of the species, however, are "nonmodel" mimics that match a general search image for flower visiting insects.[1,3,4] Other orchid species imitate sites for oviposition[3] or sleeping holes for solitary bees.[6]

The most spectacular case of floral mimicry is sexual deception, also called pseudo-copulation, a pollination mechanism that is not known outside the Orchidaceae.[1,4,5]

In sexually deceptive orchids, the flowers mimic in shape, color, and odor females of their pollinators, thereby attracting males for pollination. Such reproductive mimicry has been termed "Pouyannian mimicry,"[7] since Pouyanne was the first to describe the behavior of *Campsoscolia ciliata* (Scoliidae) and *Andrena* (Apidae) males on *Ophrys* flowers as "pseudocopulation" (Figure 10.1).[8,9] Almost at the same time, Edith Coleman, an Australian naturalist, published in a series of papers that pollination of *Cryptostylis* was achieved by male *Lissopimpla excelsa* wasps (Ichneumonidae) during pseudocopulation with the flower (see Coleman[10,11] and references therein). Later pseudocopulation was subsequently studied intensively by Kullenberg,[12] who postulated "chemical mimicry," that *Ophrys* flowers mimic the odor of hymenopteran (and other insect) females to dupe males into visiting their flowers and thereby pollinating them.

Meanwhile, sexually deceptive orchids have been documented in Asia,[13] Australia,[14,15] South Africa,[16] South and Central America,[17–20] and Europe.[6,21,22] Since the orchid genera involved are unrelated, a polyphyletic origin of this pollination syndrome can be assumed.[23–25]

Male aculeate Hymenoptera are mostly involved in pollination of sexually deceptive orchids, but also males of some sawflies, Diptera, and scarabaeid beetles.[6,15,16,26–28] Besides visual and tactile cues, olfactory stimuli have been shown to be crucial for the attraction of pollinators in the European genus *Ophrys*[12,29–31] as well as in southern Australian orchid species,[6,32,33] and release innate sexual behavior in male pollinators. In the early studies, similarity in floral shape and color were demonstrated between flowers and females of the pollinators.[12,26,34] Recent investigations, however, have clearly demonstrated that odor signals are of primary importance for the attraction of the pollinating males.[26,29–31,33,35,36]

Here we review odor communication in conjunction with pollinator attraction in sexually deceptive orchids. The focus is on more recent reports and review papers rather than giving a complete survey of the literature. Therefore this review is far from complete. The chemical structures involved in pollinator attraction and in stimulating pseudocopulatory behavior in males have only been identified in sexually deceptive orchids from Australia, Europe, and the Neotropics.

FIGURE 10.1 (See color insert following page 178.) Males of *C. ciliata* pseudocopulating with a flower of *Ophrys speculum* (head pollination). Photo by H.F. Paulus.

10.2 POLLINATOR ATTRACTION IN SEXUALLY DECEPTIVE ORCHIDS IN AUSTRALIA

The widest diversity of sexually deceptive orchids can be observed in Australia. More than 150 species of orchids representing nine genera of terrestrial Diuridae attract their pollinators by sexual deception.[15] The males of five different subfamilies of Hymenoptera are deceived and exploited as pollinators (Figure 10.2). One hundred species of orchids in the genera *Arthrochilus*, *Caladenia*, *Chiloglottis*, *Drakaea*, *Spiculaea*, and *Paracaleana* are pollinated by male thynnine wasps (Tiphiidae: Thynninae).[37] Various wasps (Ichneumonidae and Scoliidae) pollinate *Cryptostylis* and *Calochilus*, sawflies (Pergidae) pollinate *Caleana*,[37] and male ants (Myrmecidae) pollinate *Leporella*.[15,38]

According to phylogenetic studies, sexual deception evolved independently in several taxa of Australian orchids.[6,39] Molecular studies indicate at least one origin in the thynnine wasp-pollinated *Caladenia* and up to four separate origins in the remaining four orchid taxa *Cryptostylis*, *Leporella*, *Calochilus*, and *Caleana*.[39]

In Australian orchids, the flower labellum of many sexually deceptive species mimics the color of the female body of their pollinating insect.[40,41] This mimicry varies between species from being barely recognizable in the genus *Caladenia* to remarkably insectlike forms in some species of the genus *Chiloglottis*.[42] The speculum may show ultraviolet reflection, but this feature is absent in most of the pseudocopulatory orchids of southern Australia pollinated by tiphiid wasps.[6] However, in Australian orchids as well as in European *Ophrys*, odor is the most important factor in attracting pollinators and stimulating copulatory behavior. In *Caladenia*, these odors are emitted by glands on the orchid labellum or the tips of the petals or sepals,[37] and the pollinating wasps are attracted to these areas.

The chemistry of sexual deception has only recently been identified in the Australian orchid genus *Chiloglottis* (Diuridae), which relies exclusively on sexual deception for pollination.[28] A single compound or two shared compounds have been identified to account for pollinator attraction by means of electrophysiological investigations (gas chromatography-electroantennographic detection [GC-EAD]) in

FIGURE 10.2 (See color insert following page 178.) *Chiloglottis trapeziformis* attracts males of its pollinator species, the thynnine wasp *Neozeleboria cryptoides*. Photo by F.P. Schiestl.

Chiloglottis trapeziformis, *Chiloglottis valida*, *Chiloglottis trilabra*, and *Chiloglottis seminuda*, as well as in the morphologically and phylogenetically distinct sympatric Australian orchids *Cryptostylis erecta* and *Cryptostylis subulata*.[35,43] The latter two species are pollinated by the same wasp species, *L. excelsa*. The chemical structure of the odor signal has been identified in *C. trapeziformis*, which attracts males of its pollinator species, the thynnine wasp *Neozeleboria cryptoides*, by 2-ethyl-5-propylcyclohexan-1,3-dione. The same compound is also produced by female wasps as a male-attracting sex pheromone (Figure 10.2).[33] Chiloglottone represents a new chemical class of compounds. Chemical analyses aimed at identifying the structures of the other compounds involved in male attraction are in progress.[44]

10.3   POLLINATOR ATTRACTION IN *OPHRYS* ORCHIDS

With the exception of *Ophrys apifera*, which is frequently self-pollinated, all *Ophrys* species are pollinated by sexual deception.[45] The majority of *Ophrys* species occur in the Mediterranean region.[18,22,45] A few species grow as far north as Sweden and Norway, to the west in England and the Canary Islands, and to the east as far as the Crimea. The pollinators are mostly bees (Andrenidae, Colletidae, Megachilidae, and Apidae). Sphecid and scoliid wasps and beetles (Elateridae, Scarabaeidae) are involved in a few cases.[12,26,46] *Ophrys* flowers act mainly through odors, mimicking virgin females of their pollinators. Male insects are lured to the orchid by volatile semiochemicals and visual cues. At close range, chemical signals from the flowers elicit sexual behavior in males, which try to copulate with the flower labellum, and respond in the same manner as if in the presence of female sex pheromones. Thus the male touches the gynostemium and the pollinia may become attached to his head or, in some species, the tip of his abdomen. His copulatory attempts with another flower ensure that the pollinia is transferred to the flower's stigmatic surface and pollination occurs.

In addition to the selective attraction of pollinators via odors, visual cues play a more or less important role, depending on the species.[47] The morphological similarities between *O. speculum* flowers and *C. ciliata* females are obvious.[45] The indumentum on the labellum mimics the hairs of a female, the lateral lobes correspond to the legs, and the shining speculum looks like the mirrored wings. In *Ophrys* species pollinated by *Andrena*, visual cues seem to be less important.[48] In all species investigated so far, copulation attempts of the males have only been elicited by scent.[29,31,48]

Ophrys species have been intensively studied by many investigators.[12,21,26,49,50] Through chemical analysis, it was found that *Ophrys* flowers produce complex species-specific mixtures of more than 100 compounds,[26,30,31,51] mainly aliphatics and terpenoids; aromatic compounds are present in minor amounts. *Ophrys* species mimic the sex pheromones of their pollinators. Floral fragrances produced by flowers contain chemical compounds identical to those of the pheromonal secretions of the respective female insects.[26]

Early published behavioral experiments with synthetic compounds demonstrated that those identified only in insect glandular secretions had a greater capacity to attract males compared to volatiles identified in flowers. Borg-Karlson[26] therefore concluded that *Ophrys* flowers produce a set of "second class attractivity compounds" that successfully attract only a certain fraction of the male population, which, at that particular moment, has a low threshold for sexual attraction. However, more recent investigations have shown that *Ophrys* flowers produce the same compounds in more or less similar relative amounts to the sex pheromones of females of their pollinators.[29–31,36] Behavioral experiments with synthetic copies of the compounds produced by *Ophrys* flowers have shown that only certain volatiles are active in stimulating mating behavior in the males.[26,29,30,36] Among 108 compounds produced by *Ophrys sphegodes*,[51] 14 saturated and unsaturated hydrocarbons (C_{21}–C_{29}) triggered EAG responses in male antennae, and synthetic copies of these hydrocarbons, blended in the relative amounts found in the odor samples of females and the orchid flower, elicited copulatory attempts in males.[29] Almost the same alkanes and alkenes were found to be biologically active in several allopatric and sympatric species of the *Ophrys fusca* group and in one species of the *O. sphegodes/mammosa* group pollinated by either *Andrena nigroaenea* or *Andrena flavipes*.[36]

Most of the hydrocarbons are abundant in cuticular waxes of plants and are said to have the primary function of protecting the plant from dehydration.[52] Many of the compounds identified in *O. sphegodes* labellum extracts can also be found in other *Ophrys* species[26] and could therefore have such a function. Compounds that are not used to attract pollinators may act as a solvent for the male-attracting volatiles or they may repel other bee species.[53] A further selective pressure for the production of a wide array of compounds in *Ophrys* orchids might have been a frequent adaptation to new pollinator species. Over longer periods, it may be a more economical evolutionary strategy to use a broad spectrum of components and to alter the pattern of compounds that already exist than to synthesize new compounds that may involve major changes in the biosynthetic pathways.

In contrast to the *Andrena*-pollinated species, investigations in the mirror orchid (*O. speculum*) have shown that pollinator attraction in sexually deceptive orchids may also be based on a few specific chemical compounds.[31] *O. speculum* flowers produce many volatiles, including trace amounts of (ω1)-hydroxy- and (ω1)-oxo acids, especially 9-hydroxydecanoic acid. These compounds, which are novel in plants, proved to be the major components of the female sex pheromone in the scoliid wasp *C. ciliata*, and stimulate male copulatory behavior in this pollinator species (Figure 10.1). The specificity of the signal depends primarily on the structure and enantiomeric composition of the oxygenated acids, which is the same in the wasps and orchids. The overall composition of the blend differed significantly between the orchid and its pollinator and is of secondary importance.[31]

10.4 POLLINATOR ATTRACTION IN SEXUALLY DECEPTIVE ORCHIDS IN THE NEOTROPICS

Pseudocopulation has been reported for a few epidendroid epiphytic orchid genera of the neotropics.[17] Flowers of the orchid genera *Trichoceros*, *Telipogon*, and *Stellilabium* (all Ornithocephalinae) mimic female tachinid flies and attract males for pollination.[17] Most species occur at higher elevations in South and Central America. Females of the pollinators, willing to mate, land on a leaf or on flowers and open and close their genital orifice when male flies pass by.[17] This behavior serves as a signal

FIGURE 10.3 (See color insert following page 178.) Pseudocopulating drones of *N. testaceicornis* on a flower of *M. ringens*. Photo by R.B. Singer.

to the males searching for females. They approach the females rapidly and copulation lasts for a few moments. It is thought that the flowers attract males strictly on the basis of visual stimuli and that fragrances do not play a role in male attraction.[54,55] However, behavioral experiments to prove this assumption are missing.

Sexual deception has been suggested to occur in the genera *Cyrtidium*, *Mormolyca*, and *Trigonidium* (Maxillariinae)[17] and in the genus *Lepanthes* (Pleurothallidinae).[56] However, many of the early reports[57] were thought to be speculative[58] because the behavior of the male stingless bees that was described reminds one of the behavior of euglossine bees on orchids. In more recent literature,[19,20] the pollination process is documented and described in greater detail. *Trigonidium* is a genus of about 13 species, ranging from Mexico to southern Brazil.[58] *Trigonidium obtusum* (Epidendroidae), a species with traplike flowers, is pollinated by sexually excited drones of *Plebeia droryana* (Apidae: Meliponinae) that try to copulate with glandular apices of the petals or sepals.[17,19] In another species, *Mormolyca ringens*, the pollination process is similar to the European *Ophrys* system, in which the flowers resemble females of the pollinators (Figure 10.3)[20] and drones of *Nannotrigona testaceicornis* and *Scaptotrigona* sp. (Meliponinae). Chemical analyses in both orchid species revealed that the floral scent of *T. obtusum* is almost exclusively composed of pentadecane,[59] whereas the fragrance of *M. ringens* contained 31 compounds that were identified as fatty acid derivatives, aromatic compounds, and mono- and sesquiterpenes. However, thus far the male-attracting compounds have not been identified by means of behavioral experiments. Recently the pollination of *Lepanthes glicensteinii* by sexually deceived male fungus gnats has been documented.[56] Olfactory signals seem to play a role as a long-distance attractant; however, again the chemicals have not been identified. In contrast to the nonneotropical taxa of sexually deceptive orchids, in *Lepanthes* and Maxillariinae, flowers can be found for most of the year and many generations of pollinators are produced in a year.

10.5 SPECIFICITY OF THE MALE-ATTRACTING SIGNALS

There is a strong specialization of the sexually deceptive orchids to male pollinators because sex pheromones in bees are highly species specific.[60] Behavioral tests have shown that many European *Ophrys* species are visited and pollinated by males of usually only one or only a few pollinator species.[12,27,46,61] Similar species specificity is seen in the Australian orchid genera *Caladenia*[37] and *Chiloglottis*[42] and their pollinators. As many orchids are usually interfertile,[62] species-specific attraction of pollinators is important in maintaining reproductive isolation. In contrast to the clades mentioned above, pollination of the Australian genus *Cryptostylis* differs. Many species are morphologically disparate, but use the same pollinator, *L. excelsa*, an ichneumonid wasp.[63] GC-EAD analyses with floral extracts of *C. erecta* and *C. subulata* revealed that *Lissopimpla* males react to the same single compound.[35] Since many of the species are sympatric at any one site and no hybrids are known, reproductive isolation among *Cryptostylis* species may be achieved by some form of genetic incompatibility.[64] In a more detailed study producing artificial crosses between *C. erecta* and *C. subulata*, however, fruits have been produced.[65] Whether seeds are viable remains to been shown in further research.

In insects, the specificity of a chemical signal in a "noisy chemical background" may be achieved either by specific blends of simple and often ubiquitous compounds, or by the presence of unique compounds, including specific compositions of stereoisomers.[66] The European *Ophrys* orchids pollinator *A. nigroaenea* belong to the first scenario, where flowers and virgin females use almost identical bouquets of common saturated and unsaturated hydrocarbons to elicit mating behavior in males.[29,36] Interestingly, the same hydrocarbons are also major volatiles in *O. speculum* flowers and *C. ciliata* females, making up almost identical blends both with respect to chain length and double-bond positions,[67] but these neither stimulate *C. ciliata* male antennae nor induce mating behavior. The *O. speculum–C. ciliata* interaction, however, represents the second possibility, with both the flower and the pollinators' females producing highly specific oxygenated carboxylic acids as the signal. In this case the various hydrocarbons could potentially act as solvents for the male-attracting compounds, serve as tactile cues, or, given their physicochemical properties, have other functions as surface lipids.[52] A further example of the second possibility is represented by Australian orchids in which the male-attracting chemical signal often consists of only one unique compound.[33,35]

The existence of two very different plant-pollinator chemical communication systems even within various *Ophrys* species raises the following question: Why do *C. ciliata* females and mimicking *O. speculum* flowers use complex and unusual compounds rather than a species-specific blend of common compounds, especially since the orchid possesses the same set of common hydrocarbons used by sympatric *Ophrys* species to attract their pollinators?[29] Ecological and phylogenetic factors as well as physiological constraints play a fundamental role in the evolution of chemical communication systems. Many *Andrena* species occur sympatrically with closely related species.[68] Heterospecific mating is avoided by the use of species-specific blends of the same hydrocarbons[69] and by accompanying pattern recognition.

In contrast, *C. ciliata* neither temporally nor spatially overlaps with other scoliid wasps,[70] and the male-attracting scent consists of a few specific compounds that perhaps have arisen by intraspecific sexual selection, and that impart great species specificity.

In *Ophrys* orchids, reproductive isolation is linked to prepollination mechanisms, that is, attraction of and pollination by males of only one pollinator species.[12,27] Therefore it is an advantage if an *Ophrys* species selects a pollinator species that is reproductively isolated from other sympatrically occurring species, such as *A. nigroaenea*[69] or *C. ciliata*,[70] and does not visit flowers of sympatrically occurring *Ophrys* species. Mimicking the sex pheromone of a pollinator species that is reproductively isolated may prevent hybridization and pollen loss, which may lead to an increase in reproductive success. This is particularly important in sexually deceptive orchids, since they have a low rate of pollinator visitation and since the pollen, as in most orchids, is packed in pollinia.[51,71]

10.6 ARE ORCHID FLOWERS MORE ATTRACTIVE THAN POLLINATOR FEMALES?

In many aculeate insects males usually emerge before the females in order to maximize mating success. During this time period, when male competition is high, pseudocopulatory orchids start to bloom.[1,72] Flowers obviously exploit the naiveté of the male insects.[6] In contrast to the hypothesis that *Ophrys* flowers are neglected once the pollinator females are present,[26] in dual-choice experiments, *Ophrys* flowers were significantly more attractive to *C. ciliata* males than their own females.[31] Since *O. speculum* flowers and *C. ciliata* females attract males with the same compounds (see above), the greater attractiveness of *Ophrys* flowers can be explained by a greater total amount of male-attracting scent produced by the flowers as compared to the wasps (Figure 10.4).[48] Comparable results exist for the Australian orchid *C. trapeziformis*.[73] Females in the presence of orchids elicited fewer male approaches and no copulation attempts, and orchids produced 10 times more pheromone than female wasps.[35] Generally, calling sexually active hymenopteran females often produce minute quantities of sex attractant,[60,74] presumably only detectable over short distances. The risk of producing long-distance sex pheromones and thus attracting predators and brood parasites may be great for a female calling close to her nesting or oviposition site. Orchid flowers are exempt from this risk and can produce larger amounts of male-attracting scent, limited only by physiological constraints.

Both systems prove that males are unable to distinguish between the chemical signals of the model (female) and the mimic (orchid). Furthermore, several studies have shown that male pollinators habituate rapidly to the presence of orchids.[51,73,75] Orchids can therefore negatively affect their pollinators by making the attraction of mates more difficult for female wasps, and selection would be expected to favor males that are capable of discriminating between flowers and females.[76] Consequently flowers that are able to perfectly mimic the signal of the model should have a selective advantage. Whether this hypothesis is valid and female populations are harmed by the presence of orchids has to be investigated in future studies.

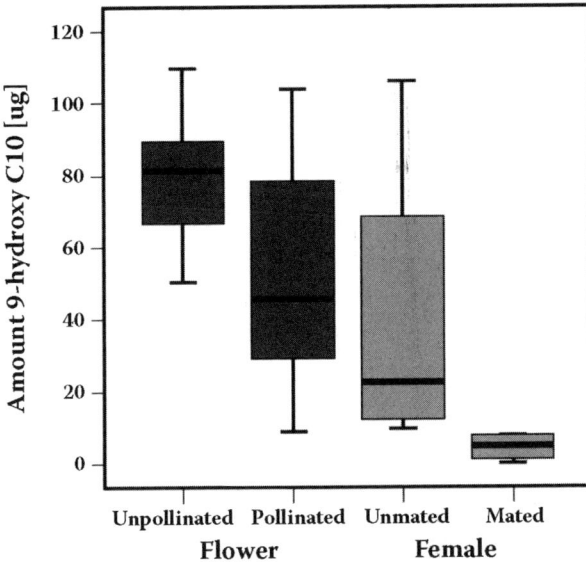

FIGURE 10.4 Comparison of the total amount of 9-hydroxydecanoic acid in unpollinated and pollinated *O. speculum* flowers as well as in unmated and mated *C. ciliata* females. There is a statistically significant difference among *C. ciliata* female groups and among the flower groups (*U*-tests, $P < 0.001$).

10.7   VARIATION OF SCENT AND REPRODUCTIVE SUCCESS

10.7.1   FLOWER-SPECIFIC VARIATION OF ODOR SIGNALS

Flower visits in deceptive pollination systems are often rare and brief,[32] and deceptive orchids often show low levels of fruit set.[77] A strong specialization of sexually deceptive orchids in attracting only males of one pollinator species and the highly sophisticated learning capabilities of bees[78] results in a low visitation rate,[71] as in other deceptively pollinated orchids,[79] resulting in a low number of fruits produced. In a population of *O. sphegodes* near Illmitz, Austria, only 4.9% of 887 plants were visited by a pollinator, as revealed by either pollinia removal, pollination, or both.[71,75] As a consequence, mechanisms have evolved in *Ophrys* orchids to increase reproductive success. Flowers are relatively long lived and the opportunity for pollination is maximally extended.[6] Furthermore, pollinated flowers produce a large number of seeds.[18] A low pollination frequency led to the evolution of strategies that increase the chance that males visit more than one flower on the same plant, thus increasing the number of pollination events and therefore the number of seeds produced.[51]

Characteristic variation in the proportions of the chemical compounds of flowers between and within an inflorescence utilizes the learning abilities of the male pollinators, thus raising the chance of more than one flower being visited by the same male bee. In the orchid *O. sphegodes*, hydrocarbons are responsible for eliciting male copulatory behavior,[29] whereas this was not achieved by synthetic blends of esters and aldehydes,[44] although they are electrophysiologically active (GC-EAD) and can be perceived by the males.[30] Ayasse et al.[51] reported an alternative function for these compounds. The GC-EAD active esters and aldehydes showed a characteristic variation among flowers of different stem positions within an inflorescence. In behavioral field tests, male bees learned the ester and aldehyde bouquets of individual flowers during mating attempts and recognized them in further encounters. Bees thereby avoid trying to mate with flowers they have visited previously, but do not avoid other flowers either from a different or from the same plant. Therefore *Ophrys* plants may take advantage of the learning abilities of the pollinators and influence flower visitation behavior. Sixty-seven percent of the males that visited one flower in an inflorescence returned to visit a second flower in the same inflorescence (Figure 10.5). However, geitonogamy is prevented and the likelihood of cross-fertilization is enhanced by the time required for the pollinium deposited on the pollinator to complete its bending movement, which is necessary for pollination to occur. Cross-fertilization is also enhanced by the high degree of odor variation between plants. This variation minimizes learned avoidance of the flowers and increases the likelihood that a given pollinator will visit several to many different plants within a population.

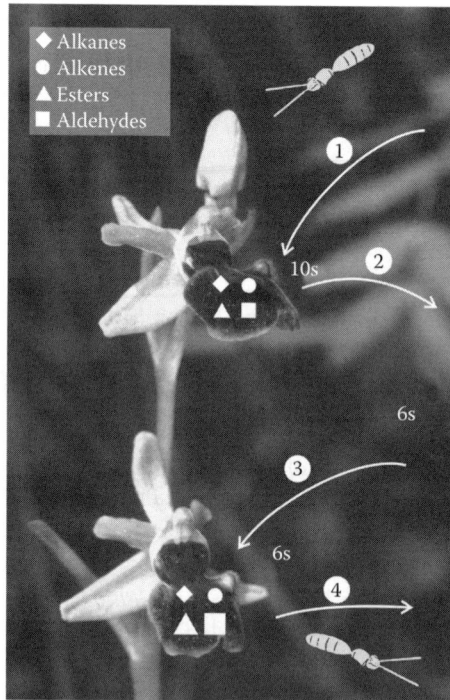

FIGURE 10.5 *Andrena nigroaenea* males learn and recognize odor bouquets of individual *Ophrys sphegodes* flowers. First the males are attracted by a mixture of unsaturated and saturated hydrocarbons, which is the same in different flowers of an inflorescence. By flower-specific variation of esters and aldehydes, a plant takes advantage of the learning abilities of the males and redirects them to flowers that have not yet been visited. Thereby, the reproductive success of a plant is increased. Males first visit one flower of an inflorescence (1), remove or add pollinia, disappear (2), and afterward visit a second flower of the same inflorescence (3,4).

10.7.2 SCENT VARIATION IN UNPOLLINATED AND POLLINATED FLOWERS

As an adaptation to highly specialized pollination mechanisms and rare pollinator visits, orchid flowers are usually long lived.[80] Under conditions of a low flower visitation rate of pollinators in sexually deceptive orchids, strategies evolved in order to increase the number of pollinated flowers of a plant and therefore the reproductive success of a plant.[75,81] Pollinated flowers often show a rapid change in color, and cessation of scent production can be observed[82] and seems to be widespread in orchids.[80,83] Postpollination alterations in pollinator-attracting signals may have different functions. A first function is to save energy, since flower maintenance and fragrance production are energy consuming.[80,84] Pollinated flowers with removed pollinia may gain only low additional reproductive success by prolonged maintenance of pollinated flowers. Therefore a second function is to direct pollinators to unpollinated flowers of an inflorescence, as was shown in the sexually deceptive orchid *O. sphegodes*. Floral scent was found to be altered after pollination, resulting in a decreased attractiveness of these flowers.[82] These postpollination changes in the production of scent were shown to guide pollinators to the unpollinated flowers of an inflorescence,[81,82] thereby increasing the number of pollinated flowers. In *O. sphegodes*, the changes involved a decrease in the total amount of scent produced as well as an alteration of the odor bouquet. (*E,E*)-farnesyl hexanoate is the major component of the Dufour's gland secretion in breeding (mated) females of the pollinator bee *A. nigroaenea*,[85] and is produced by *O. sphegodes* flowers after pollination. Behavioral tests showed that (*E,E*)-farnesyl hexanoate has a key function as a repellent signal of pollinated flowers.[81] Therefore flowers not only imitate the receptive females to attract pollinators, but also imitate mated females in order to guide pollinators to unpollinated flowers of the inflorescence. Thus the reproductive success of a plant is maximized (Figure 10.6).

Ophrys plants retain the attractiveness of the whole inflorescence after the pollination of single flowers. This is achieved by two different mechanisms.[81] First, the amount of farnesyl hexanoate produced by pollinated flowers is minute (about 1000 times less than in mated females). Second, the pseudocopulation-eliciting odor compounds (alkanes and alkenes) are still present. Furthermore, because of the low volatility of these compounds, they probably remain on the flower surface for a considerably long time after pollination.

In *O. speculum* plants, a different mechanism evolved in order to direct pollinators to unpollinated flowers of an inflorescence. Mated females as well as pollinated flowers decrease the amount of male-attracting compounds, mainly hydroxy acids (Figure 10.4).[48] *O. speculum* flowers produce hydroxy acids in trace amounts only, and these compounds are more volatile compared to the hydrocarbons used for pollinator attraction in *O. sphegodes*. A cessation of the production of hydroxy acids in pollinated flowers would quickly result in a smaller amount on the flower surface and consequently a lower attractiveness of pollinated flowers.

However, revisiting a plant may lead to geitonogamous pollination, causing a fitness disadvantage. The genetic consequence of self-pollination is inbreeding depression.[86] Therefore in orchids, mechanisms have evolved to prevent autogamy and geitonogamy[18] and to support allogamy. In *O. sphegodes*, the number of fertile seeds with embryos in seed capsules is significantly higher in artificially cross-pollinated

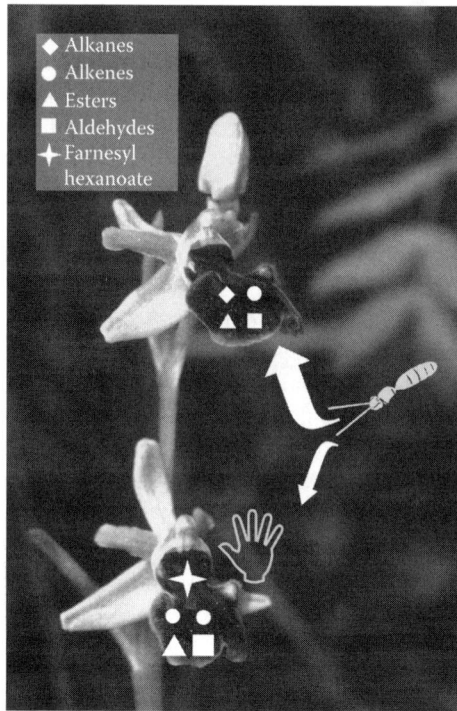

FIGURE 10.6 Pollinated *O. sphegodes* flowers produce *(E,E)*-farnesyl hexanoate, a compound that is produced by mated females of their pollinator. This guides pollinators to unpollinated flowers of the inflorescence. Thus the reproductive success of a plant is maximized. For the function of the other chemical compounds see Figure 10.5.

flowers compared to self-pollinated flowers (Figure 10.7). Therefore self-pollinated flowers should remain attractive for pollinators in order to receive pollen or gametes from flowers of other plants within a population. The question of whether male pollinators can discriminate between self-pollinated and cross-pollinated flowers was investigated by a dual-choice experiment that showed self-pollinated flowers remain attractive for pollinators (Figure 10.8). Chemical analyses of the scent of self-pollinated versus cross-pollinated flowers showed a nonsignificant lower amount of *(E,E)*-farnesyl hexanoate in self-pollinated flowers (self-pollinated: mean ± SE = 7.2 ± 2.04 ng, $n = 15$; cross-pollinated: mean ± SE = 12.3 ± 3.40 ng, $n = 16$; Mann-Whitney U-test, $P > 0.05$) which was responsible for the greater attractiveness in the dual-choice experiment.

10.8   SCENT VARIATION AND SPECIATION

10.8.1   Reproductive Isolation and Scent Variation

Isolation mechanisms in orchids fall into two types: prezygotic mechanisms, which work before fertilization, and postzygotic mechanisms, which work afterward.[6,45] In *Ophrys* orchids, ecological and ethological barriers are of greatest importance,

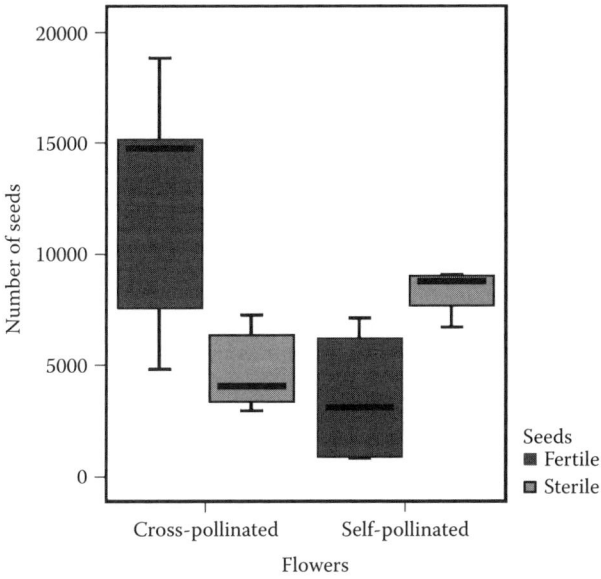

FIGURE 10.7 The number of fertile and sterile seeds in seed capsules of cross-pollinated and self-pollinated *O. sphegodes* flowers. Cross-pollinated flowers produced significantly more fertile seeds and significantly fewer sterile seeds than self-pollinated flowers (*t*-test, $P < 0.05$).

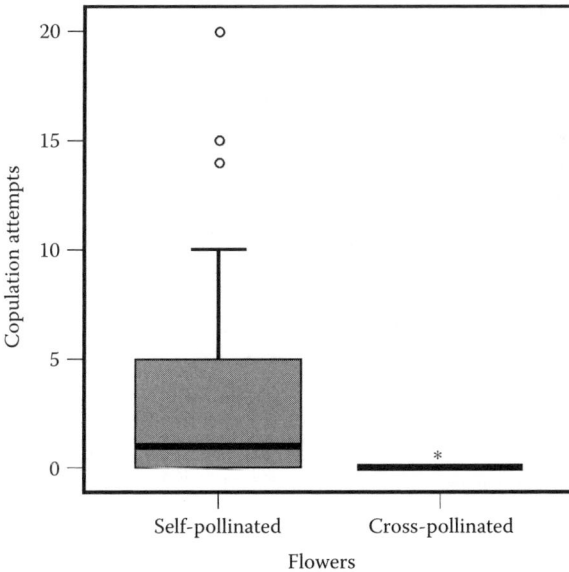

FIGURE 10.8 Mean attractiveness of self-pollinated and cross-pollinated *O. sphegodes* flowers. Self-pollinated flowers were significantly more attractive for pollinating males than cross-pollinated flowers (*t*-test, $P < 0.05$).

whereas in the orchid genera of southern Australia, flower morphology and genetic barriers play a more important role.[6]

The highly specific *Ophrys*-pollinator relationship represents the main mechanism of reproductive isolation between the often interfertile *Ophrys* species[62] and the species-specific scent is mainly responsible for prezygotic isolation.[69] In addition to the selective attraction of pollinators via odors, the location of attachment of the pollinia to the male provides a mechanical component to the isolating mechanism between sympatric *Ophrys* species with the same pollinator. *A. nigroaenea* males are the pollinators of *O. sphegodes*, *O. sitiaca*, and *O. fusca* flowers.[47] However, while the pollinia are deposited on the head of the male bee in *O. sphegodes*, in *O. sitiaca* and *O. fusca*, males reverse into a flower and receive the pollinia on the tip of their abdomen (Figure 10.9). Thus an effective mechanical isolating mechanism between sympatrically occurring species is achieved and the loss of pollinia to heterospecific orchids is minimized.

In the sympatric, closely related orchid species *O. fusca* and *O. bilunulata*, which are pollinated by *A. nigroaenea* and *A. flavipes*, Schiestl and Ayasse[69] identified and compared the pollinator-attracting scent. They found that the difference in scent between both species is rather small. Among the biologically active compounds, the pattern of alkanes was mostly the same, whereas the relative proportions of alkenes differed.[69] Based on behavioral experiments Schiestl and Ayasse[69] found that the pattern of alkenes has an important function in selectively attracting pollinators of both species, a result that was later supported by investigations by Stökl et al.[36] Using electrophysiology (GC-EAD) and chemical analyses, they studied the odor bouquets of several allopatric and sympatric species of the *O. fusca* group and one species of the *O. sphegodes/mammosa* group, all pollinated by either *A. nigroaenea* or *A. flavipes*. A comparison of the investigated species based on the proportions of all GC-EAD active compounds revealed that allopatrically occurring *Ophrys* species with the same pollinator, independent of their phylogenetic relationship, use the same odor compounds for pollinator attraction (Figure 10.10). Therefore there is a convergent evolution of pollinator-attracting volatiles in *Ophrys* orchids. Differences between the *Ophrys* species pollinated either by *A. nigroaenea* or by *A. flavipes* mainly involve different odor bouquets. In congruence with previous results,[69] Stökl et al.[36] found that alkenes with a certain double-bond position are most important for pollinator attraction.

Consequently the odor bouquet of an *Ophrys* species should be under strong stabilizing selection to remain attractive to their pollinators. If the scent of a plant deviates too much from the species-specific sex pheromone of the pollinator, it will result in lower reproductive success. However, negative frequency-dependent selection in response to odor learning in deceptive systems may favor variability of the pollinator-attracting signal within orchid populations.[75,87,88] Ayasse and Schiestl[75] found considerable odor variation within *O. sphegodes* populations. This variation minimizes learned avoidance of the flowers and increases the likelihood that a given pollinator will visit several to many different plants within a population (see above).

Interestingly, in *O. sphegodes*, intraspecific variation of the behaviorally active compounds is smaller compared to the nonactive compounds.[51] One explanation

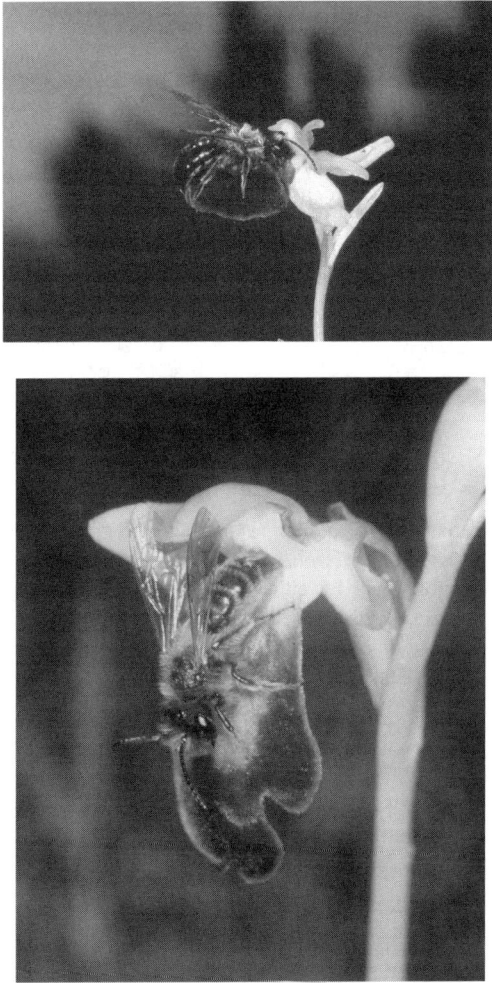

FIGURE 10.9 (See color insert following page 178.) In the sympatrically occurring *O. sphegodes* and *O. fusca*, the location of attachment of the pollinia to the male provides a mechanical component to the isolating mechanism via odor. *A. nigroaenea* males are the pollinators of both species. However, while the pollinia are deposited on the head of the male bee in *O. sphegodes* (top), males visiting *O. fusca* back into the flower and receive pollinia on the tip of the abdomen (bottom).

for the smaller intraspecific variation of the behavior-modulating compounds may be a higher selective pressure on the pollinator-attracting communication signal. Sex pheromones of female insects consist of highly species-specific mixtures of compounds with a fairly low degree of variability.[89,90] Therefore the male-attracting volatiles of *Ophrys* flowers should also show a fairly low degree of variability in order to match the signal males are able to perceive.

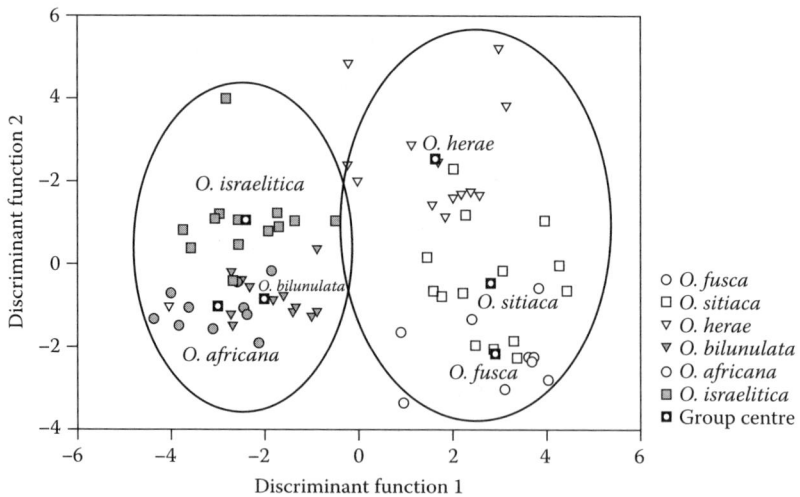

FIGURE 10.10 Comparison of the pollinator attracting bouquets of hydrocarbons of allopatric and sympatric species of the *Ophrys fusca* group and one species of the *O. sphegodes/mammosa* group, pollinated by either *Andrena nigroaenea* (symbols white) or *A. flavipes* (symbols dark gray). *Ophrys* species with the same pollinator, independent of their phylogenetic relationship, use the same odor compounds in a similar compound composition for pollinator attraction. For details, see Ayasse et al.[51]

10.8.2 HYBRIDIZATION AND SPECIATION

Hybridization and introgression is thought to be an important mechanism for speciation in many plants.[62] Data on naturally occurring plant hybrids are quite extensive.[91,92] In the Mediterranean region, sympatric speciation has been suggested in the genus *Ophrys*.[21] A possible mechanism for speciation may be hybridization and introgression, which may be common within *Ophrys*[45,93,94] for the following reasons: First, the genus contains many closely related species that are sympatric to a great extent. Second, the chromosome number of all species is the same, and hybrids are fertile (homodiploidy). Anderson[95] stated that the effects of introgression are often so subtle that they cannot be detected. Hybrid speciation is conceivable, if the same pollinator is attracted by two *Ophrys* taxa that occur in sympatry.[47] Odor changes as a result of genetic mutations that affect the genes involved in odor production and hybridization can be the driving force for speciation, since they may become the cause for reproductive isolation between the interfertile species.[27,62] Formation of hybrids in *Ophrys* may be stabilized, thus providing the potential for speciation, since the attraction of a new pollinator by the hybrids may act as a prezygotic isolation barrier. Thus the odor produced by hybrid plants is essential for the reproductive isolation of hybrids.

In the *O. fusca* group, speciation may be brought about by changing the patterns of alkenes, which leads to the attraction of a different pollinator and reproductive isolation.[69] Such plants would successfully reproduce, being reproductively isolated from other sympatrically occurring plants, and the new odor would be established

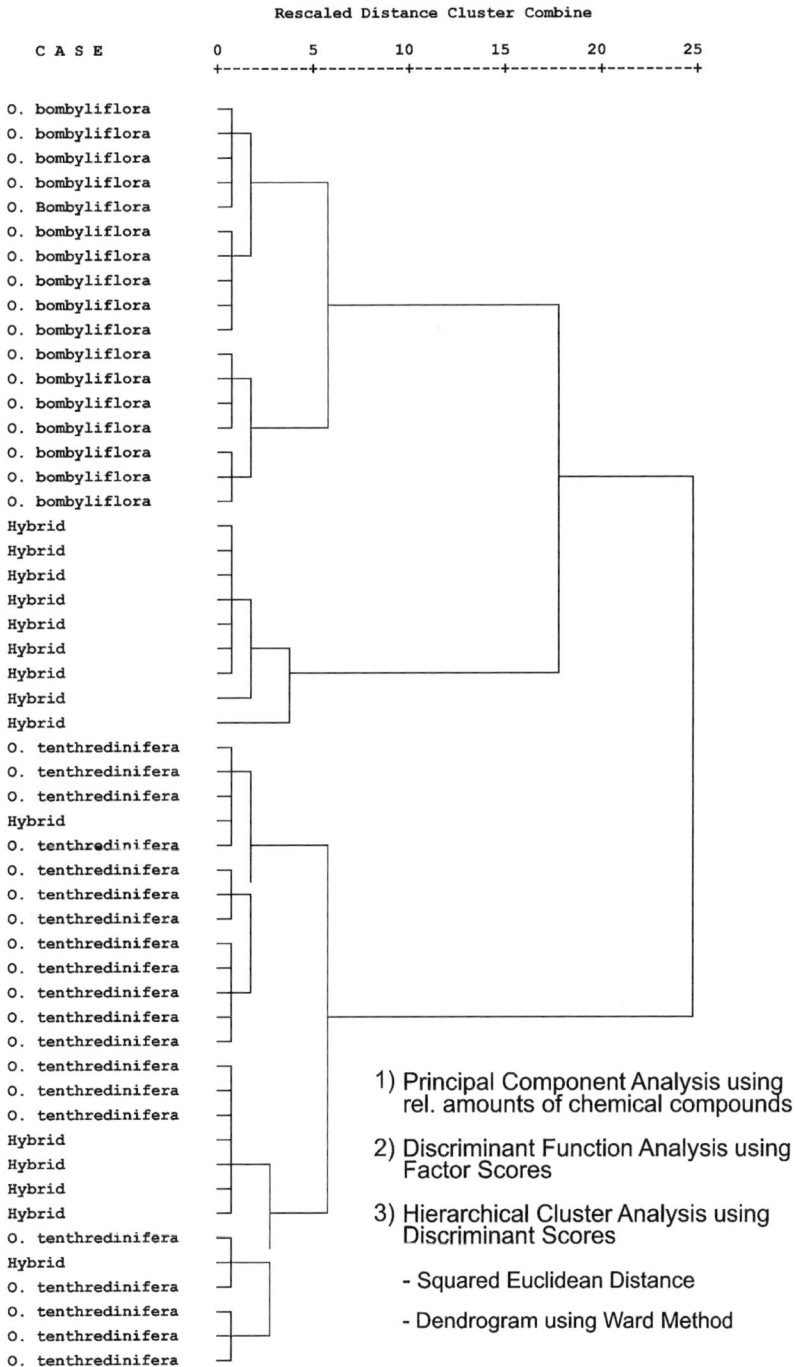

FIGURE 10.11 Dendrogram using floral scent of *Ophrys tenthredinifera*, *Ophrys bombyliflora* and *O. tenthredinifera* × *O. bombyliflora*. Most hybrids form a separate cluster, indicating an intermediate scent. Several hybrids cluster together with *O. tenthredinifera*.

by stabilizing selection. However, if a mutant attracts the pollinators of sympatrically occurring species, hybridization may take place.

Data on the floral scents of hybrid plants have not yet been published for orchids. In hybrids from two species of *Salix*, secondary chemistry was intermediate, suggesting additive inheritance of the chemicals investigated.[96] The floral scents of *Ophrys* hybrids support the finding that hybrids may produce intermediate scents (Figure 10.11).[48] *O. tenthredinifera* and *O. bombyliflora* are both pollinated by *Eucera* bees and different degrees of cross-attraction can be observed,[48] which may lead to hybrid formation. In a comparison of the odor bouquets of hybrids identified by morphological traits and both parent species, several specimens of the hybrid swarm clustered together with one of the parent species (Figure 10.11). If the specimens that produce intermediate scent by chance attract a new pollinator, speciation may take place. Hybridization may favorably occur in *Ophrys* species that are visited by pollinators of the same genus of bees, such as in the *O. fusca* group, where most species are pollinated by *Andrena* males.[45,47] Future investigations combining molecular techniques, chemical analyses, behavioral experiments, and electrophysiology will help to clarify the potential hybrid origin of *Ophrys* populations and may demonstrate processes of sympatric speciation.

ACKNOWLEDGMENTS

I thank Hannes Paulus, Florian Schiestl, and Rodrigo Singer for providing me with photos, and Gerhard Gottsberger, Stefan Jarau, Rodrigo Singer, and Johannes Stökl for helpful comments on the manuscript. Parts of this research review were supported by a grant from the Fonds zur Förderung der Wissenschaftlichen Forschung (FWF Austria; Austrian Science Foundation) (P12275-BIO) and the Deutsche Forschungs-gemeinschaft (DFG; German Research Foundation) (AY 12/1-1).

REFERENCES

1. Nilsson, L.A., Orchid pollination biology, *Trends Ecol. Evol.* 7, 255, 1992.
2. Darwin, C., *On the Various Contrivances by Which Orchids Are Fertilized by Insects*, John Murray, London, 1885.
3. Dafni, A., Mimicry and deception in pollination, *Annu. Rev. Ecol. Syst.* 15, 259, 1984.
4. Dafni, A., Floral mimicry: mutualism and undirectional exploitation of insects by plants, in *Plant Surface and Insects*, Arnold, E., Ed., Richard Clay, Ltd., London, 1986, p. 81.
5. Ackermann, J.D., Mechanisms and evolution of food-deceptive pollination systems in orchids, *Lindelyana* 1, 108, 1986.
6. Dafni, A. and Bernhardt, P., Pollination of terrestrial orchids of southern Australia and the Mediterranean region, in *Evolutionary Biology*, Hecht, M.K., Wallace, B., and Macintyre, R.J., Eds., Plenum Press, New York, 1990, p. 193.
7. Pasteur, G., A classificatory review of mimicry systems, *Annu. Rev. Ecol. Syst.* 13, 169, 1982.
8. Correvon, H. and Pouyanne, A., Un curieux cas de mimétisme chez les Ophrydées, *J. Soc. Nat. Hort. France* 17, 29, 1916.

9. Pouyanne, A., Le fecondation des *Ophrys* par les insectes, *Bull. Soc. Hist. Nat. Afr. Nord.* 8, 6, 1917.

10. Coleman, E., Pollination of an Australian orchid by male ichneumonid *Lissopimpla semipunctata* Kirby, *Trans. R. Entomol. Soc. Lond.* 76, 533, 1928.

11. Coleman, E., Further observations on the pseudocopulation of the male *Lissopimpla semipunctata* Kirby (*Hymenoptera parasitica*) with the Australian orchid *Cryptostylis leptochila* F.v.M., *Proc. R. Entomol. Soc. Lond.* 13, 82, 1938.

12. Kullenberg, B., Studies in *Ophrys* pollination, *Zool. Bidrag. (Uppsala)* 34, 1, 1961.

13. Sasaki, M., Ono, M., Asada, S., and Yoshida, T., Oriental orchid (*Cymbidium pumilum*) attracts drones of the Japanese honeybee (*Apis cerana japonica*) as pollinator, *Experientia* 47, 1229, 1991.

14. Peakall, R., The unique pollination of *Leporella fimbriata* (Orchidaceae): pollination by pseudocopulating male ants (*Myrmecia urens*, Formicidae), *Plant Syst. Evol.* 167, 137, 1989.

15. Peakall, R., Responses of male *Zaspilothynnus trilobatus* Turner wasps to females and the sexually deceptive orchid it pollinates, *Funct. Ecol.* 4, 159, 1990.

16. Steiner, K.E., Whitehead, V.B., and Johnson, S.D., Floral and pollinator divergence in two sexually deceptive South African orchids, *Am. J. Bot.* 81, 185, 1994.

17. van der Pijl, L. and Dodson, C.H., *Orchid Flowers: Their Pollination and Evolution*, University of Miami Press, Coral Gables, FL, 1966.

18. van der Cingel, N.A., An *Atlas of Orchid Pollination: European Orchids*, A.A. Balkema, Rotterdam, 1995.

19. Singer, R.B., The pollination mechanism in *Trigonidium obtusum* Lindl (Orchidaceae: Maxillariinae): sexual mimicry and trap-flowers, *Ann. Bot. Lond.* 89, 157, 2002.

20. Singer, R.B., Flach, A., Koehler, S., Marsaioli, A.J., and Amaral, M.C.E., Sexual mimicry in *Mormolyca ringens* (Lindl.) Schltr. (Orchidaceae: Maxillariinae), *Ann. Bot. Lond.* 93, 755, 2004.

21. Kullenberg, B. and Bergström, G., The pollination of *Ophrys* orchids, *Bot. Notiser* 129, 11, 1976.

22. Paulus, H.F. and Gack, C., Pollination of *Ophrys* (Orchidaceae) in Cyprus, *Plant Syst. Evol.* 169, 177, 1990.

23. Ames, O., Pollination of orchids through pseudocopulation, *Bot. Mus. Leafl. Harv. Univ.* 5, 1, 1937.

24. Baker, H.G. and Hurd, P.D., Jr., Intrafloral ecology, *Annu. Rev. Entomol.* 13, 385, 1968.

25. Dressler, R.L., *The Orchid: Natural History and Classification*, Harvard University Press, Cambridge, MA, 1981.

26. Borg-Karlson, A.-K., Chemical and ethological studies of pollination in the genus *Ophrys* (Orchidaceae), *Phytochemistry* 29, 1359, 1990.

27. Paulus, H.F. and Gack, C., Pollinators as prepollinating isolation factors: evolution and speciation in *Ophrys* (Orchidaceae), *Isr. J. Bot.* 39, 43, 1990.

28. Bower, C.C., Demonstration of pollinator-mediated reproductive isolation in sexually deceptive species of *Chiloglottis* (Orchidaceae: Caladeniinae), *Aust. J. Bot.* 44, 15, 1996.

29. Schiestl, F.P., Ayasse, M., Paulus, H.F., Löfstedt, C., Hansson, B.S., Ibarra, F., and Francke, W., Orchid pollination by sexual swindle, *Nature* 399, 421, 1999.

30. Schiestl, F.P., Ayasse, M., Paulus, H.F., Löfstedt, C., Hansson, B.S., Ibarra, F., and Francke, W., Sex pheromone mimicry in the early spider orchid (*Ophrys sphegodes*): patterns of hydrocarbons as the key mechanism for pollination by sexual deception, *J. Comp. Physiol. A* 186, 567, 2000.

31. Ayasse, M., Schiestl, F.P., Paulus, H.F., Ibarra, F., and Francke, W., Pollinator attraction in a sexually deceptive orchid by means of unconventional chemicals, *Proc. R. Soc. Lond. B* 270, 517, 2003.

32. Peakall, R. and Beattie, A.J., Ecological and genetic consequences of pollination by sexual deception in the orchid *Caladenia tentactulata*, *Evolution* 50, 2207, 1996.

33. Schiestl, F.P., Peakall, R., Mant, J.G., Ibarra, F., Schulz, C., Franke, S., and Francke, W., The chemistry of sexual deception in an orchid-wasp pollination system, *Science* 302, 437, 2003.

34. Ågren, L., Kullenberg, B., and Sensenbaugh, T., Congruences in pilosity between three species of *Ophrys* (Orchidaceae) and their hymenopteran pollinators, *Nova Acta Reg. Soc. Sci. Ups. Ser. V:C* 3, 15, 1984.

35. Schiestl, F.P., Peakall, R., and Mant, J., Chemical communication in the sexually deceptive orchid genus *Cryptostylis*, *Bot. J. Linn. Soc.* 144, 199, 2004.

36. Stökl, J., Paulus, H., Dafni, A., Schulz, C., and Francke, W., Ayasse M Pollinator attracting odour signals in sexually deceptive orchids of the *Ophrys fusca* group, *Plant Syst. Evol.* 254, 105, 2005.

37. Stoutamire, W.P., Wasp pollination species of *Caladenia* (Orchidacea) in southwestern Australia, *Aust. J. Bot.* 31, 383, 1983.

38. Peakall, R., Beattie, A.J., and James, S.J., Pseudocopulation of an orchid by male ants: a test of two hypotheses accounting for the rarity of ant pollination, *Oecologia* 73, 522, 1987.

39. Kores, P.J., Molvray, M., Weston, P.H., Hopper, S.D., Brown, A.P., Cameron, K.M., and Chase, M.W., A phylogenetic analysis of Diurideae (Orchideaceae) based on plastid DNA sequence data, *Am. J. Bot.* 88, 1903, 2001.

40. Stoutamire, W.P., Pollination studies in Australian terrestrial orchids, *Natl. Geog. Soc. Res. Rep.* 13, 591, 1981.

41. Wallace, B.J., On *Cryptostylis* pollination and pseudocopulation, *Orchadian* 5, 168, 1978.

42. Bower, C.C., The use of pollinators in the taxonomy of sexually deceptive orchids in the subtribe Caladeniinae (Orchidaceae), *Orchadean* 10, 331, 1992.

43. Mant, J.G., Schiestl, F.P., Peakall, R., and Weston, P.H., A phylogenetic study of pollinator conservatism among sexually deceptive orchids, *Evolution* 56, 888, 2002.

44. Schiestl, F.P., personal communication, 2005.

45. Delforge, P., *Guide des Orchidées d'Europe, d'Afrique du Nord et du Proche-Orient*, Delachaux et Niestlé S.A., Lausanne-Paris, 2001.

46. Paulus, H.F. and Gack, C., Signalfälschung als Bestäubungsstrategie in der mediterranen Orchideengattung *Ophrys* — Probleme der Artbildung und der Artabgrenzung, in *International Symposium on European Orchids, Eurorchis*, vol. 92, Brederoo, P. and Kapteyn den Boumeester, D.W., Eds., Stichting Uitgeverij Koninkliijke Nederlandse Natuuhistorische Vereniging in cooperation with the Stichting Europese orchideen van de KNNV, Utrecht/Haarlem, 1994, p. 45.

47. Paulus, H.F., Signale in der Bestäuberanlockung: Weibchenimitation als Bestäubungsprinzip bei der mediterranen Orchideengattung *Ophrys*, *Verh. Zool.-Bot. Ges. Österreich.* 134, 133, 1997.

48. Ayasse, M., unpublished data.

49. Bergström, G., Role of volatile chemicals in *Ophrys*-pollinator interactions, in *Biochemical Aspects of Plant and Animal Coevolution*, Harborne, G., Ed., Academic Press, London, 1978, p. 207.

50. Tengö, J., Odour-released behaviour in *Andrena* male bees (Apoidea, Hymenoptera), *Zoon* 7, 15, 1979.

51. Ayasse, M., Schiestl, F.P., Paulus, H.F., Lofstedt, C., Hansson, B., Ibarra, F., and Francke, W., Evolution of reproductive strategies in the sexually deceptive orchid *Ophrys sphegodes*: how does flower-specific variation of odor signals influence reproductive success?, *Evolution* 54, 1995, 2000.

52. Eigenbrode, S.D. and Espelie, K.E., Effects of plant epicuticular lipids on insect herbivores, *Annu. Rev. Entomol.* 40, 171, 1995.

53. Francke, W., Convergency and diversity in multicomponent insect pheromones, in *Advances in Invertebrate Reproduction*, Porchet, M., Andries, J.-C., and Dhainaut, A., Eds., Elsevier Science, Amsterdam, 1986, p. 327.

54. Dodson, C.H., The importance of pollination in the evolution of the orchids of tropical America, *Am. Orchid Soc. Bull.* 31, 525, 1963.

55. Dodson, C.H. and Escobar, R.E., The Telipogons of Costa Rica (I), *Orquideologia* 17, 3, 1987.

56. Blanco, M.A. and Barboza, G., Pseudocopulatory copulation in *Lepanthes* (Orchidaceae: Pleurothallidinae) by fungus gnats, *Ann. Bot.* 95, 763, 2005.

57. Kerr, W.E. and Lopez, C.R., Biologia da reproducao de *Trigona* (*Plebeia*) *droryana* F. Smith, *Rev. Bras. Biol.* 22, 335, 1962.

58. Van der Cingel, N.A., *An Atlas of Orchid Pollination: America, Africa, Asia and Australia*, A.A. Balkema, Rotterdam, 2001.

59. Flach, A., Dondon, R.C., Singer, R.B., Koehler, S., Amaral, M.C.E., and Marsaioli, A.J., The chemistry of pollination in selected Brazilian Maxillariinae orchids: floral rewards and fragrance, *J. Chem. Ecol.* 30, 1039, 2004.

60. Ayasse, M., Paxton, R.J., and Tengö, J., Mating behavior and chemical communication in the Hymenoptera, *Annu. Rev. Entomol.* 46, 31, 2001.

61. Paulus, H.F. and Gack, G., Neue Befunde zur Pseudokopulation und Bestäuberspezifität in der Gattung *Ophrys*—Untersuchungen in Kreta, Süditalien und Israel, *Die Orchidee Sonderheft* 39, 48, 1986.

62. Ehrendorfer, F., Hybridisierung, Polyploidie und Evolution bei europäisch-mediterranene Orchideen, *Die Orchidee Sonderheft*, 33, 15, 1980.

63. Jones, D.L., *Native Orchids of Australia*, Frenchs Forest, Australia, Reed, 1988.

64. Stoutamire, W.P., Pseudocopulation in Australian terrestrial orchids, *Am. Orchid Soc. Bull.* 44, 226, 1975.

65. Lloyd, G., The pollination biology of *Cryptostylis erecta* and *Cryptostylis subulata* (Orchidaceae) and the maintenance of species integrity, unpublished B.Sc. Honours Thesis, School of Biological Sciences, Sydney, Australia, 2003.

66. Roelofs, W.L., The chemistry of sex attraction, in *Chemical Ecology: The Chemistry of Biotic Interaction*, Eisner, T. and Meinwald, J., Eds., National Academy Press, Washington, DC, 1995, p. 103.

67. Erdmann, D.H., Identifizierung und Synthese flüchtiger Signalstoffe aus Insekten und ihren Wirtspflanzen, PhD dissertation, University of Hamburg, Hamburg, Germany, 1996.

68. Westrich, P., *Die Wildbienen Baden Württembergs I*, Verlag Eugen Ulmer, Stuttgart, Germany, 1989.

69. Schiestl, F.P. and Ayasse, M., Do changes in floral odor cause sympatric speciation in sexually deceptive orchids?, *Plant Syst. Evol.* 234, 111, 2002.

70. Osten, T., Die Scoliiden des Mittelmeer-Gebietes und angrenzender Regionen (Hymenoptera) Ein Bestimmungsschlüssel, *Linzer Biol. Beitr.* 32, 537, 2000.

71. Ayasse, M., Schiestl, F., Paulus, H.F., Erdmann, D., and Francke, W., Chemical communication in the reproductive biology of *Ophrys sphegodes*, *Mitt. Dtsch. Ges. Allg. Angew. Angw. Ent.* 11, 473, 1997.

72. Paulus, H.F., Co-Evolution und einseitige Anpassung in Blüten-Bestäuber-Systemen. Bestäuber als Schrittmacher in der Blütenevolution, *Verh. Dt. Zool. Gesellsch.* 81, 25, 1988.

73. Wong, B.B.M. and Schiestl, F.P., How an orchid harms its pollinator, *Proc. R. Soc. Lond.* 269, 1529, 2002.

74. Ayasse, M., Engels, W., Lübke, G., and Francke, W., Mating expenditures reduced via female sex pheromone modulation in the primitively eusocial halictine bee, *Lasioglossum (Evylaeus) malachurum* (Hymenoptera: Halictidae), *Behav. Ecol. Sociobiol.* 45, 95, 1999.

75. Ayasse, M. and Schiestl, F.P., Evolution of reproductive strategies in the sexually deceptive orchid *Ophrys sphegodes*: how does variation of floral scent emission after pollination influence pollinator behaviour, *Zoology* 103, 39, 2000.

76. Stowe, M.K., Chemical mimicry, in *Chemical Mediation of Coevolution*, Spencer, K.C., Ed., Academic Press, Chicago, 1988, p. 513.

77. Neiland, M.R.M. and Wilcock, C.C., Maximisation of reproductive success by European Orchidaceae under conditions of infrequent pollination, *Photoplasma* 187, 39, 1995.

78. Smith, B.H. and Ayasse, M., Kin-based male mating preferences in two species of halictine bees, *Behav. Ecol. Sociobiol.* 20, 313, 1987.

79. Fritz, A.-L. and Nilsson, L.A., Reproductive success and gender variation in deceit-pollinated orchids, in *Floral Biology: Studies on Floral Evolution in Animal-Pollinated Plants*, Lloyd, D.G. and Barrett, S.C.H., Eds., Chapman & Hall, New York, 1996, p. 319.

80. Arditti, J., Aspects of the physiology of orchids, in *Advances in Botanical Research*, vol. 7, Woolhouse, H.W., Ed., Academic Press, London, 1979, p. 422.

81. Schiestl, F.P. and Ayasse, M., Post-pollination emission of a repellent compound in a sexually deceptive orchid: a new mechanism for maximising reproductive success? *Oecologia* 126, 531, 2001.

82. Schiestl, F.P., Ayasse, M., Paulus, H.F., Erdmann, D., and Francke, W., Variation of floral scent emission and postpollination changes in individual flowers of *Ophrys sphegodes* subsp. *sphegodes*, *J. Chem. Ecol.* 23, 2881, 1997.

83. Arditti, J., *Fundamentals of Orchid Biology*, John Wiley & Sons, New York, 1992.

84. Vogel, S., *The Role of Scent Glands in Pollination*, Smithsonian Institution Libraries, Washington DC, 1990.

85. Schiestl, F.P. and Ayasse, M., Post mating odor in females of the solitary bee, *Andrena nigroaenea* (Apoidea, Andrenidae), inhibits male mating behavior, *Behav. Ecol. Sociobiol.* 48, 303, 2000.

86. Proctor, M., Yeo, P., and Lack, A., *The Natural History of Pollination*, Timber Press, Portland, OR, 1996.

87. Moya, S. and Ackerman, J.D., Variation in the floral fragrance of *Epidendrum ciliare* (Orchidaceae), *Nord. J. Bot.* 13, 41, 1993.

88. Gigord, L.D.B., Macnair, M.R., and Smithson, A., Negative frequency-dependent selection maintains a dramatic flower color polymorphism in the rewardless orchid *Dactylorhiza sambucina* (L.) Soó, *Proc. Natl. Acad. Sci. USA* 98, 6253, 2001.

89. Baker, T.C., Sex pheromone communication in the Lepidoptera: new research progress, *Experientia* 45, 248, 1989.

90. Löfstedt, C., Herrebout, W.M., and Menken, J., Sex pheromones and their potential role in the evolution of reproductive isolation in small ermine moth (Yponomeutidae), *Chemoecology* 2, 20, 1991.

91. Rieseberg, L.H., Hybrid origins of plant species, *Annu. Rev. Ecol. Syst.* 28, 359, 1997.

92. Knobloch, I.W., Intergeneric hybridization in flowering plants, *Taxon* 21, 97, 1971.
93. Stebbins, G.L. and Ferlan, L., Population variability, hybridization, and introgression in some species of *Ophrys*, *Evolution* 10, 32, 1956.
94. Nelson, E., *Gestaltwandel und Artbildung, erörtert am Beispiel der Orchidaceen Europas und der Mittelmeerländer, insbesondere der Gattung Ophrys*, Eigenverlag, Chernex-Montreux, 1962.
95. Anderson, E., Introgressive hybridisation, *Biol. Rev.* 28, 280, 1953.
96. Orians, C.M. and Fritz, R.S., Secondary chemistry of hybrid and parental willows: phenoloic glycosides and condensed tannins in *Salix sericea*, *S. eriocephala*, and their hybrids, *J. Chem. Ecol.* 21, 1245, 1995.

11 Detection and Coding of Flower Volatiles in Nectar-Foraging Insects

Mikael A. Carlsson and Bill S. Hansson

CONTENTS

11.1 INTRODUCTION

Unlike us, insects live in an odor world where an ability to accurately monitor the chemical environment is essential for survival. For example, mates are often located and identified by intersexual signals and pheromones, and oviposition sites with high competition are avoided by deterring compounds. In addition, nectar-foraging insects, like honeybees and moths, use olfactory cues emitted by flowers to find the food source. The importance of floral scent in pollination systems is discussed throughout this book. Flower-emitted compounds may trigger both innate behavioral responses and be involved in associative learning processes. In this chapter we will discuss the neuronal mechanisms that underlie these cognitive processes. Both peripheral detection mechanisms of flower molecules and central representation and coding of this information will be discussed.

11.2 DETECTION OF FLOWER MOLECULES

The major function of the olfactory organs (antennae and mouthparts; Figure 11.1a) is to provide the central nervous system with information about the identity and abundance of odor molecules in the environment. To accomplish this task, specific cells sense the presence of a chemical stimulus and transform it into changes in membrane potentials that can quickly and reliably send information to the target cells in the brain.

The main olfactory organs in insects are the antennae. On the third flagellum of the antenna there are numerous cuticular formations—sensilla—containing the sensory cells (Figure 11.1b). Each sensillum normally houses 2 to 5 olfactory receptor neurons (ORNs), but rarely more than 100.[1] The ORNs are bipolar cells and connect directly to the brain without any peripheral synapses. From the cell somata at the sensillar base, the dendritic end extends into an aqueous fluid, the sensillar lymph, which acts as the interface between neuron and environment. Odor molecules enter the sensilla through pores in the cuticular walls.[2] As most odorous volatiles are lipophilic, the transfer from the pores to the receptor sites on the ORNs is believed to be facilitated by docking to so-called odorant binding proteins (OBPs).[3]

In contrast to other sensory modalities, the olfactory system has to recognize and discriminate stimuli that are multidimensional with respect to physical properties. In the visual system, for example, thousands of colors can be discriminated by less than a handful of differentially tuned but overlapping receptors. The solution to how the olfactory sense deals with this problem came when the Nobel laureates Linda Buck and Richard Axel[4] discovered a multigene family coding for odorant

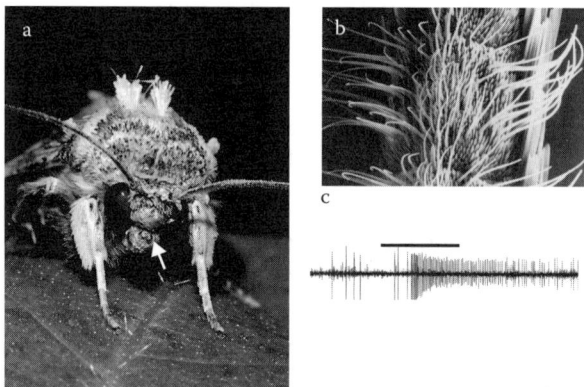

FIGURE 11.1 (a) A male *Spodoptera littoralis* (Photo by Dr. Richard Ignell). Olfactory receptor cells are located on the antenna and the mouthparts. The arrow shows the location of the labial palp pit organs, which, in addition to the antennae, houses ORNs. (b) A scanning electron micrograph (SEM) image of a single segment from the male antenna of *S. littoralis* showing different types of olfactory sensilla housing the receptor cells (Photo by Håkan Ljungberg). (c) A typical response to a flower volatile (β-caryophyllene) measured extracellularly from a single sensillum. Shortly after the onset of the stimulus, a train of action potential is evoked. The bar represents the stimulus period (500 msec).

receptor proteins in rats. Since this historic landmark, putative odorant receptor proteins have been identified in a number of organisms, including insects.[5-8] The sizes of the gene families coding for these receptors are remarkable and the number of different receptors expressed in olfactory tissues can be as large as 1300 in the mouse.[9] Even though the number is far lower in insects (about 40 in *Drosophila melanogaster*[10] and about 80 in *Anopheles gambiae*[7]), the gene family still covers a substantial part of the genome. All known odorant receptors are G-protein coupled 7-transmembrane proteins, but they show little homology between phylogenetically divergent groups of organisms.[11] Most importantly, with a few exceptions, each insect ORN expresses only a single receptor type[5,6,12] (but see Goldman et al.[13]).

Binding of an odor molecule to a receptor protein triggers a second messenger cascade. The major pathway in insects involves generation of inositol 1,4,5,-triphosphate (IP_3), which causes an influx of calcium ions into the dendrite.[13] The calcium then activates nonspecific cation channels. An inflow of cations changes the membrane potential. If the depolarization exceeds a certain threshold, an action potential is evoked at the initiation site near the soma. Action potentials carry information along the axons into the primary olfactory center of the brain, the antennal lobe (AL). The frequency of the evoked action potentials in a neuron is proportional to the concentration of the stimulus.

Since the pioneering work of Schneider,[15] a multitude of electrophysiological studies of insect ORNs have been performed. Figure 11.1c shows an example of a typical response from a recording using a tungsten electrode to pick up extracellular signals from the ORNs within a sensillum. An overwhelming majority of these studies have focused on the detection of pheromone-related compounds. The pheromone-detecting ORNs generally display a very narrow tuning. Even a minor deviation from the "required" structure, such as the movement of double bonds or the replacement of functional groups, may result in a substantially reduced response.[16-18] Such high selectivity has, in a few cases, been recorded in ORNs tuned to flower and other plant-related odors.[19-22] However, even though certain plant compounds activate highly selective ORNs, the same compounds are often also detected by other less discriminative ORNs.

The molecular receptive range of ORNs primarily tuned to floral odorants is exemplified by an extensive study performed by Shields and Hildebrand[23] in the female hawkmoth (*Manduca sexta*). They used a large panel of volatiles (more than 100 different compounds) known to be emitted by preferred flowers such as *Datura wrightii*.[24] First, two major groups of ORNs could be discerned. The first group of neurons responded to a narrow range of terpene compounds, whereas the second group responded to aromatic compounds. These groups were basically nonoverlapping. Within each group, an ORN responded to a limited range of generally similar molecules. For instance, one type of cell responded to the monoterpene geraniol and the sesquiterpene farnesol. Another type of ORN responded to geraniol, but not to farnesol. Instead, it responded strongly to linalool, a monoterpene, which structurally resembles geraniol. In addition, a third population of ORNs was more broadly tuned, responding to structurally different chemicals. In parallel, Kalinová et al.[25] recorded from plant odor, specific ORNs in both male and female *M. sexta*. Despite a less extensive stimulus panel, the results of this study agreed well with what was found by Shields and Hildebrand[23] and also revealed a very high similarity between the sexes. In about 100 neurons investigated, males and females displayed more or less the same proportions of neurons responding to the different compounds. Studies of ORN responses to plant- and flower-related compounds in other species converge on the conclusion that both broad and narrow tuning occur.[19–21,26,27] As several different types of ORNs can be activated to a different degree by the same compound, the identity of the stimulus is likely contained in an "across-neuron" pattern.

The often broader response spectrum, compared to the pheromone-detecting ORNs, of neurons tuned to plant-related compounds does not necessarily mean that the neurons are "sloppy." Rather, the neurons are likely highly selective to molecular features shared by several compounds.[28] Due to the limited number of receptor proteins, an insect with exclusively specific receptors (i.e., tuned to a single compound like the pheromone receptors) will have a low coding capacity. On the other hand, such specific receptors may be beneficial in order to recognize stimuli that are particularly important and highly predictable over time (e.g., sexual pheromones). However, important odorous sources are rarely predictable. Even the composition of volatiles emitted from the same flower changes over time.[29] Therefore it should be beneficial to have receptors with broader tuning. Broad and overlapping tuning allows detection and discrimination of an almost infinite number of potential odors. As a comparison, it has been estimated that humans possess about 300 different functional receptor proteins.[30] Still, we can recognize more than 400,000 different odorous molecules.[31]

Raguso and Willis[32] proposed that nectar-feeding insects used carbon dioxide (CO_2) as an additional cue to locate nectar sources. It was recently demonstrated that the CO_2 level was correlated with the secretion of nectar in the flower *D. wrightii*.[33] Thus CO_2 may act as an indicator of food abundance to the insect. The unique structure of the CO_2 suggests that information regarding this compound uses a specialized pathway. Indeed, ORNs tuned to CO_2 in moths are not located on the antenna, but in the labial palp pit organ (Figure 11.1a).[34] This organ houses more than 2000 ORNs in *M. sexta*[34] and responds specifically to CO_2.[35,36] Thus reception of CO_2 resembles that of sexual pheromones by traveling along a labeled neural line.

As different insects and other organisms seem to have highly divergent genes coding for receptor proteins,[11,37] the ability to detect odor molecules likely differs. We expect that evolutionary pressures led to the development of receptors adapted to the requirement of the animal. As a consequence, different animals have different perceptions of the odor world.

In summary, a number of plant odor tuned ORNs are as selective as the pheromone-detecting neurons. Other populations of ORNs are less specific and overlap with similarly tuned ORNs. Such a mechanism may be the best of two worlds. High selectivity enables the insect to be prepared for especially important and predictable stimuli. Broad and overlapping tuning, on the other hand, increases the coding capacity immensely and prepares the animal for an unpredictable and ever-changing odor world.

11.3   THE AL: ORGANIZATION, PROCESSING, AND ENCODING

When a floral scent plume hits the antenna, a large number of differentially tuned and overlapping populations of ORNs are activated. The different types of ORNs are more or less randomly dispersed across the antenna. So how does the brain "know" which receptors have been activated? Moreover, how does the brain encode concentration differences and volatiles presented in complex mixtures? And how does experience affect the encoding scheme?

11.3.1   THE UBIQUITOUS GLOMERULUS

The primary center for olfactory processing in insects is the AL, which is the anatomical and functional analogue of the vertebrate olfactory bulb (Figure 11.2a). The AL consists of spheroidal neuropilar structures—glomeruli—where synaptic contacts between ORN axons and AL interneurons take place. In most insect species, the AL contains 40 to 200 glomeruli arranged in one or two layers around a central fibrous core. Within a glomerulus, the axons of ORNs synapse with higher order neurons, which are first, the output neurons, the projection neurons (PNs) (Figure 11.2b). Most PNs have dendritic uniglomerular arborizations and the axonal projections leave the AL through several different antennocerebral tracts.[38] In addition, there are several different morphological types of amacrine cells, called local interneurons

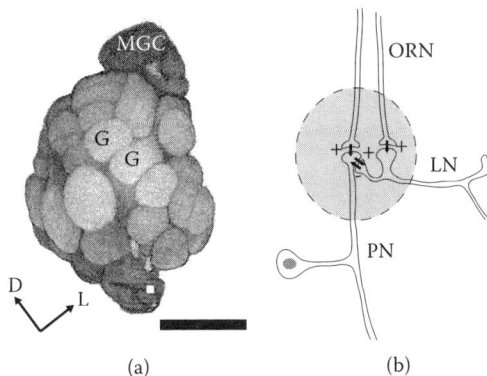

(a)   (b)

FIGURE 11.2 (a) A surface reconstruction of a male AL of the moth *S. littoralis*. The brain was immunostained with synapsin antibody and optically sectioned using a confocal microscope. Stacks of images were integrated with the software Imaris 2.7 (Bitplane AG, Switzerland) on a Silicon Graphics workstation to obtain surface projections of the lobe. The macroglomerular complex (MGC) is located close to the entrance of the antennal nerve. All other glomeruli (G) are sexually isomorphic. L, lateral; D, dorsal; scale bar, 100 μm. (modified from Carlsson et al.[62]). (b) Synaptic organization of the major types of AL neurons. Sensory neurons (ORNs) make uniglomerular synapses both directly with projection neurons (PNs) and indirectly via local interneurons (LNs). In addition, LNs innervate several glomeruli and generally make inhibitory synapses. Cell bodies of PNs and LNs are located within the AL. The arrows show the synaptic pathways. –, inhibitory synapse; +, excitatory synapse.

(LNs) (Figure 11.2b). The LNs convey information from the ORNs to the PNs.[39–43] They also make reciprocal synapses with PNs and feedback contact with ORNs. A major part of the LNs have been found to be GABAnergic (gamma-aminobutyric acid).[44,45] Thus the LNs have been proposed to have a role in an inhibitory, possibly contrast-enhancing network.

In several insect species there is a sexual dimorphic organization, in that a specific cluster of glomeruli in one of the sexes is devoted to processing of information about sexual pheromones (e.g., the male-specific macroglomerular complex [MGC] in moths). However, the major part of the glomerular population in nectar-foraging insects is sexually isomorphic and receives input from ORNs tuned to plant-associated compounds. There is a growing body of evidence that glomeruli act as functional units in insects. For example, a strict topology of physiologically characterized input as well as output neurons from the MGC has been demonstrated in several insect species.[46–53] Another piece of evidence comes from tracing ORNs expressing specific gene products in fruit flies. Receptor neurons expressing a specific putative receptor protein send axons to the same one or two glomeruli in *D. melanogaster*.[54,55] These target glomeruli have stereotyped locations across individuals. Thus there appears to be a highly ordered targeting of ORNs onto the glomeruli. However, physiological evidence for a glomerular organization of food odor-detecting ORNs with similar tuning came with the development of optical imaging techniques in insects.[56,57] In most of these experiments, a Ca^{2+}-sensitive dye was applied to the brain. This means a nonselective staining of AL neurons, but gives a reasonable estimation of input signals to the AL.[58]

11.3.2 Organization of Input to the AL

11.3.2.1 How Is the Identity of Flower Molecules Represented in the AL?

A number of Ca^{2+} imaging studies in honeybees and moths have shown that flower and other plant compounds activate specific combinations of glomeruli.[56,59–63] The spatial distribution of the glomerular patterns evoked by these odorants depends on the stimulus and is similar across animals. A typical flower molecule activates several glomeruli and a specific glomerulus is activated by several compounds (Figure 11.3). However, each combination of responding glomeruli is unique. For example, in *M. sexta*, a panel of terpenes and aromatic flower compounds were tested.[63] These compounds evoked patterns of clustered activated glomeruli. A medial cluster was activated by aromatic compounds and a lateral cluster by the terpenes. The clusters were more or less nonoverlapping. However, each single compound still evoked a unique activity pattern within the respective cluster. These results suggest a hierarchical organization with a rough classification according to primary structures (e.g., cyclic and noncyclic structure) and a fine-scale division within each cluster. Similar organizations were found in males and females. The results further confirm the study by Shields and Hildebrand,[23] who showed specific responses from receptor neurons to a number of terpene and aromatic compounds. The responses to the two groups were more or less exclusive, so that one group of neurons responded to aromatics

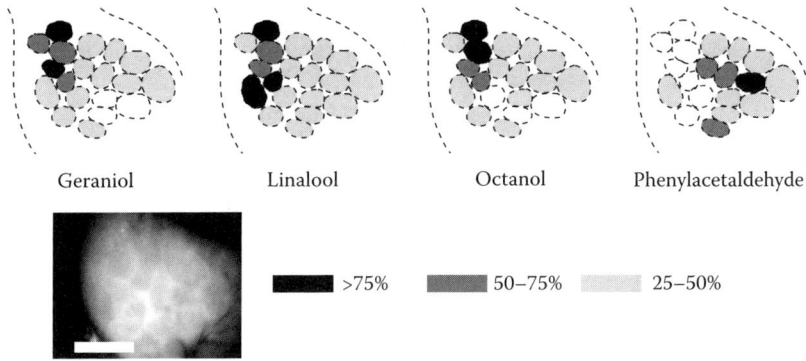

FIGURE 11.3 Glomerular Ca^{2+} responses to different plant-related compounds in a female *S. littoralis*. The glomerular borderlines were obtained from an anatomical poststaining (below left). Each glomerulus was colored according to the optically recorded odor-evoked changes in light intensity. For each odor, the strongest responding glomerulus was set to 100% and the rest scaled accordingly. Scale bar, 100 μm.

and one to terpenes. Also, in the moths *Spodoptera littoralis* and *Helicoverpa zea*, there appears to be a similar spatial separation of glomerular clusters of glomeruli responding preferentially to either terpenes or aromatics.[62,64] A single study showing the projections of physiologically characterized ORNs tuned to floral compounds was conducted in the moth *Trichoplusia ni*.[65] Here was shown a longitudinal partitioning of axonal projections, in that ORNs responding to aromatic flower volatiles terminated deeper in the lobe (i.e., in the medial region) than neurons responding to the monoterpene linalool. Thus there seems to be a coarse spatial division of labor in the ALs that is conserved across phylogenetically diverse groups of moths.

In both honeybees and moths, more systematic structure response studies have been performed.[60,66] Series of aliphatic compounds were used and each molecule in a series differed by a single carbon atom. Thus, within a series, the structural difference was reduced to a single dimension. In both insects it was demonstrated that individual glomeruli were activated by a range of stimuli, but showed optimal responses to one or a few neighboring chain lengths, irrespective of functional group (Figure 11.4). Different glomeruli had different optimal chain length preferences, but overlapped with each other. The correlation of response patterns decreased as the distance in chain length between compounds increased (i.e., the more structural similarity, the more similar the response patterns). Behavioral experiments have shown that *M. sexta* can discriminate even structurally similar compounds.[67] However, the ability to discriminate is a function of the structural similarity of the odorants (e.g., the chain length). These results support the idea that the spatial patterns may be part of a code that could be parsed by downstream brain regions. Furthermore, glomeruli that demonstrated similar response profiles in Ca^{2+} -imaging were often located close to each other (Figure 11.4).[60,66] If such architecture of neighboring glomeruli with similar response profiles reflects an underlying physiological requirement (e.g., facilitation of mechanisms such as lateral inhibition) is not known.

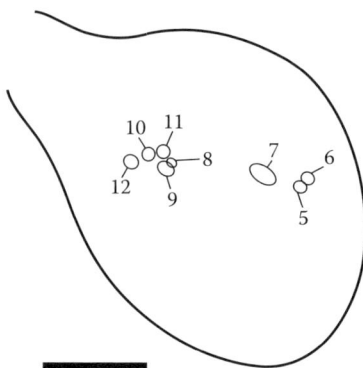

FIGURE 11.4 Structure dependency of AL representations measured with Ca^{2+} imaging. A standardized AL shows the foci of highest activity for a series of homologous aldehydes (C5–C12) in females of the moth *S. littoralis*. Due to differing vapor pressures of the tested compounds, different absolute concentrations were used. The coordinates are averaged across eight animals and corrections were made for differences in AL size. The ovals represent averaged coordinates ± SEM. Scale bar, 100 μm.

From the labial palp pit organ (Figure 11.1a), where ORNs are selectively tuned to CO_2, axons innervate a uniquely identified glomerulus in the medial region of the AL of several moth species.[34,35,68] Thus the CO_2 innervation pattern resembles that of the pheromone-tuned ORNs, in that a single type of ORN is activated, and as a consequence also a single glomerulus.

In summary, all ORNs belonging to a specific type converge on a specific glomerulus. As most flower volatiles activate a specific combination of ORNs, each compound evokes a unique pattern of activated glomeruli. These patterns are genetically determined, and ORNs expressing receptors with overlapping tuning appear to innervate neighboring glomeruli. The unique molecule of CO_2, however, constitutes an exception, in that it activates a specific type of ORN, innervating a single glomerulus.

11.3.2.2 How Is the Abundance of Flower Molecules Represented in the AL?

Nectar-foraging insects are likely exposed to a very wide range of odor concentrations. Small amounts of flower molecules have to be recognized when trying to locate a source from a long distance. On the other hand, while actually feeding, surrounded by the petals, the insect should be exposed to an enormous abundance of volatiles. This fact puts extra pressure on the olfactory system to correctly recognize the identity of the stimulus. A concentration-invariant code would solve the problem. On the other hand, the insect should also be able to discriminate between concentrations. The olfactory system must not only be able to detect and process information about the quality of a stimulus, but also accurately code changes in concentration. At the peripheral level, insect ORNs can recognize differences in stimulus concentration over several orders of magnitude.[69]

In honeybee ALs, Ca^{2+} activity patterns evoked at high and low odor concentrations showed nearly identical spatial distribution, though with a difference in signal intensity.[56] Furthermore, the receptive range of glomeruli was not dependent on the stimulus concentration over a range of two decadic steps.[60] Using a wider range of concentrations, Carlsson and Hansson[70] demonstrated that the number of activated glomeruli in *S. littoralis* increased with odor concentration. This was likely due to a recruitment of ORNs with lower affinity. However, intensity did not increase uniformly in all glomeruli and saturation level was reached at different concentrations. As a consequence, the qualitative patterns changed with concentration. This is illustrated in Figure 11.5, which shows that the focus of activity was not located in the same glomerulus at different concentrations. The fact that this movement of activity foci is generally restricted to adjacent glomeruli further indicates that ORNs expressing related receptors target proximal glomeruli. The correlation between odor maps evoked by different compounds was calculated and it was shown that structurally similar compounds evoked more correlated patterns at high than at low concentrations, possibly because these compounds utilized the same input channels. If a spatial olfactory map plays a role in the recognition and discrimination of odors, then we would expect concentration-dependent variations of odor maps to have an impact on the perceived odor quality. Stimulus concentration can indeed influence the perception of odor quality and behavior in insects. For example, olfactory responses of the fruit fly shifted from attraction to repulsion as the concentration increased.[71,72]

11.3.3 PROCESSING AND CODING OF FLOWER ODORS IN THE AL

The input information is thus organized into the glomeruli so that each glomerulus represents activity from a specific population of ORNs. But do downstream brain

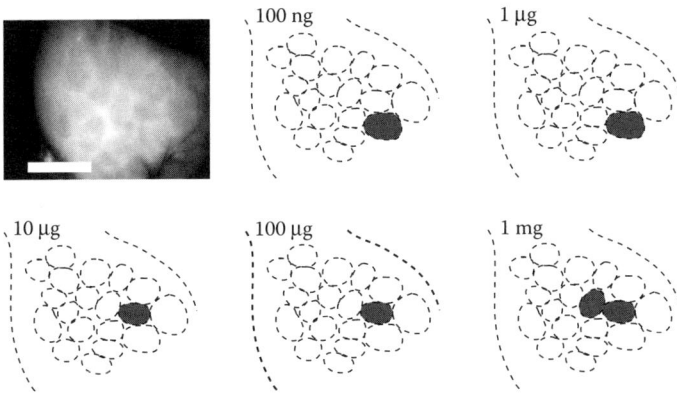

FIGURE 11.5 The effect of concentration on glomerular Ca^{2+} responses in the moth *S. littoralis*. Responses to a concentration series of the aromatic flower compound phenylacetaldehyde. Only the principal glomeruli (the most strongly responding) are shaded, but these are not identical across concentrations. Movement of the highest activity takes place between neighboring glomeruli. At the highest concentration, two different glomeruli are almost equally activated. The AL image shows the stained glomeruli from which the outlines have been constructed. Scale bar, 100 μm.

regions use the spatial maps of responding glomeruli to analyze the identity and abundance of odors? A spatial arrangement is very likely an important part of the code, but in addition there are temporal features carrying odor-specific information. The AL network with its amacrine cells proposes that the information flow from ORNs to PNs is not simply isomorphic. In fact, responses in single PNs are often temporally complex with multiphasic patterns. These slow temporal patterns are not driven directly by afferent input, as ORNs normally spike in a rather noncomplex fashion (Figure 11.1c). The complex PN patterns are both odor and neuron specific.[73] Despite the odor specificity, the patterns may not constitute an olfactory code per se.[74] Rather, slow temporal patterns contribute to optimize the code over time. This means that the ability to discriminate between odors may be time dependent (see below).

In addition to slow temporal patterns, it has long been known that odors can evoke oscillations in different brain areas.[75,76] Odor-evoked oscillations have also been observed in a number of insects by measuring field potentials.[73] The frequency of the oscillations is species, but not odor specific. Regular oscillations result from large populations of AL interneurons spiking synchronously. Coincident spiking has not been observed among ORNs, which means that afferent input cannot be the driving force behind the PN synchrony. Instead, the reciprocal coupling of LNs and PNs gives rise to subthreshold oscillations in the PNs.[77] As a consequence, action potentials often appear periodically. During a cycle of an oscillation, a number of PNs spike in synchrony.[78,79] Other neurons may spike as well, but are not phase locked to any cycles. The population of neurons acting in synchrony depends on the stimulus identity and concentration. Thus, a different odor evokes synchronous activity in another (but often overlapping) subset of PNs. Coincident spiking activity is abolished by application of picrotoxin (PTX), which selectively blocks GABA-mediated Cl⁻ channels.[80,81] However, PTX treatment had no effect on the slow temporal patterns, which suggests at least two independent inhibitory networks.

Calcium imaging of a selected population of PNs in the honeybee showed that the output patterns were significantly more restricted than the input patterns.[82] This sharpening is likely mediated by a mechanism that resembles the contrast enhancement mediated by lateral inhibition in the visual system.[83] Moreover, Sachse and Galizia[82] provided evidence for a glomerulus-specific network that selectively inhibited glomeruli with overlapping response profiles. This network was insensitive to PTX treatment and may thus be mediated by transmitters other than GABA. In another study using PN selective imaging, Carlsson et al.[84] showed that glomerular activity patterns in *S. littoralis* systematically changed with time (Figure 11.6), which was most likely caused by different slow temporal patterns in the individual PNs. Moreover, Carlsson et al.[84] showed that responses evoked by structurally similar flower compounds were initially highly correlated. However, during a sustained period of exposure (1 second), the patterns became successively less similar. Also, responses to dissimilar compounds decorrelated over time. Similarly Stopfer et al.[85] showed that the classification performance of a population of PNs in the locust was optimized 200 to 300 msec after the onset of the response. Furthermore, the classification success of concentrations also increased with time. Behavioral support for a temporal optimization of the discrimination ability was recently demonstrated in the honeybee.[86] The honeybees could easily discriminate between structurally

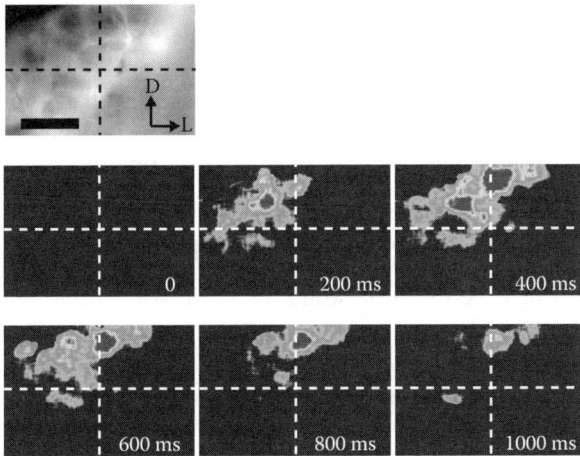

FIGURE 11.6 The spatial activity pattern evoked by an odor (linalool 50 µg) is dynamic over time. Here a selected population of PNs in a male *S. littoralis* are stained with a Ca^{2+}-sensitive dye. Every second false color-coded image during the period of stimulation is shown. The focus of activity is moving from a glomerulus in the medial region to a glomerulus more laterally located. The image of the AL is from the same recordings. D, dorsal; L, lateral. Scale bar, 100 µm. Time after stimulus onset is indicated.

similar compounds. Correct discrimination, however, required about 700 msec and was independent of the similarity of the stimuli.

Due to the limited temporal resolution of calcium imaging, it cannot be used to study fast temporal events like synchronized activity. Instead, a different methodological approach was adapted to the moth.[87] Multiprobe electrodes were used, which allowed recordings from up to 16 AL neurons simultaneously. Using this method in *M. sexta*, Lei et al.[88] found that it was difficult to discriminate between, for example, the two structurally similar floral molecules, nerol and linalool, simply by comparing spatial patterns of responding PNs. However, the two compounds showed very different patterns of coincident spiking. Thus synchronous spiking appears to be involved in fine odor discrimination. This was indeed demonstrated by Stopfer et al.[89] Using the proboscis extension reflex paradigm,[90,91] bees were trained to respond to one odor and were then tested with the same odor, a novel similar odor, and a novel dissimilar odor. The animals received either an application of PTX, which abolished the odor-evoked oscillations, or saline prior to conditioning. Both groups learned the conditioned stimulus equally well and had no problems discriminating this from a structurally different odor. However, the group of bees that was treated with PTX was unable to discriminate between the learned odor and a structurally similar odor.

In addition to the discussed coding mechanisms, the AL organization likely provides a mechanism to increase the signal:noise ratio of the afferent signals due to the large number of ORNs synapsing on each PN. This convergence may allow a spatial summation mechanism. The sensitivity in AL interneurons has been reported to be several orders of magnitude larger than the corresponding ORNs.[92]

In conclusion, the olfactory code appears to consist of spatial as well as temporal ingredients. The spatial organization of PNs of different tuning may be sufficient for a crude classification of odors. However, adding the feature of synchronously spiking PNs enables the animal to distinguish between similar stimuli. Moreover, the code is dynamic and the slow temporal patterning optimizes this code and increases the capacity to recognize and discriminate odors with time.

11.3.4 THE OLFACTORY CODE MAY BE MODIFIED BY EXPERIENCE

Since the pioneering experiments by von Frisch,[93] it is well established that honeybees and other insects learn their odor cues from visited flowers. Learning enables the insect to more efficiently utilize resources. The fact that a sensory-induced behavior changes can have many different ultimate explanations. All different neural levels in the sensory systems are theoretically open to be influenced by neuroactive processes. In early experiments in the honeybee, it was shown that short-term memory was affected when the AL was subjected to cooling, narcosis, or electrical stimulation.[94,95] In addition, local injection of octopamine in the AL, as a substitute for an unconditioned stimulus, established a stable memory of the coupled odor.[96] Long-term morphological changes in the AL have been observed in flies and honeybees. Sigg et al.[97] demonstrated that the volume of a certain glomerulus increased due to a shift in foraging behavior in the honeybee. The increased size was due to an increased density of synapses.[98] In contrast, the volume of specific glomeruli in the AL of *D. melanogaster* was reduced after long-term odor exposure.[99] The reductions in glomerular volume were odor specific, in that different odors affected the size of different glomeruli. Thus the AL anatomy may change continuously during the lives of insects due to different experiences.

Honeybees can easily be trained to respond to odorants using the proboscis extension reflex paradigm.[90,91] By comparing odor representations in the ALs before and after such training, Faber et al.[100] showed, using Ca^{2+} imaging, that the response to a rewarded odor, but not to an unrewarded odor, significantly increased in activated glomeruli. Furthermore, the correlation between odor representations of a rewarded and an unrewarded odor decreased after conditioning. Thus it seems that the ALs work alongside the mushroom bodies and other higher brain areas as centers for odor memory formation, either as primary learning centers or secondary by feedback from higher brain centers.

Sensitization of honeybees by sucrose stimulation led to an overall increase of Ca^{2+} signals in PNs.[101] The PNs were selectively labeled with a calcium-sensitive dye and the response in activated glomeruli was significantly stronger than in control animals. Hence neural correlates of both associative and nonassociative processes are formed already at the level of the AL. A sensitized animal will be in a state more ready to pay attention to environmental stimuli, thereby optimizing, for example, nectar search.

Hence, in addition to higher brain regions like the mushroom bodies, the AL appears to be a site for learning mechanisms and memory formation. Both anatomical and physiological modifications are elicited by different experiences. This means that "olfactory code" is dynamic and should differ between animals living in different chemical environments.

11.3.5   How Are Multimolecular Stimuli Encoded?

Imagine a flowering meadow: How can a nectar-foraging insect distinguish the smell of a specific flower, which may consist of hundreds of odorous volatiles, intermingled among hundreds of other odor-emitting flower species? Under natural conditions, odors rarely occur as single compounds. Instead, odors are often mixtures of hundreds of components. Whereas the mechanisms behind detection, coding, and discrimination of monomolecular volatiles are well investigated, in only a few studies have blends of compounds been tested. A mixture may be represented in the ALs either as an addition of the individual components or as a novel configurational representation. A configurational representation is the result if the components of a mixture interact with each other in an unpredictable manner. In honeybees, Ca^{2+} imaging responses to binary mixtures could not always be predicted based on responses to the individual components.[56,102] Activity in individual glomeruli in response to a mixture either equaled the sum of the components or was weaker than predicted. The power of the mixture suppressions increased when adding components to the blend.[56] Mixture interactions were both odor and glomerulus specific. It is likely that suppressed glomerular activity is the result of interglomerular inhibition mediated by LNs or by inhibitory feedback to the ORN terminals. However, it cannot be excluded that odor molecules interact with each other at the receptor site. A few research groups have reported peripheral mixture interactions in insects as either suppression or synergistic effects.[103–107]

Antennal lobe neurons sensitive to specific blends of floral compounds have not been reported. However, such neurons are common in the MGC[108] and may also exist for interneurons tuned to nonpheromonal compounds.

Future studies will very likely focus much more on how the olfactory system recognizes and discriminates odor blends. Moreover, the question of how the olfactory system filters out important olfactory information from a background of odor noise should be addressed. It is indeed amazing how a complex multimolecular blend like that emitted from a specific flower can be distinguished from a molecular soup of volatiles emitted from flowers with overlapping chemical compositions.

11.4   CONCLUSION

Floral scent is composed of a wide variety of molecules. These compounds are detected by ORNs mainly located on the antennae. Each type of receptor cell responds to several structurally similar compounds and each of these compounds activates several types of receptor cells. Since many insects obviously can behaviorally discriminate between these compounds, a combinatorial, across-neuron mechanism is required. Likely the ORNs are tuned to a molecular feature shared by several different compounds and each compound possesses several of these features and thus activates different receptors. All ORNs expressing the same receptor protein converge on the same glomerulus in the AL. As a consequence, a pattern of activated glomeruli is formed. The identity of floral compounds is represented as unique combinations of activated glomeruli. These activity patterns depend on the odor identity, the odor abundance, and on previous experience. In addition, molecules in a blend are generally not linearly represented. The input signals are then processed and modified by collateral inhibitory LNs. As a consequence, the output signals in the PNs are temporally more complex than the input signals. Patterns of coincidently spiking across the AL clearly carry odor- and concentration-specific information. Slow temporal patterns, on the other hand, dynamically optimize this code over time. A fundamental question is of course how this output information is parsed by downstream brain regions and how a meaningful image of the odor world is created. And finally, how do these higher brain regions incorporate information from other sensory modalities? Clearly odors do not work alone as floral attractors of insects, but rather they work in conjunction with other (e.g., visual) cues.[32,109]

REFERENCES

1. Keil, T.A., Morphology and development of the peripheral olfactory organs, in *Insect Olfaction*, Hansson, B.S., Ed., Springer, Berlin, 1999, p. 5.
2. Steinbrecht, R.A., Pore structures in insect olfactory sensilla: a review of data and concepts, *Int. J. Insect Morphol. Embryol.* 26, 229, 1997.
3. Vogt, R.G. and Riddiford, L.M., Pheromone binding and inactivation by moth antennae, *Nature* 193, 161, 1981.
4. Buck, L. and Axel, R., A novel multigene family may encode odorant receptors: a molecular basis for odor recognition, *Cell* 65, 175, 1991.
5. Clyne, P.J., Warr, C.G., Freeman, M.R., Lessing, D., Kim, J., and Carlson, J.R., A novel family of divergent seven-transmembrane proteins: candidate odorant receptors in *Drosophila*, *Neuron* 22, 327, 1999.
6. Vosshall, L.B., Amrein, H., Morozov, P.S., Rzhetsky, A., and Axel, R., A spatial map of olfactory receptor expression in the *Drosophila* antenna, *Cell* 96, 725, 1999.
7. Hill, C.A., Fox, A.N., Pitts, R.J., Kent, L.B., Tan, P.L., Chrystal, M.A., Cravchik, A., Collins, F.H., Robertson, H.M., and Zwiebel, L.J., G protein-coupled receptors in *Anopheles gambiae*, *Science* 298, 176, 2002.
8. Krieger, J., Raming, K., Dewer, Y.M., Bette, S., Conzelmann, S., and Breer, H., A divergent gene family encoding candidate olfactory receptors of the moth *Heliothis virescens*, *Eur. J. Neurosci.* 16, 619, 2002.
9. Zhang, X. and Firestein, S., The olfactory receptor gene superfamily of the mouse, *Nat. Neurosci.* 5, 124, 2002.
10. Vosshall, L.B., The molecular logic of olfaction in *Drosophila*, *Chem. Senses* 26, 207, 2001.
11. Mombaerts, P., Seven transmembrane proteins as odorant and chemosensory receptors, *Science* 286, 707, 1999.
12. Dobritsa, A.A., van der Goes van Naters, W., Warr, C.G., Steinbrecht, R.A., and Carlson, J.R., Integrating the molecular and cellular basis of odor coding in the *Drosophila* antenna, *Neuron* 6, 827, 2003.
13. Goldman, A.L., van der Goes van Naters, W., Lessing, D., Warr, C.G., and Carlson, J.R., Coexpression of two functional odor receptors in one neuron, *Neuron* 45, 661, 2005.

14. Stengl, M. et al. Perireceptor events and transduction mechanisms in insect olfaction, in *Insect Olfaction*, Hansson, B.S., Ed., Springer, Berlin, 1999, p. 49.
15. Schneider, D., Elektrophysiologische Untersuchungen von Chemo- und Mechanorezeptoren der Antenne des Seidenspinners *Bombyx mori* L., *Z. Vergl. Physiol.* 40, 8, 1957.
16. Liljefors, T., Thelin, B., and van der Pers, J.N.C., Structure-activity relationships between stimulus molecules and response of a pheromone receptor in turnip moth, *Agrotis segetum*: modifications of the acetate group, *J. Chem. Ecol.* 10, 1661, 1984.
17. Liljefors, T., Thelin, B., van der Pers, J.N.C., and Löfstedt, C., Chain-elongated analogues of a pheromone component of the turnip moth, *Agrotis segetum*. A structure-activity study using molecular mechanisms, *J. Chem. Soc. Perkin. Trans.* II, 1957, 1985.
18. Liljefors, T., Bengtsson, M., and Hansson, B.S., Effects of double-bond configuration on interaction between a moth sex pheromone component and its receptor. A receptor-interaction model based on molecular mechanisms, *J. Chem. Ecol.* 13, 2023, 1987.
19. Hansson, B.S., Larsson, M.C., and Leal, W.S., Green leaf volatile-detecting olfactory receptor neurons display very high sensitivity and specificity in a scarab beetle, *Physiol. Entomol.* 24, 121, 1999.
20. Larsson, M.C., Leal, W.S., and Hansson, B.S., Olfactory receptor neurons detecting plant odours and male volatiles in *Anomala cuprea* beetles (Coleoptera: Scarabidae). *J. Insect Physiol.* 47, 1065, 2001.
21. Stensmyr, M.C., Larsson, M.C., Bice, S.B., and Hansson, B.S., Detection of fruit- and flower-emitted volatiles by olfactory receptor neurons in the polyphagous fruit chafer *Pachnoda marginata* (Coleoptera: Cetoniinae), *J. Comp. Physiol. A* 187, 509, 2001.
22. Röstelien, T., Borg-Karlson, A.-K., Faldt, J., Jacobsson, U., and Mustaparta, H., The plant sesquiterpene germacrene D specifically activates a major type of antennal receptor neuron of the tobacco budworm moth *Heliothis virescens*, *Chem. Senses* 25, 141, 2000.
23. Shields, V.D. and Hildebrand, J.C., Responses of a population of antennal olfactory receptor cells in the female moth *Manduca sexta* to plant associated volatile organic compounds, *J. Comp. Physiol. A* 186, 1135, 2001.
24. Raguso, R.A. and Willis, M.A., Floral scent and its role(s) in hawkmoth attraction, *Chem. Senses* 22, 774, 1997.
25. Kalinová, B., Hoskovec, M., Liblikas, I., Unelius, C.R., and Hansson, B.S., Detection of sex pheromone components in *Manduca sexta* (L.), *Chem. Senses* 26, 1175, 2001.
26. Anderson, P., Hansson, B.S., and Löfqvist, J., Plant-odour-specific receptor neurons on the antennae of the female and male *Spodoptera littoralis*, *Physiol. Entomol.* 20, 189, 1995.
27. Jönsson, M. and Anderson, P., Electrophysiological response to herbivore-induced host plant volatiles in the moth *Spodoptera littoralis*, *Physiol. Entomol.* 24, 377, 1999.
28. Araneda, R.C., Kini, A.D., and Firestein, S., The molecular receptive range of an odorant receptor, *Nat. Neurosci.* 3, 1248, 2000.
29. Kaiser, R., *The Scent of Orchids*, Elsevier Science, Amsterdam, 1993.
30. Mombaerts, P., The human repertoire of odorant receptor genes and pseudogenes, *Annu. Rev. Genom. Hum. Genet.* 2, 493, 2001.
31. Mori, K., Grouping of odorant receptors: odour maps in the mammalian olfactory bulb, *Biochem. Soc. Trans.* 31, 134, 2003.
32. Raguso, R.A. and Willis, M.A., Synergy between visual and olfactory cues in nectar feeding by naïve hawkmoths, *Manduca sexta*, *Anim. Behav.* 63, 685, 2002.

33. Guerenstein, P.G., Yepez, A., Van Haren, J., Williams, D.G., and Hildebrand, J.G., Floral CO_2 emission may indicate food abundance to nectar-feeding moths, *Naturwissenschaften* 91, 329, 2004.

34. Kent, K.S., Harrow, I.D., Quartararo, P., and Hildebrand, J.G., An accessory olfactory pathway in Lepidoptera: the labial pit organ and its central projections in *Manduca sexta* and certain other sphinx moths and silk moths, *Cell Tissue Res.* 245, 237, 1986.

35. Bogner, F., Boppré, M., Ernst, K.D., and Boeckh, J., CO_2 sensitive receptors on labial palps of *Rhodogastria* moths (Lepidoptera: Arctiidae): physiology, fine structure and central projection, *J. Comp. Physiol. A* 158, 741, 1986.

36. Guerenstein, P.G., Christensen, T.A., and Hildebrand, J.G., Sensory processing of environmental CO_2 information in the moth nervous system, *Chem. Senses* 27, 661, 2002.

37. Krieger, J., Klink, O., Mohl, C., Raming, K., and Breer, H., A candidate olfactory receptor subtype highly conserved across different insect orders, *J. Comp. Physiol. A* 189, 519, 2003.

38. Anton, S. and Homberg, U., Antennal lobe structure, in *Insect Olfaction*, Hansson, B.S., Ed., Springer, Berlin, 1999, p. 97.

39. Boeckh, J. and Tolbert, L.P., Synaptic organization and development of the antennal lobe in insects, *Microsc. Res. Tech.* 24, 260, 1993.

40. Distler, P.G. and Boeckh, J., Synaptic connection between olfactory receptor cells and uniglomerular projection neurons in the antennal lobe of the American cockroach, *Periplaneta americana*, *J. Comp. Neurol.* 370, 35, 1996.

41. Distler, P.G. and Boeckh, J., Synaptic connections between identified neuron types in the antennal lobe glomeruli of the cockroach, *Periplaneta americana.* II. Local multiglomerular interneurons, *J. Comp. Neurol.* 383, 529, 1997.

42. Malun, D., Synaptic relationships between GABA-immunoreactive neurons and an identified uniglomerular projection neuron in the antennal lobe of *Periplaneta americana*: a double-labeling electron microscopic study, *Histochemistry* 96, 197, 1991.

43. Sun, X.J., Tolbert, L.P., and Hildebrand, J.G., Synaptic organisation of the uniglomerular projection neurons of the antennal lobe of the moth *Manduca sexta*: a laser scanning confocal and electron microscopic study, *J. Comp. Neurol.* 379, 2, 1997.

44. Hoskins, S.G., Homberg, U., Kingan, T.G., Christensen, T.A., and Hildebrand, J.G., Immunocytochemistry of GABA in the antennal lobes of the sphinx moth *Manduca sexta*, *Cell Tissue Res.* 244, 243, 1986.

45. Homberg, U., Christensen, T.A., and Hildebrand, J.G., Structure and function of the deutocerebrum in insects, *Annu. Rev. Entomol.* 34, 477, 1989.

46. Hansson, B.S., Christensen, T.A., and Hildebrand, J.G., Functionally distinct subdivisions of the macroglomerular complex in the antennal lobe of the male sphinx moth *Manduca sexta*, *J. Comp. Neurol.* 312, 264, 1991.

47. Hansson, B.S., Ljungberg, H., Hallberg, E., and Löfstedt, C., Functional specialization of olfactory glomeruli in a moth, *Science* 256, 1313, 1992.

48. Hansson, B.S., Almaas, T.J., and Anton, S., Chemical communication in heliothine moths. V. Antennal lobe projection patterns of pheromone-detecting olfactory receptor neurons in the male *Heliothis virescens* (Lepidoptera: Noctuidae), *J. Comp. Physiol. A* 177, 535, 1995.

49. Anton, S. and Hansson, B.S., Sex pheromone and plant-associated odour processing in antennal lobe interneurons of male *Spodoptera littoralis* (Lepidoptera: Noctuidae), *J. Comp. Physiol. A* 176, 773, 1995.

50. Anton, S., Löfstedt, C., and Hansson, B.S., Central nervous processing of sex pheromones in two strains of the European corn borer, *Ostrinia nubilalis* (Lepidoptera: Pyralidae), *Exp. Biol.* 200, 1073, 1997.

51. Ochieng, S.A., Anderson, P., and Hansson, B.S., Antennal lobe projection patterns of olfactory receptor neurons involved in sex pheromone detection in *Spodoptera littoralis* (Lepidoptera: Noctuidae), *Tissue Cell* 27, 221, 1995.

52. Vickers, N.J., Christensen, T.A., and Hildebrand, J.G., Combinatorial odor discrimination in the brain: attractive and antagonist odor blends are represented in distinct combinations of uniquely identifiable glomeruli, *J. Comp. Neurol.* 400, 35, 1998.

53. Berg, B, G., Almaas, T.J., Bjaalie, J., and Mustaparta, H., The macroglomerular complex of the antennal lobe in the tobacco budworm moth *Heliothis virescens*: specified subdivision in four compartments according to information about biologically significant compounds, *J. Comp. Physiol. A* 183, 669, 1998.

54. Gao, Q., Yuan, B., and Chess, A., Convergent projections of *Drosophila* olfactory neurons to specific glomeruli in the antennal lobe, *Nat. Neurosci.* 3, 780, 2000.

55. Vosshall, L.B., Wong, A.M., and Axel, R., An olfactory sensory map in the fly brain, *Cell* 102, 147, 2000.

56. Joerges, J., Küttner, A., Galizia, C.G., and Menzel, R., Representations of odours and odour mixtures visualized in the honeybee brain, *Nature* 387, 285, 1997.

57. Galizia, G.C., Joerges, J., Kuettner, A., Faber, T., and Menzel, R., A semi-in-vivo preparation for optical recording of the insect brain, *J. Neurosci. Methods* 76, 61, 1997.

58. Galizia, C.G., Nägler, K., Hölldobler, B., and Menzel, R., Odour coding is bilaterally symmetrical in the antennal lobes of honeybees (*Apis mellifera*), *Eur. J. Neurosci.* 10, 2964, 1998.

59. Galizia, C.G., Sachse, S., Rappert, A., and Menzel, R., The glomerular code for odor representation is species specific in the honeybee *Apis mellifera*, *Nat. Neurosci.* 2, 473, 1999.

60. Sachse, S., Rappert, A., and Galizia, C.G., The spatial representation of chemical structures in the antennal lobe of honeybees: steps toward the olfactory code, *Eur. J. Neurosci.* 11, 3970, 1999.

61. Galizia, C.G., Sachse, S., and Mustaparta, H., Calcium responses to pheromones and plant odours in the antennal lobe of the male and female moth *Heliothis virescens*, *J. Comp. Physiol. A* 186, 1049, 2000.

62. Carlsson, M.A., Galizia, C.G., and Hansson, B.S., Spatial representation of odours in the antennal lobe of the moth *Spodoptera littoralis* (Lepidoptera: Noctuidae), *Chem. Senses* 27, 231, 2002.

63. Hansson, B.S., Carlsson, M.A., and Kalinová, B., Olfactory activation patterns in the antennal lobe of the sphinx moth, *Manduca sexta*, *J. Comp. Physiol. A* 189, 301, 2003.

64. Carlsson, M.A., Baker, T.C., and Hansson, B.S., unpublished observation.

65. Todd, J.L. and Baker, T.C., Antennal lobe partitioning of behaviorally active odors in female cabbage looper moths, *Naturwissenschaften* 83, 324, 1996.

66. Meijerink, J., Carlsson, M.A., and Hansson, B.S., Spatial representation of odorant structure in the moth antennal lobe: a study of structure-response relationships at low doses, *J. Comp. Neurol.* 467, 11, 2003.

67. Daly, K.C., Chandra, S., Durtschi, M.L., and Smith, B.H., The generalization of an olfactory-based conditioned response reveals unique but overlapping odour representations in the moth *Manduca sexta*, *J. Exp. Biol.* 204, 3085, 2001.

68. Lee, J.K. and Altner, H., Primary sensory projections of the labial palp-pit organ of *Pieris rapae* L. (Lepidoptera: Pieridae), *Int. J. Insect Morphol. Embryol.* 15, 439, 1986.

69. Todd, J.L. and Baker, T.C., Function of peripheral olfactory organs, in *Insect Olfaction*, Hansson, B.S., Ed., Springer, Berlin, 1999, p. 67.

70. Carlsson, M.A. and Hansson, B.S., Dose-response characteristics of glomerular activity in the moth antennal lobe, *Chem. Senses* 28, 269, 2003.

71. Siddiqi, O., Olfactory neurogenetics of *Drosophila*, in *Genetics: New Frontiers*, vol. III, Chopra, V.L., Joshi, B.C., Sharma, R.P., and Bawal, H.C., Eds., Oxford University Press & IBH, London, 1983, p. 243.

72. Stensmyr, M.C., Dekker, T., and Hansson, B.S., Evolution of the olfactory code in the *Drosophila melanogaster* subgroup, *Proc. R. Soc. Lond. B Biol. Sci.* 270, 2333, 2003.

73. Stopfer, M., Wehr, M., MacLeod, K., and Laurent, G., Neural dynamics, oscillatory synchronization, and odour codes, in *Insect Olfaction*, Hansson, B.S., Ed., Springer, Berlin, 1999, p. 163.

74. Laurent, G., Olfactory network dynamics and the coding of multidimensional signals, *Nat. Rev. Neurosci.* 3, 884, 2002.

75. Adrian, E.D., The electric activity of the olfactory bulb, *Electroencephalogr. Clin. Neurophysiol.* 2, 377, 1950.

76. Adrian, E.D., Sensory messages and sensation. The response of the olfactory organ to different smells, *Acta Physiol. Scand.* 29, 5, 1953.

77. Wehr, M. and Laurent, G., Relationship between afferent and central temporal patterns in the locust olfactory system, *J. Neurosci.* 19, 381, 1999.

78. Laurent, G. and Davidowitz, H., Encoding of olfactory information with oscillating neuronal assemblies, *Science* 265, 1872, 1994.

79. Wehr, M. and Laurent, G., Odour encoding by temporal sequences of firing in oscillating neural assemblies, *Nature* 384, 162, 1996.

80. MacLeod, K. and Laurent, G., Distinct mechanisms for synchronization and temporal patterning of odor-encoding neural assemblies, *Science* 274, 976, 1996.

81. MacLeod, K., Backer, A., and Laurent, G., Who reads temporal information contained across synchronized and oscillatory spike trains?, *Nature* 395, 693, 1998.

82. Sachse, S. and Galizia, C.G., Role of inhibition for temporal and spatial odor representation in olfactory output neurons: a calcium imaging study, *J. Neurophysiol.* 87, 1106, 2002.

83. Hartline, H.K., Wagner, H.G., and MacNichol, E.F., The peripheral origin of nervous activity in the visual system, *Cold Spring Harbor Symp. Quant. Biol.* 17, 125, 1952.

84. Carlsson, M.A., Knüsel, P., Verschure, P.F.M.J., and Hansson, B.S., Spatio-temporal Ca^{2+} dynamics of moth olfactory projection neurons, *Eur J Neurosci.* 22, 647–657, 2005.

85. Stopfer, M., Jayaraman, V., and Laurent, G., Intensity versus identity coding in an olfactory system, *Neuron* 39, 991, 2003.

86. Ditzen, M., Evers, J.-F., and Galizia, C.G., Odor similarity does not influence the time needed for odor processing, *Chem. Senses* 28, 781, 2003.

87. Christensen, T.A., Pawlowski, V.M., Lei, H., and Hildebrand, J.G., Multi-unit recordings reveal context-dependent modulation of synchrony in odor-specific neural ensembles, *Nat. Neurosci.* 3, 927, 2000.

88. Lei, H., Christensen, T.A., and Hildebrand, J.G., Spatial and temporal organization of ensemble representations for different odor classes in the moth antennal lobe, *J. Neurosci.* 24, 11108, 2004.

89. Stopfer, M., Bhagavan, S., Smith, B.H., and Laurent, G., Impaired odour discrimination on desynchronization of odour-encoding neural assemblies, *Nature* 390, 70, 1997.

90. Kuwabara, M., Bildung des bedingten Reflexes von Pavlovs Typus bei der Honigbiene, *Apis mellifica, J. Fac. Sci. Hokkaido Univ. Ser. V Zool.* 13, 458, 1957.

91. Bitterman, M.E., Menzel, R., Fietz, A., and Schäfer, S., Classical conditioning of proboscis extension in honeybees (*Apis mellifera*), *J. Comp. Psychol.* 97, 107, 1983.

92. Hansson, B.S. and Christensen, T.A., Functional characteristics of the antennal lobe, in *Insect Olfaction*, Hansson, B.S., Ed., Springer, Berlin, 1999, p. 125.

93. von Frisch, K., The dance language and orientation of bees, Harvard University Press, Cambridge, MA, 1967.

94. Menzel, R., Erber, J., and Masuhr, T., Learning and memory in the honeybee, in *Experimental Analysis of Insect Behaviour*, Barton-Browne, L., Ed., Springer, Berlin, Germany, 1974, p. 195.

95. Erber, J., Masuhr, T., and Menzel, R., Localization of short-term memory in the brain of the bee, *Apis mellifera*, *Physiol. Entomol.* 5, 343, 1980.

96. Hammer, M. and Menzel, R., Multiple sites of associative odor learning as revealed by local brain microinjections of octopamine in honeybees, *Learn. Mem.* 5, 146, 1998.

97. Sigg, D., Thompson, C.M., and Mercer, A.R., Activity-dependent changes to the brain and behavior of the honey bee, *Apis mellifera* (L.), *J. Neurosci.* 17, 7148, 1997.

98. Brown, S.M., Napper, R.M., Thompson, C.M., and Mercer, A.R., Stereological analysis reveals striking differences in the structural plasticity of two readily identifiable glomeruli in the antennal lobes of the adult worker honeybee, *J. Neurosci.* 22, 8514, 2002.

99. Devaud, J.-M., Acebes, A., and Ferrus, A., Odor exposure causes central adaptation and morphological changes in selected olfactory glomeruli in *Drosophila*, *J. Neurosci.* 21, 6274, 2001.

100. Faber, T., Joerges, J., and Menzel, R., Associative learning modifies neural representations of odors in the insect brain, *Nat. Neurosci.* 2, 74, 1999.

101. Weidert, M., Galizia, C.G., and Menzel, R., Sensitization increases odor-evoked Ca^{2+}-signals in projection neurons of the honeybee, *Apis mellifera*, in *Proceedings of the 28th Göttingen Neurobiology Conference*, vol. 1, Elsner, N. and Kreutzberg, G.W., Eds., George Thieme, Stuttgart, 2001, p. 177.

102. Galizia, C.G., Kuttner, A., Joerges, J., and Menzel, R., Odour representation in the honeybee olfactory glomeruli shows slow temporal dynamics: an optical recording study using a voltage-sensitive dye, *J. Insect Physiol.* 46, 877, 2000.

103. Den Otter, C.J., Schuil, H.A., and Sander-van Oosten, A., Reception of host-plant odors and female sex pheromone in *Adoxophyes orana* (Lepidoptera:Tortricidae): electrophysiology and morphology, *Entomol. Exp. Appl.* 24, 370, 1978.

104. De Jong, R. and Visser, J.H., Specificity-related suppression of responses to binary mixtures in the olfactory receptors of the Colorado potato beetle, *Brain Res.* 447, 18, 1988.

105. Akers, R.P. and Getz, W.M., Response of olfactory receptor neurons in honeybees to odorants and their binary mixtures, *J. Comp. Physiol. A* 173, 169, 1993.

106. Getz, W.M. and Akers, R.P., Response of American cockroach (*Periplaneta americana*) olfactory receptors to selected alcohol odorants and their binary combinations, *J. Comp. Physiol. A* 180, 701, 1997.

107. Ochieng, S.A., Park, K.C., and Baker, T.C., Host plant volatiles synergize responses of sex pheromone-specific olfactory receptor neurons in male *Helicoverpa zea*, *J. Comp. Physiol. A* 188, 325, 2002.

108. Christensen, T.A., Mustaparta, H., and Hildebrand, J.G., Chemical communication in heliothine moths. II. Central processing of intraspecific and interspecific olfactory messages in the male corn earworm moth. *Helicoverpa zea*, *J. Comp. Physiol. A* 169, 259, 1991.

109. Raguso, R.A. and Willis, M.A., Synergy between visual and olfactory cues in nectar feeding by wild hawkmoths, *Manduca sexta*, *Anim. Behav.* 69, 407, 2005.

12 Learning-Based Recognition and Discrimination of Floral Odors

Brian H. Smith, Geraldine A. Wright, and Kevin C. Daly

CONTENTS

12.1 INTRODUCTION

A foraging moth or bee visits up to a few dozen to more than a hundred flowers on a foraging trip,[1] and it can make many such trips in a single day.[2] During these visits it associates floral stimuli—color,[3–6] shape,[7,8] texture,[9,10] odor[11,12]—with nectar and pollen rewards produced by flowers to attract pollinators.[13] Based on these experiences, the honeybee's memory is continuously updated with current information about the nature and distribution of reward associated with a species of flower.[14] This memory influences ongoing decisions about staying or leaving a resource patch[15] and whether to specialize on a particular species of flower.[1,16]

The purpose of this chapter is to review learning mechanisms that can influence the consolidation and updating of the memory that foraging honeybees form about the association of floral odors with nectar and pollen rewards given off by flowers. We will review information from the moth (*Manduca sexta*) and the honeybee (*Apis mellifera*). But the majority of data will focus on the honeybee for two reasons: first, it is a generalist pollinator that is capable of learning about many different floral odors; second, the nonassociative and associative learning mechanisms it uses to analyze floral odors have been extensively studied, beginning with the pioneering work of Karl von Frisch almost 100 years ago.[13,17] Thus studies of the honeybee can serve as a model for extension of these types of analyses to both generalist and specialist pollinators.[18] The vast majority of pollinators have not been well studied, and many species may be threatened with extinction.[19] Finally, we will argue that an understanding of how animals learn about and analyze floral odors may motivate the development of new sampling strategies for chemical analyses.

A honeybee's behavior toward a floral odor is guided by a mixture of innate and learned components. They are innately curious about floral stimuli. A honeybee scout may be attracted to the color, shape, and odor given off by a flower that has just come into bloom. This "orienting response" to a novel stimulus, which occurs before any learning and allows animals to bring appropriate sensory receptors to bear on the stimulus, is a necessary and important part of subsequent learning about the association of that stimulus with reinforcement.[20,21] Once attracted to a flower, further response is shaped by whether or not the nectar or pollen reward exceeds or fails to exceed that offered by other available floral species.[2] Thus the attractiveness of a reward is not simply an absolute quantity associated with the stimulus,[22] such as the concentration or amount of nectar. Attractiveness is instead one or both of those qualities relative to other floral resources available in the animal's environment, which is a psychological phenomenon known as "incentive contrast."

The protocols for studying these processes provide experimenters with the ability to ask animals questions about what they know about a stimulus, how they know it, and how that learning is translated into behavior.[23] Can they detect stimuli? Can they discriminate stimuli? Can they form categories among discriminable stimuli based on associations of those stimuli with a common meaning? We will show how learning paradigms have been adapted to evaluate questions such as these for the sensory processing of odors. The first sections are meant for researchers from a variety of backgrounds as an entrée to what can seem to be complicated experimental protocols. It is not meant to be an extensive review of learning mechanisms. More extensive discussion can be found in a number of excellent recent reviews.[23-25] For the same reason, we only briefly review the extensive literature on the physiology underlying insect olfaction.[26,27] All of the behavioral mechanisms we review below provide testable hypotheses about how they are implemented in the central nervous system. Finally, our discussion focuses on the use of insects as models for studies of olfaction. This is because of the vast literature on insect learning about flowers. Nevertheless, all of the olfactory learning mechanisms discussed below have been identified in vertebrates as well. Thus we feel our conclusions will have broader applicability to how many animals, including vertebrates, perceive floral odors.

12.2 LEARNING MECHANISMS AS "TOOLS" FOR ANALYZING FLORAL ODORS

Several different learning protocols have been employed to analyze how insects such as moths and honeybees learn about floral odors. We review these basic mechanisms here and in the next section. We view the mechanisms that underlie performance in these protocols as tools that influence and shape what insects learn about natural, more complex floral odors.

12.2.1 SOME BASIC DEFINITIONS: CONDITIONED STIMULUS, UNCONDITIONED STIMULUS, AND ASSOCIATIVE CONDITIONING

It is helpful to begin any discussion of more complex learning phenomena with basic definitions of the types of stimuli and learning mechanisms.[25] Stimuli are divided into two types based on the nature of innate responses to those stimuli. Prior to conditioning, a conditioned stimulus (CS) elicits little or no response beyond an initial orienting response typically shown toward all novel stimuli introduced into an animal's environment.[20] Odors, as we use them in most of the learning protocols below, are CSs. An unconditioned stimulus (US) elicits powerful innate responses associated with, for example, feeding, sex, or escape behaviors. When a CS is properly associated with a US, the response to the CS changes in a predictable way that prepares animals for occurrence of the US. The conditioned response to the CS is the index of learning and may, or may not, resemble the response to the US.

Olfactory conditioning of honeybees has been extensively studied in free-flying situations, in which foragers are induced to visit a scented feeder that contains a sucrose-water solution,[13,17,28] as well as in honeybees restrained in harnesses that allow them to freely move antennae and mouthparts.[17,29,30] In the latter procedure, called proboscis extension response (PER) conditioning, a sucrose-water solution (US) applied to sucrose-sensitive sensillae on the antennae elicits extension of a honeybee's mouthparts (proboscis). An odor (CS) typically does not elicit proboscis extension at high frequency in naïve honeybees, although a honeybee may show appetitive antennal movements in the presence of odor.[31] After one or a few associations with the US, the odor comes to elicit proboscis extension without the US, which is the conditioned response.

Proboscis extension to odor can be quantified through a variety of means, such as a binary response (respond or not) or in terms of the latency and duration of proboscis extension.[32–34]

The probability that the CS will come to elicit proboscis extension depends on the way it is associated with the US.[25,30] In "forward pairing," the CS precedes the US and is thus capable of predicting its imminent occurrence. There is a time interval between the onset of the CS and onset of the US that is optimal, but the exact interval depends on the species, on the conditioning protocol, and on the interval between conditioning (CS-US pairing) trials.[25] For PER conditioning in the honeybee, this interval is 1 to 5 sec.[30] Longer and shorter intervals, such that the CS and US overlap or the US actually precedes the CS (simultaneous and backward pairing, respectively), produce poorer conditioning performance. Furthermore, regular experience with backward pairing can produce inhibitory learning of the odor CS.[35]

Appropriate forward pairing of an odor with an appetitive (e.g., sucrose[30,36]) or aversive (e.g., shock[37]) US produces rapid excitatory conditioning (Figure 12.1). In PER conditioning, a single CS-US association will significantly increase a honeybee's response to the odor CS.[17,30] Frequently only three to six such trials are required to successfully condition about 90% of the honeybees in a test population, which usually consists of 20 to 30 honeybees for most of the experiments reviewed below. Once honeybees have been conditioned, the memory lasts for several days as it consolidates through several phases that have different underlying biochemical and molecular bases.[17]

Because of the pairing specificity, the mechanisms just described refer to associative (classical/Pavlovian) conditioning. Other associative learning mechanisms exist, such as operant/instrumental conditioning.[24,38] They may be important for understanding how honeybees and other pollinators learn about floral stimuli. However, they are beyond the scope of the current review.

12.2.2 NONASSOCIATIVE MODIFICATION OF RESPONSES TO ODOR

Some odors elicit powerful responses without associative conditioning. In this respect, they are more like a US (e.g., sucrose to a honeybee). These are typically species-specific sex or alarm pheromones (see Chapter 11),[39] or they are plant-derived substances (such as kairmones, allomones, and synomones) that have innate meaning for an insect because, for example, the odors may be part of plant defenses.[40] Males of many species of insects show upwind anemotaxis in wind currents carrying the pheromone of a conspecific female.[41,42] Anemotaxis is expressed through upwind surges when an odor filament is detected. Surges are interspersed by crosswind casting when a sufficient period has passed without detection of a filament. The location of the source of the plume—a female—elicits further courtship behavior.

Repeated, regular exposure to a US produces habituation, which is a nonassociative form of learning that is characterized in part by a decline in the response to the US over repeated exposures.[25] The sex pheromone-mediated responses of male moths (*Heliothis virescens*)[43] (Figure 12.2) and sweat bees (*Lasioglossum zephyrum*)[44] habituate. Both studies showed that the pheromone-mediated response declines over repeated exposure. Other criteria must be shown in order to prove

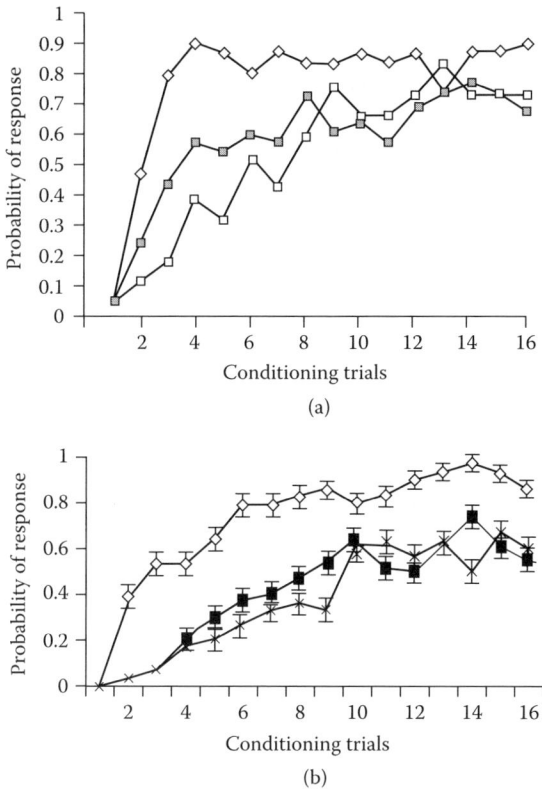

FIGURE 12.1 Acquisition curves for honeybees learning to associate an odor CS with a sucrose US. (a) Odor concentration affected the rate of learning. Honeybees were conditioned over 16 trials to perceptually similar odorants (1-hexanol, 1-heptanol, 1-octanol) ($n = 88$) or to perceptually dissimilar odorants (1-hexanol, 2-octanone, geraniol) ($n = 89$). Data were combined for both sets of odorants. Individuals trained with the high concentration (2.0 M; open diamonds) learned to associate odor with sucrose faster than those trained with either the intermediate (0.02 M; closed boxes) or the low (0.0002 M; open boxes) concentration of odorant (logistic regression, $n = 175$, concentration × trial number, $P = 0.014$). The level of response reached by the 16th trial was not significantly different for each of the concentrations (logistic regression, df = 2, $P = 0.196$) or for each of the odorants (logistic regression, df = 2, $P = 0.293$). (b) The low concentration odorant (0.0002 M; open diamonds) is a more salient stimulus than a stimulus containing only hexane or a stimulus with neither hexane nor odorant (air). Individuals conditioned with the low-concentration odorant over 16 trials reach a greater level of association than subjects conditioned only with hexane or air (for odorant: $n = 28$; for hexane: $n = 30$; for air: $n = 30$; logistic regression: odor versus hexane: df = 1, $P = 0.007$). (From Wright, G.A. and Smith, B.H., *Chemical Senses* 29, 127–135, 2004, with permission.)

habituation.[45,46] Several of these other criteria were met in the Daly and Figueredo[43] study.

Another example of habituation is provided by studies of orchid bees. Orchid flowers mimic species-specific sex pheromones in an effort to attract males for pollination (see Chapter 10).[47] In this case, the "pheromone" produced by the plant,

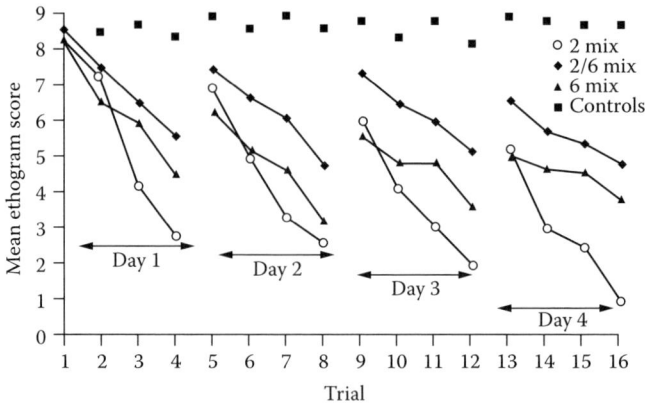

FIGURE 12.2 Habituation of upwind-oriented flight of male *H. virescens* moths to sex pheromone components in a wind tunnel. Symbols represent the mean response strength calculated from an ethogram of flight behavior (see Daly and Figueredo[43] for details and statistical analyses) for three treatment groups and pooled controls. Four trials were performed on each of four days. Control animals (black squares: $n = 9$ per mean) were not given repeated exposure. Their response strength continues to be strong both within days and between days. Experimental groups were repeatedly exposed to a pheromone blend (see Daly and Figueredo[43] for details of the blends). Experimental groups show a consistent pattern of decreasing response strength within days and recovery between days. Also notice that the group receiving the natural blend—2MIX (open circle)—clearly habituates at a much higher rate with more spontaneous recovery over days. (From Daly, K.C. and Figueredo, A.J., *Physiological Entomology* 25, 180–190, 2000, with permission.)

to the detriment of the insect, is called an "allomone." Males attempt to copulate with the flowers, and in the process they pick up pollen, which they carry to another flower and accomplish fertilization of the flower. Males quickly learn not respond to flowers after such a misguided copulation attempt, and it is likely that this occurs through habituation. It is possible that males habituate to the specific mixture of odors given off by a flower and thereafter avoid that odor.

Little is known about the biological role of habituation to sex pheromones or allomones. It may be very important in several ecological situations. For example, habituation could potentially affect diversification of pheromones in overlapping populations of moths using similar compounds in their pheromones.[48]

A few studies have used pheromones or plant-derived kairmones as an odor CS for PER conditioning. Honeybees easily learn to extend their mouthparts to components of their Nasonov pheromone during PER conditioning.[32,33] That is perhaps not surprising because the Nasonov pheromone is a multipurpose pheromone. Among other uses, it is frequently released as an attractant to a food or water source. The same study found that honeybees learned much more slowly when the CS was a component of their sting (alarm) pheromone (isoamyl acetate). Similarly, moths (*M. sexta*) can be conditioned to respond to a variety of floral odors.[49,50] Yet Daly et al.[51] found it difficult to condition them to methyl jasmonate. Methyl jasmonate is produced by the host plant of *Manduca*, *Nicotiana tabacum*, when its leaves are being eaten by *Manduca* larvae, and it has been shown to deter oviposition by gravid *Manduca* females.[52,53] Associative conditioning in these cases has not been investigated further. It is possible that the difficulty in learning arose from interference derived from the US-like quality of the odor. The odors may not interact strongly enough with associative pathways in the brain,[54] or they may invoke competing motor pathways that interfere with acquisition or the expression of learning.[55]

12.2.3 GENERALIZATION AND DISCRIMINATION

Once conditioned, honeybees and moths can be tested in various ways to evaluate hypotheses about the perceptual similarity between test odorants and the odorant used as the CS. Typically these tests involve manipulation of a molecular feature, or determinant,[56] such as carbon chain length or the type of functional group present in the molecule. Using test odors that have physical differences such as these makes it possible to examine specific hypotheses about the way that physical features of a stimulus influence perception.

Two types of tests that are frequently used are generalization[32,33] and discrimination.[36] Generalization tests involve one or more odorants or mixtures presented in randomized order and without reinforcement. Test odors vary in a systematic way from the CS. For example, Daly et al.[50] (Figure 12.3) conditioned moths (*M. sexta*) to aliphatic alcohols (1-hexanol or 1-decanol) or to a ketone (2-hexanone). After conditioning, each moth was tested with its CS odorant as well as with a series of aliphatic alcohols and ketones containing 7, 8, or 10 carbons in the straight chain. As the difference in chain length increased or decreased relative to the CS, subjects responded gradually less to the test odorant. For any given difference in chain length, if the functional group of the test odorant differed from the CS, there was a more substantial decrease in response. These two physical parameters produced additive decrements in the response, which indicates that they may constitute independent coding dimensions.

Furthermore, the results of these studies with the moth corresponded to earlier behavioral studies with the honeybee[32,33] and to those with rats.[57,58] This growing body of data provides a strong indication that simpler features of molecules such as chain length and functional group represent relevant and perhaps independent neural coding dimensions in these diverse species.

One can predict from these data that the molecular features of an odorant give rise to a pattern of neural activity in the brain that is characteristic for that odorant molecule. This neural pattern then changes in a graded fashion as the molecular features are systematically changed. Daly et al.[59] tested this prediction and found that the behavioral results corresponded well to recordings of neural activity in the moth antennal lobe when moths were tested with the same odorants. Indeed the spatiotemporal activity patterns of ensembles changed in a graded manner that was predicted from the behavioral studies. Odorants between which there was a low probability of generalization of the conditioned response in behavioral studies were statistically more different in physiological analyses. Conversely, odorants that elicited a high degree of behavioral generalization produced statistically more similar patterns of neural activity. However, these results should not be taken as an indication

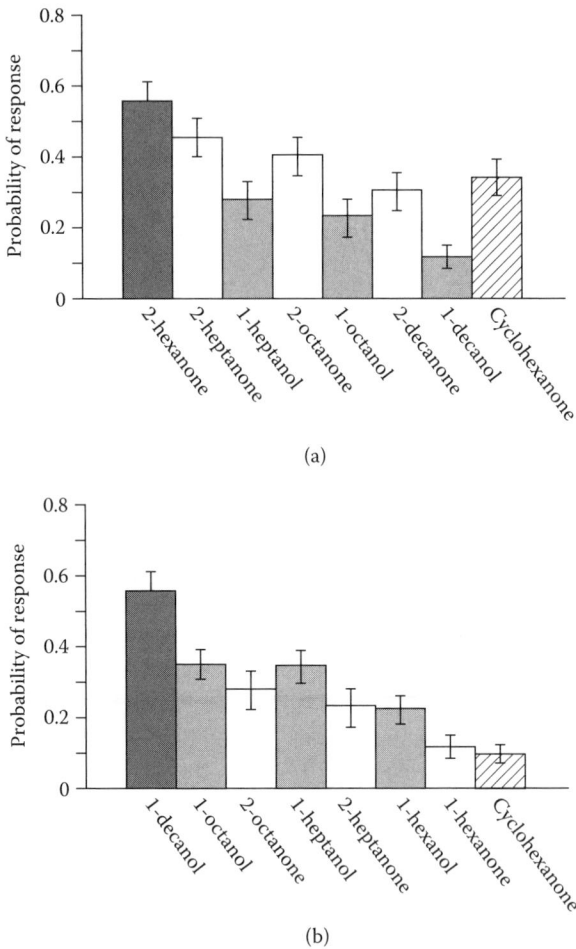

FIGURE 12.3 Response probabilities of *M. sexta* moths for the post-test generalized response to the conditioning odor (black, CS) and to test odors, color coded by functional group: gray, alcohols; white, ketones; striped, cyclohexanone. Moths were conditioned to either 1-hexanol (A; $N = 0$) or to 1-decanol (B; $N = 80$). Statistical analysis revealed that with each incremental increase in chain length there is a corresponding 9% decrease in the probability of a feeding response. See Daly et al.[50] for the complete statistical analysis. (From Daly K.C., Chandra, S.B.C., Durtschi, M.L., and Smith, B.H., *Journal of Experimental Biology* 204, 3085–3095, 2001, with permission.)

that molecular features will always perfectly predict perceptual/neural similarity. Given constraints on coding (see below), one might expect nonlinear relationships to emerge.

Discrimination conditioning differs from generalization in that two odorants are simultaneously conditioned. One odorant is reinforced (A+; a "+" indicates sucrose reinforcement), as described above. However, a second type of trial is pseudorandomly interspersed[37] with reinforced trials. The pseudorandom sequence of trials ensures that subjects cannot predict which stimulus will follow on the next trial. They therefore must respond only as a function of the properties of the CS. These latter trials involve identical presentations of an odorant, but either without sucrose reinforcement or in association with punishment (X–, where a "–" indicates a lack of reinforcement or punishment).[37] If the two odorants (A and X) are discriminable then animals will eventually respond to the reinforced odorant, but not to the unreinforced odorant (Figure 12.4a).

Several measures of discriminability are possible. The strictest criterion involves conditioning until almost complete discrimination occurs in an attempt to discover pairs of odorants that animals fail to discriminate. Vareschi[60] found that honeybees

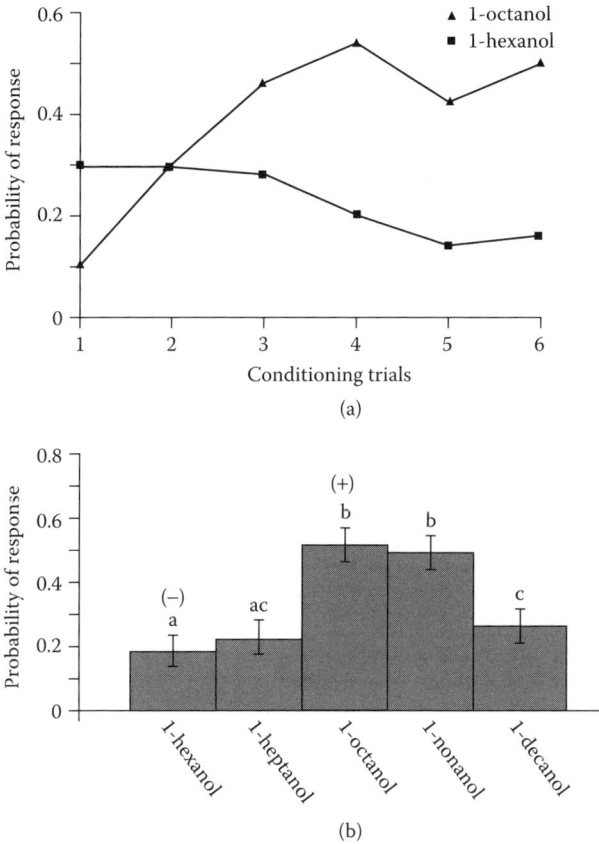

(a)

(b)

FIGURE 12.4 Acquisition curves and test-phase response probabilities for the differential conditioning experiments with moths. (a) Differential conditioning of 1-octanol (reinforced) versus 1-hexanol (not reinforced) ($N = 50$). (b) Generalization across a range of alcohols (1-hexanol through 1-decanol). A "+" or "–" indicates reinforced or nonreinforced odorants, respectively. Lowercase letters above values denote results of one-tailed t-tests between means ($P < 0.05$); like letters indicate nonsignificance. (From Daly K.C., Chandra, S.B.C., Durtschi, M.L., and Smith, B.H., *Journal of Experimental Biology* 204, 3085–3095, 2001, with permission.)

could be relatively easily conditioned to discriminate a very large number of odor pairs in the PER procedure. The few pairs that were not easily discriminable could be discriminated using a larger number of conditioning trials. Similar results were found more recently using a free-flying conditioning procedure.[61] Alternatively, the number of trials to reach a set criterion (e.g., a set number of correct responses in a row) or a correlate of it[37] could be used as a more sensitive measure of similarity.

Generalization as we use it here is a special case of discrimination.[33] It is meant as a measure of perceptual similarity. Thus we would expect that indices of discriminability, especially the more sensitive trials-to-criterion measure, would correlate to generalization. Either type of procedure is adequate for evaluation of perceptual similarity, but they might be used to test somewhat different hypotheses. Discrimination conditioning might be expected to highlight the slightest perceptual differences between two odors. Absolute failure to discriminate is rare in species studied to date. When such a failure occurs, it is very informative for neural hypotheses about how those odorants are represented in the brain. Such a failure may inform ecological hypotheses about those odorants. Generalization might be better for revealing more subtle similarities and establishing graded response patterns with changing molecular features, as we discussed above. In this view, complete generalization (equal response to the CS and to a test odor) would not necessarily mean that odorants are not discriminable. It would only be an indication that the odors are perceptually very similar CSs and that the two odorants may share many features used for predicting the US.

This use of generalization differs from "higher order," more cognitive interpretations.[62] By association with a common meaning, such as reward, easily discriminable objects might be grouped into the same category (e.g., all brightly colored roughly circular objects are flowers). Thus any new flower might be recognized as such even though it is easily distinguishable from all other objects in that category. This type of generalization might also affect perception of floral odors. This type of generalization has been implied in a recent study of variability in odor stimuli.[63]

We retain the first usage of generalization (i.e., from a single stimulus to additional stimuli during test conditions designed to examine the perceptual similarity of odors) throughout the remainder of this text in the interest of consistency with previously published studies.[32,33,50,63,64] We recommend that any future use of the term explicitly recognize which sense is being used.[33]

12.2.4 Learning without Reinforcement

Under natural foraging conditions in the field, flowers will sometimes fail to contain a nectar reward. The frequency of nonrewarding flowers has been shown to dramatically affect foraging decisions,[65] and nonrewarding trials affect performance in the PER procedure.[66] In an analog to this foraging problem, insects can learn about an odor CS over several presentations when it has not been reinforced with a US.[67] Under normal circumstances, it is difficult to resolve changes in responsiveness to an unreinforced odor. Nevertheless, it can be revealed that honeybees learn something about the unreinforced odor by subsequently pairing it with sucrose in a way that would normally produce a robust conditioned response. After presentation of eight or more unreinforced exposures to odor in the PER paradigm, honeybees are slow to learn about that odor when it is subsequently paired with reinforcement.[67] After 25 or more unreinforced exposures, learning is very slow relative to appropriate control groups. Thus when an odor is not followed by anything of consequence (i.e., a US), honeybees explicitly learn not to respond to that odor.

During discrimination conditioning, one odor is either not reinforced (Figure 12.4a) or is punished.[37] The presence of an unreinforced odor influenced the shape of the generalization gradient shown in Figure 12.4b.[50] After differential conditioning, when moths were tested for generalization to odorants of the same functional group (alcohols, ketones, etc.), but different carbon chain lengths as the CS+, the distribution was skewed away from the unreinforced (CS) odorant. This pattern is called a "peak shift."[38] The skew might result from the additive interaction of two gradients. An excitatory gradient would slope gradually away from the reinforced odorant, such that the neural representations of test odorants become progressively less similar to that of the CS odorant as a molecular feature of the test odorant is changed to be less like that of the CS. Likewise, some form of an inhibitory gradient would be expected to slope away from the unreinforced odorant. Odorants with features where the two gradients overlap—the intermediate chain lengths in this example—thus elicit less of a response than would be expected given only excitatory conditioning.

This provides support for two propositions. First, it provides a further indication that some form of learning occurs during unreinforced presentations. Second, it further supports the proposal that the physical dimensions of odorant molecules— functional group and chain length—are represented by odor codes in the brain because the interaction of excitation and inhibition can move the response along this dimension.[38] The existence of one or more forms of inhibition implies that how an animal perceives an odorant or a mixture depends on the associative context of other, similar odorants and mixtures. This could have important implications for studies of olfactory neural representations in the brain.

12.3   DISCRIMINATION OF INTENSITY

Floral scents, and the compounds found in floral scents, occur over many orders of magnitude in concentration.[68] The concentration, or intensity, of a scent is therefore likely to be an important feature of naturally occurring odors that pollinators use to identify floral scents. Several studies of honeybees have recently shown that the intensity of floral scent influences the rate of learning, the ability to discriminate among odors, and the perceptual qualities of odors.[63,64]

The salience of an odorant as a CS, which we define as perceived intensity, is a function of the concentration. In this sense, salience and intensity refer to perceptual properties, and concentration is the physical dimension to which they are correlated. The relationship between these parameters can be revealed by conditioning honeybees to different concentrations of the same odorant. Wright and Smith[64] showed that the rate of acquisition, which is measured as the rate of increase in the probability of responding across trials, was proportionately lower at lower concentrations (Figure 12.1). This indicates that detectability influences how quickly an odor is learned.

However, all of the groups ultimately reached the same final "asymptotic" level of response, when 80 to 90% of the honeybees responded to the odor CS. Therefore it would appear that more experience over a larger number of conditioning trials can overcome limitations imposed by low concentration.

Perhaps a more surprising finding from this study was that the discriminability of odors is influenced by intensity. After conditioning to the highest concentration (Figure 12.5),[64] honeybees showed only low levels of generalization to novel test odorants that were dissimilar in molecular structure (i.e., different chain lengths, shapes, and functional groups) but tested at the same concentration as the odorant CS. Thus at this concentration these odorants were easily discriminable. As the conditioning and test concentrations decreased, animals responded equally well to the odorant CS. But the discriminability of the CS and test odorants decreased, which is revealed by progressively flatter generalization gradients. At the lowest concentration, animals responded equally to all test odors. As noted above, this does not mean that the odorants at the lowest concentration were not discriminable, it only means that the odorants became less discriminable at lower test concentrations.

Therefore the perceptual qualities of odors that pollinators use for the identification of floral scents may be affected by odor concentration. These data indicate that the neural code for quality (identity) is not independent of intensity. At the lowest detectable intensities, animals behave as though they can detect the presence of an odor, but they have more difficulty in determining its identity.

The results fit studies of concentration coding in primary sensory cells.[69] Olfactory sensory cells respond to their "best" odorants at low concentrations.[70] The same sensory cells respond to other "nonbest" odorants, but only at higher concentrations. Thus the cells become more broadly tuned at higher concentrations. This implies

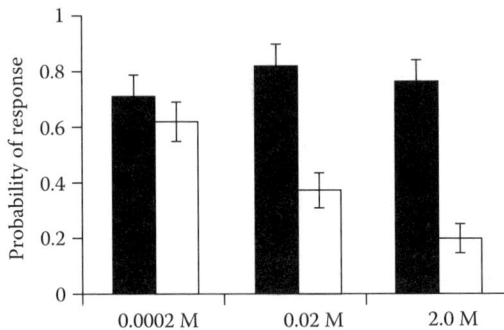

FIGURE 12.5 Generalization from the conditioned odorant (filled columns) to different test odorants (open columns) at different conditioning/test concentrations. After conditioning honeybees for 16 trials with an odorant at a fixed concentration, we tested each honeybee ($n = 88$ animals) with perceptually dissimilar odorants (i.e., easily discriminable) at the same concentration as the conditioned odorant. Dissimilar odorant compounds became progressively easier for honeybees to discriminate with increasing concentration. Generalization response levels to novel odors (open columns) are shown next to the response to the conditioned odor (shaded column) at a given training/testing concentration. (From Wright, G.A. and Smith, B.H., *Chemical Senses* 29, 127–135, 2004, with permission.)

that more sensory cells respond to any given odorant at higher concentrations, which would provide more information that could improve discriminability of that odorant from others. Furthermore, in the first layers of synaptic processing in the brain— the antennal lobes of insects and olfactory bulbs of vertebrates[26]—patterns of activation tend to become more distinct at higher concentrations. Stopfer et al.[71] showed in the locust that odorants are still statistically distinct at low concentrations, but the distinctiveness improves with concentration. Recently Sachse and Galizia[72] showed in the honeybee that, across a broad range of concentrations, the neural activity patterns change, so the neural code for lower concentrations may differ from that for higher concentrations, although they show that local processing in the antennal lobes contributes to making the code less variant across narrower concentration ranges.

The fact that an odorant can have different perceptual qualities across a wide range of concentrations raises an interesting issue. Insects detect odors, and they show upwind-oriented movement, when they are still far away from the source (positive anemotaxis).[41,42] Odorants are carried downwind in turbulent plumes that rapidly separate into eddies and filaments. Thus when following a plume, insects are exposed to rapidly fluctuating intensities, which include long periods of very low or zero concentrations punctuated by periods of high concentration as a filament is crossed.

How might it be possible for an animal to recognize the identity of a source from a distance if odor quality depends on concentration? The answer to this question must await further research. If insects average concentration over long time periods, they may perceive a low concentration that is qualitatively different from the source. Alternatively, some filaments, even far away from the source, may expose insects to brief concentrations that approximate the source concentration.[73] Filaments at the appropriate concentration might become encountered more frequently as the distance to the source is reduced. We propose that insects might simply keep track of the time since the last capture of a filament that approximated the source (learned) concentration. This would require integrating concentration over much shorter intervals; for example, it might be integrated over the average width of a filament. This mechanism might even provide a means for determining odor intensity and quality when still far away from the source.

12.4 LEARNING ABOUT ODOR MIXTURES

Floral odors typically arise from mixtures of many odorants. Therefore, any discussion about learning of floral odors must include a discussion of learning about mixtures. A common finding from studies of how animals perceive odor mixtures is that the whole stimulus (AX) is treated differently from, and thus is not simply the sum of, its parts (A and X).[74,75] This has been observed in mixtures of only a few odorants and in more complex mixtures. In particular, one study has shown that as the number of compounds in a mixture increases, the ability of humans to identify the presence of individual compounds decreases.[76] Studies of this kind strongly suggest that mixture-unique olfactory entities arise as a result of the way the odors are perceived and encoded by the olfactory system. In addition, experience with odors appears to influence the perceptual qualities of mixtures and the way they are represented as stimuli in the brain.[76,77]

Studies of controlled mixtures have identified several behavioral mechanisms that all reflect one or another type of perceptual interaction among odorants. After introducing those behavioral studies, we will review different theoretical treatments that make different assumptions about the nature of that interaction.[38] In general, one such treatment assumes that perceptual properties of different odorants are independently learned and processed in mixtures, such that the perceptual qualities do not change when two or more odorants occur in mixtures. The other type of theoretical account assumes that a unique neural activity pattern occurs for the mixture and becomes associated with reinforcement. As we will discuss, there is support from behavioral and physiological analyses for both of these assumptions. Thus the final explanation for any of the phenomena described in this section might need to await detailed physiological analyses coupled directly to behavioral analyses.

12.4.1 COOPERATION BETWEEN MIXTURE COMPONENTS: SYNTHETIC AND CONFIGURAL PROCESSING

Several behavioral studies support the proposition that mixture components interact to produce mixture-unique cues that can be discriminated from the components. PER conditioning of honeybees to mixtures of two odorants (AX) followed by testing for generalization to the individual odorant components (A or X) reveals overshadowing (Figure 12.6),[34,78] which is characterized by lowered response to a component after mixture conditioning than after equivalent conditioning with the pure odorant. If the two components (A and X) are perceived independently of one another in the mixture, one might expect equivalent responses to the components when they are conditioned in the mixture (AX) versus when they are conditioned separately (A or X) (Figure 12.6a). Alternatively, if the mixture is perceived as a unique stimulus, different from the simple summation of A and X, then one might expect overshadowing (Figure 12.6b). Clearly Figure 12.6 shows that pairs of odorants can show either possibility. A or X are each different, and hence discriminable, from AX. However, the existence of overshadowing is not an unambiguous indication that mixtures produce unique perceptual cues. As we will see, overshadowing is also consistent with models that propose independent processing of odors (see the section on blocking).

The existence of mixture interactions is further supported by studies of positive and negative patterning.[74,75] Honeybees can be conditioned to discriminate a reinforced mixture from unreinforced components (positive patterning: AX+/A/X), and they can be conditioned to discriminate reinforced components from the unreinforced mixture (negative patterning: A+/X+/AX). Finally, biconditional discrimination, also referred to as transswitching,[79] has been reported (Figure 12.7).[74] In this protocol, honeybees can be easily conditioned to discriminate two reinforced mixtures of four components from the remaining two possible mixtures of the same components, which are unreinforced (AX+/BY+/AY/BX). This protocol is impossible to solve by attention only to components, because each component (A, B, X, and Y) is equally often reinforced as unreinforced.

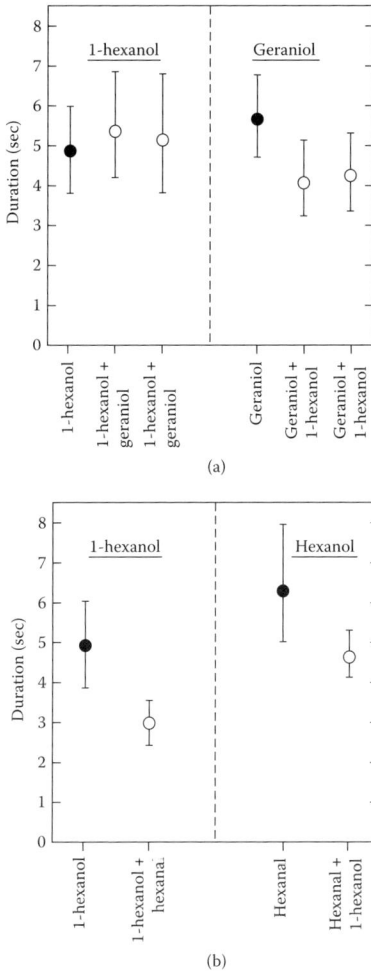

FIGURE 12.6 Interaction (overshadowing) between 1-hexanol and hexanal but not between 1-hexanol and geraniol. Symbols represent mean response durations during test trials with the underlined odorant. In each set, one group was conditioned over six acquisition trials to that pure odorant (filled symbols), and one or two groups were conditioned to that same odorant in a mixture with a second (open symbols). All responses were to a common pure test odorant (underlined) Each of the means shown in the figure represents an independent group of 15 subjects. (a) Reciprocal experiment for 1-hexanol and geraniol that failed to demonstrate significant interaction. Two-way analysis of variance (ANOVA) failed to reveal an effect to test odor (1 hexanol or geraniol; $F = 0.9$, nsec), conditioning treatment (mixture versus pure odor; $F = 0.3$, nsec), or crossed effects ($F = 0.8$, nsec). (b) Reciprocal experiment for 1-hexanol and hexanal that showed interaction. Two-way ANOVA revealed an effect of test odor (1-hexanol or hexanal; $F = 3.5$, $P < 0.05$), and the effect of conditioning treatment was significant (mixture versus pure odor; $F = 4.0$, $P < 0.05$), which substantiates the mixture interaction effect. That the mixture interaction between the odorants was symmetrical is revealed by the lack of a significant crossed effects term ($F = 0.3$, nsec). (From Smith, B.H., *Physiology and Behavior* 65, 397–407, 1998, with permission.)

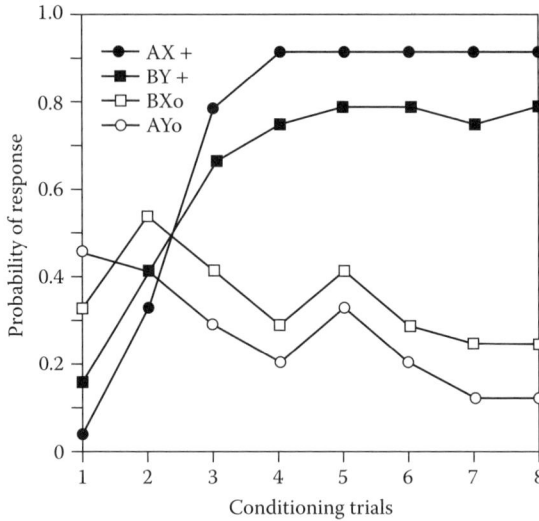

FIGURE 12.7 Learning of a transswitching problem that required honeybees ($n = 24$ animals) to discriminate four odorants (A, B, X, and Y) used to make AX+/BY+/AYo/BXo. Two of the mixtures were reinforced with sucrose (+) and two were not (o). Honeybees successfully discriminated AX+/BY+ from BXo/AYo by trial 3 ($F = 13.9$, df = 3.92, $P < 0.001$). (From Chandra, S.B.C. and Smith, B.H., *Journal of Experimental Biology* 201, 3113–3121, 1998, with permission.)

Each of these protocols—overshadowing, positive and negative patterning, and biconditional discrimination—reveals interactions of one form or another in mixtures. However, as noted above, qualitatively different theoretical accounts have been proposed for the nature of the interaction. The mixture AX may activate neural pathways in early sensory processing that are "synthetic," and hence different from A or X.[74] There may be sensory cells that fail to respond to A or X, but that respond to AX. Activation of an AX pathway might thereby diminish activation of A and X pathways due to interactions of odorant molecules at the level of transduction processes in sensory receptors. A mixture-unique cue could also arise from inhibitory interactions among A and X pathways in early synaptic processing in the brain (see below). In optical recordings from the honeybee antennal lobe, which is the first synaptic contact sensory afferents make with brain interneurons, tests with mixtures produced neural activity patterns that were similar to the components as well as patterns that are unique to the mixture.[80] However, the degree to which mixture-unique or component-like patterns arose depended on the odorants.

Thus, at least for some odorant combinations, both behavioral and physiological studies indicate that the salience of A and X is diminished by the simultaneous presence of both, which might prevent or diminish the association of A and X pathways with reinforcement. Since AX is essentially a different odor from A or X, it is not surprising that overshadowing occurs. This explanation could account for discrimination of mixtures from other mixtures during biconditional discrimination, and it could account for discrimination of mixtures or from components in or feature negative/positive discrimination.

Another theoretical account for mixture interaction assumes that odorants A and X are processed independently in early sensory pathways, which is the case for some pairs of odorants.[80] This would be the case for cues from different sensory modalities (e.g., visual, olfactory, auditory), which also show the same types of interactions described above.[28] In this interpretation, A and X together may give rise to "configural" cues that arise by associative links between A and X because of their joint occurrence with reinforcement.[79] Breakage of the configuration during testing with A or X alone may lead to a diminished response and hence account for overshadowing as well as the "not A and X" logical operation implied by negative patterning. Behavioral evidence supporting this type of inter-CS association has recently been reported for some odor mixtures in honeybees.[81]

12.4.2 COMPETITION BETWEEN COMPONENTS OF MIXTURES: BLOCKING

Overshadowing might arise because mixture components are independently processed, but compete with each other for a limited neural capacity to process sensory information or associate it with reinforcement.[23,38] The assumption made by these models is that there is a limited amount of attention or associative strength that a given CS or US can support. In a mixture, A and X would compete for this limited neural resource.

The best evidence that some form of competition can contribute to learning about odor mixtures arises from blocking experiments with honeybees (Figure 12.8).[78,82–86] One component (A) of a mixture (AX) is given a "competitive" advantage by preconditioning A with the same US used for later conditioning of the mixture. One or another type of control group of animals is identically conditioned to the AX mixture, but differs from the blocking (experimental) group in that the control is not preconditioned to A. It is important to recognize that animals in blocking and control groups each experience X in the background with A during mixture conditioning. The groups differ only in the prior association of A with reinforcement. Both groups are then tested with X. The expectation for blocking is that the response to X in the group preconditioned to A will be less than in the control group. That is, because of the increased salience from preconditioning, odor A blocks learning about X. Overshadowing, in this interpretation, can be considered to be a case of reciprocal blocking, as A and X together gain associative strength when they are paired with reinforcement.[23]

Therefore it appears that the types of competitive interactions that blocking reveals contribute to the perception of mixtures. It is clear that it consistently occurs for some pairs of odorants used as A and X,[78,83,87] but not for all pairs.[85,88] It is as yet unclear why differences in expression of blocking among different pairs of odorants should exist. Smith[89] suggested that the ecological role for blocking could be to minimize interference of changing background odors with perception of an odor that is consistently associated with reinforcement. Computational models have then suggested that blocking will be most evident when the neural representations for odors overlap,

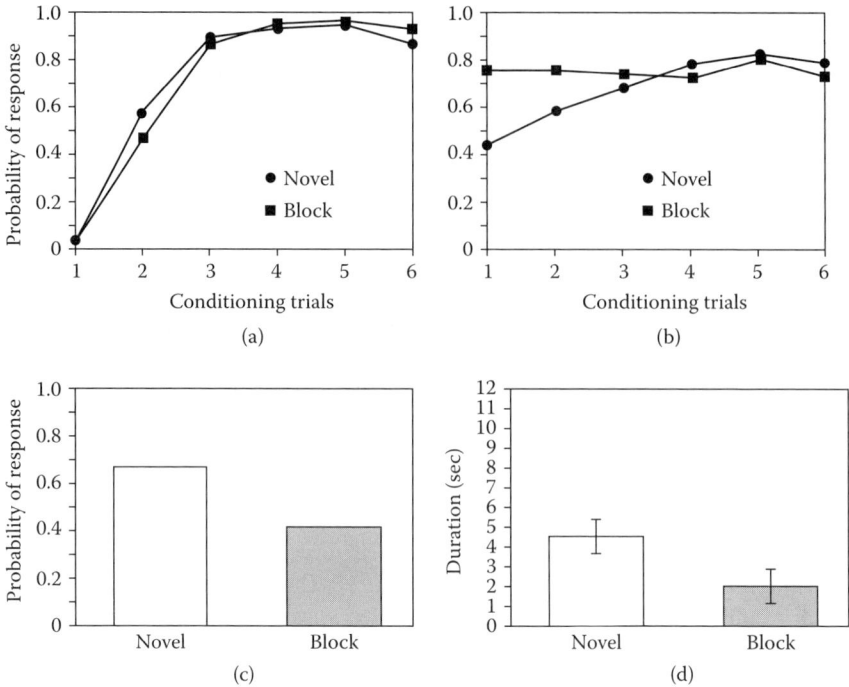

FIGURE 12.8 Summary of acquisition curves and test trials in a blocking experiment. The odorants 1-hexanol, 1-octanol, geraniol, and 2-octanone were counterbalanced as A, X, and N. (a) Subjects in the NOVEL ($n = 60$) control group and in the BLOCK ($n = 60$) group were pretrained with a novel odor N or with odor A, respectively. Subjects were conditioned over six forward-pairing trials and reinforced with 1.25 M sucrose. The figure represents the percentage of subjects that extended their mouthparts (PER) on each trial prior to presentation of the sucrose US. (b) Following pretraining, all groups received six conditioning trials with a mixture of odor A and odor X (A + X). (c) Percentage of subjects that extended their mouthparts (PER) when tested for their response to X alone after mixture conditioning. (d) This graph shows proboscis extension duration measurements of NOVEL and BLOCK groups in response to X. Duration scores were taken from videotape analysis. Subjects in the NOVEL group responded to X significantly more strongly than did subjects in the BLOCK group, whether the response was measured as the percentage of proboscis extensions (Figure 12.8c; $\chi^2 = 7.6$, $N = 120$, $P < 0.01$) or as the duration of the PER (Figure 12.8d; $F = 12.2$, df $= 1$, $P < 0.001$). (From Hosler, J.S. and Smith, B.H., *Journal of Experimental Biology* 203, 2797–2806, 2000, with permission.)

particularly in regard to lateral inhibitory interactions.[90,91] However, these models have not yet been extensively evaluated. Finally, the ability to detect blocking can be made difficult, or at least inconsistent, by the type of control group chosen.[83,85]

12.4.3   SUMMARY OF MIXTURE INTERACTIONS

It should now be evident that learning about even simpler, binary mixtures can be complex. Behavioral studies suggest that several mechanisms generally categorized here as cooperative (synthetic and configural) or competitive (blocking) can contribute to how animals perceive mixtures. Which mechanism applies in any given odor mixture might depend on the specific components and how they are represented in the central nervous system. Furthermore, some computational models synthesize two or more of these processes.[91] Future studies need to focus on unraveling the relative contributions of these mechanisms to mixture perception through coupled behavioral and physiological investigations, which will be indispensable for unraveling the basis of mixture processing.

12.5   THE IMPORTANCE OF VARIATION IN ODOR COMPOSITION

So far we have discussed the perception of intensity and quality of pure, monomolecular odorants. We then went on to show various ways those odorants interact in mixtures to produce new perceptual effects. However, there is as yet another level of complexity that needs to be added to the mixture problem. Pavlov[20] recognized that two stimuli will never be identical, even when the same stimulus is presented at two different points in time under very controlled laboratory conditions. A light stimulus, for example, might be perceived from two different angles because subjects have moved in the conditioning arena from one trial to the next.

Consider this problem for a honeybee foraging under natural conditions in a field of flowers. Every flower can be uniquely identified by its spatial location, and movement patterns of bees are designed to avoid revisits to the same location. A forager will probably deplete a flower of nectar or pollen on a single visit. Thus the forager's problem is to identify the same type of flower, but at a different spatial location; the forager must do this if it wants to avoid revisiting the same flower it just depleted. Many floral stimuli—shape, color, odor—help foragers identify the same type of flower at a new location, which is essentially the problem of stimulus generalization through either or both of the mechanisms reviewed above. However, the precise combination of color, shape, and odor can vary from one flower to the next due to a variety of factors such as environment,[92] time of day,[93] age,[94] and pollination status,[95] for example. The contribution of intraspecific genetic variability alone to differences in odor composition is substantial enough to allow for selection of different varieties of many types of flowers, which have dramatically different odor compositions.[68] Every flower can therefore give off a slightly different odor.

How does a foraging honeybee thus establish whether an odor emanating from a new flower is "similar enough" to have the same meaning, in terms of nectar or pollen, as an odor experienced a few moments or hours ago? Conversely, how does a honeybee establish whether a new odor is "too different," such that the flower may be a different pollination status, and thus present no reward, or it might be a different species altogether? This is the foraging problem, and it is certainly more complex than simply learning an association between odor and reward.

From this standpoint, it is important to understand what effect trial-to-trial variation in odor composition has on the olfactory system. We proposed above that the mechanisms we reviewed for mixtures—overshadowing, blocking, synthetic and configural learning—were "tools" that the olfactory system employs to solve olfactory problems involved in the detection and discrimination of odors. Here, we specifically propose that these mechanisms help solve the variability problem. If a subset of components in a complex mixture is less variable from flower to flower than other components, blocking might help to focus the olfactory system on those components. For example, in typical blocking protocols,[34,78] odorant A is present on 100% of the trials and the "blocked" odorant X is present on 50% of the trials. The consistency of A helps it block, or "asymmetrically overshadow," X. Subset A might consist of two or more odorant components, which might be linked through synthetic or configural processing, as described above. In the extreme sense, the memory formed of the mixture might be dominated by the perceptual qualities of the less variable subset (A). For example, in the following sequence of four-component mixtures— BAXY, CAXZ, DAXT—each of the four positions represents a different odorant component, and different letters represent different concentrations of a given odorant. If this were presented to a honeybee, we propose that the perceptual qualities of the mixture would be dominated by AX.

Wright and Smith[63] demonstrated that variance in this type of experiment has an impact on what honeybees learn about mixtures. Honeybees were conditioned using the PER protocol to mixtures of three odorant components. Subjects were then tested in random order with each component in order to assess how similar the mixture was to each component. In one experiment, one of the three components was constant across all the trials, but the other two varied from trial to trial. In situations where the odorants composing the mixture were perceptually dissimilar, honeybees responded more often to the constant component than to the variable components, even when the mean across-trial level of the constant component was lower than the means for the variable components. Thus there is some indication that the perception of the mixture was biased toward the constant component, as the model presented above would predict.

A second experiment tested for the effect of variation by allowing all three of the components to vary from trial to trial (Figure 12.9). The across-trial coefficient of variation (CV) of the mixture in this group was greater than zero (CV > 0). The subsequent responses to the components in this group were compared to a different group of honeybees that were conditioned to the same three-component mixtures, but in this group the composition was constant from trial to trial (CV = 0). Honeybees in the variable group(CV > 0) responded significantly more often when subsequently tested with the components than in the constant (CV = 0) group. The relative difference between these treatments was robust across different conditioning concentrations and types of odor mixtures. This experiment shows clearly that trial to trial variation in mixture composition significantly affects the quality of the olfactory memory for the mixture. In the variable condition, the memory was more inclusive of the qualities of the components than it was in the constant condition.

However, as we have written elsewhere in this article, expression of the effect of variation depended on the types of odorants included in the mixture. Thus the same admonition applies here: behavioral experiments will be most informative when coupled to physiological characterization of neural response patterns to the same mixtures.

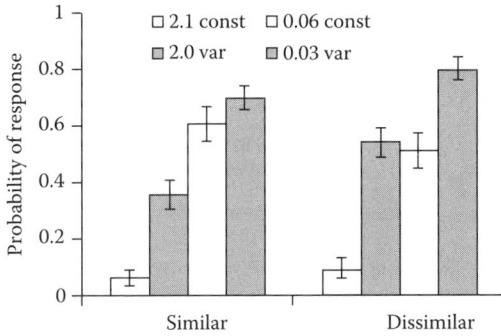

FIGURE 12.9 The effect of trial-to-trial variation on learning about a mixture. Honeybees were conditioned to chemically defined odorant mixtures that systematically varied from trial to trial. They were then tested for the response to a pure odorant component present in the mixture. This test thus evaluates generalization from the mixture to a component. One group of subjects was conditioned with no variation from trial to trial (labeled *const*); another group was conditioned with high variation from trial to trial (labeled *var*). This basic experiment was replicated with odorants that were of "similar" or "dissimilar" molecular structure. And it was also replicated for high (2.1 and 2.0) and low (0.06 and 0.03) concentrations. When subjects were conditioned with high-intensity mixtures (2.0 and 2.1 M) they responded less than subjects conditioned with low-intensity mixtures (0.03 and 0.06 M). In all four cases, subjects conditioned with highly variable mixtures (*var*) responded with a higher probability than subjects conditioned with no variation present in the mixture (*const*) (logistic regression:^{2} = 59.7, n = 229, $P < 0.001$). (From Wright, G.A. and Smith, B.H., *Proceedings of the Royal Society London B* 271, 147–152, 2004, with permission.)

12.6 PUTTING IT TOGETHER: DISCRIMINATION OF NATURAL FLORAL MIXTURES

To discriminate the scents of flowers, pollinators could potentially use all the compounds present in the scent. Conversely, it is possible that they could use key features of scent that are common among the scents of rewarding flowers. Which components of scents are extracted as dominant perceptual features depend on several factors, such as prior experience with components, mean levels of components, correlations among subsets of components, variation among components over trials, and how components are processed in the central nervous system. These factors have been identified using artificial, and hence simpler, mixtures of two or three components. A current challenge therefore is to understand how features in complex mixtures of natural odors arise and how they may be used by animals to identify the most salient fractions of odor mixtures.

Studies measuring the compounds found in floral scent have documented differences among flowers with respect to the number of compounds, the types of compounds, and their concentrations.[68] Each of these is a possible feature that could be used by pollinators to identify flowers. In fact, studies relating pollinator behavior to the compounds present in scent showed honeybees would respond to specific components found in the scent of canola when they had been conditioned to a mixture containing six compounds found in its scent.[96,97] These experiments showed that it was possible for honeybees to generalize a conditioned response from a mixture to specific compounds found in the mixture. Studies of olfactory perception and learning in other animals, however, have observed that odors composed of several compounds often have sensory properties that are not simply related to these three variables. Other aspects of a floral scent, such as its overall intensity or the diversity of the compounds present, may contribute sensory properties that arise from the mixture.

One way of understanding how the features of complex mixtures of odorant compounds arise is to examine behavior under conditions where pollinators experience the complex scent and are asked to discriminate it from other complex scents. Few studies have examined the structure of complex mixtures and have related this structure to the behavior of pollinators.[47,68,98] Ayasse et al.[47] examined the response of male bees to the scent produced by sexually deceptive orchids. Males could discriminate the scent of individual flowers; this discrimination was based, in part, on compounds that were tested in gas chromatography-electroantennogram (GC-EAG) studies and found to be "inactive." These results indicate that males used information from the entire scent of each orchid, including compounds that are not detected by EAG, to avoid mistaking the same orchid flower for a female bee. These same "inactive" compounds were also shown to be the most variable in concentration from flower to flower.

Another recent study of honeybees indicates that honeybees use not only all the compounds present in the scent to identify salient odors, they are also able to distinguish floral scents based on the ratios of the scent concentration of individual compounds.[68] In this study, Wright et al.[68] showed that four different genetic varieties (cultivars) of snapdragon had statistically different floral scents. Scents of each variety possessed the same chemical compounds, but the differences arose based on the ratios of the concentrations of compounds in the mixture. Furthermore, based on these ratios, these scents could be statistically ranked according to similarity. After conditioning, honeybees could not distinguish the scent of flowers of the same cultivar. They could, however, distinguish the scents of different cultivars based both on overall intensity and on the ratios of the compounds. Moreover, the ability of honeybees to discriminate among these scents matched the statistical differences from the chemical analyses. This correlation indicates that the olfactory system has the capacity to use information from all of the compounds in an unbiased way, when all of those components provide relevant information for discriminating among the varieties. This is analogous to the equal weighting provided to all of the components in the statistical analysis.

In comparison to this study with snapdragon varieties, Wright et al.[11] found that canola flowers containing both different odorant compounds and different concentrations of those compounds were indistinguishable to honeybees, using the same type of behavioral assay. Snapdragons and canola flowers have different compounds and also different numbers of compounds; snapdragons have 8 to 10 identified compounds and canola flowers have more than 75. It is possible that as the number of compounds in a floral scent increases, the ability to distinguish features in these also decreases.

12.7   INTEGRATION OF FLORAL ODORS WITH VISUAL CUES

Flowers provide a host of visual cues that can serve as important cues themselves or as contexts for learning about odors. Von Frisch[13] showed long ago that free-flying honeybees could be conditioned to associate colors and odors with sucrose reinforcement. Furthermore, when conditioned to a compound of color and odor, honeybees will approach the color from a long distance. Once up close, they exhibit a preference for the odor. Thus colors exert an effect over a longer distance than odors.

More recent studies have revealed that visual cues (colors) and odors can interact in nonadditive ways, much as was discussed for odor-odor interactions. In a series of studies,[28] odor was shown to overshadow color, although color potentiated learning about odor. Furthermore, compound-component discrimination (CO+/C/O) for colors (C) and odors (O) was also demonstrated. These results could be explained through an interaction between colors and odors producing a unique cue that represents the compound, perhaps through within-compound association.

Other types of associations between colors and odors have been demonstrated. Free-flying honeybees associate odors with locations around their colony where they have collected sucrose. When exposed to that odor in the colony some time later, many foragers will visit the specific site associated with the odor.[99] Finally, honeybees can learn abstract rules governing the relationships between stimuli, such as sameness and difference.[100] They can learn that an odor or a visual stimulus presented at one point in time indicates a correct choice at a later point, which is called delayed match to sample. More important for this discussion, they can then transfer this concept from odors to visual cues, or vice versa.

To conclude, honeybees form associations across modalities such as olfaction and vision, and information from one modality influences recall in the other. As concluded above, further understanding of these processes must await more detailed analysis of neural mechanisms.

12.8   NEURAL MECHANISMS OF OLFACTORY PROCESSING

12.8.1   WHY IS OLFACTORY PROCESSING SO COMPLEX?

Neural solutions for any ecological problem, like the detection and discrimination of odors, must be implemented within the constraints dictated by the computational capacity of the peripheral and central nervous systems.[26,27] Being generalist floral visitors, honeybees must be capable of detecting an essentially infinite number of floral odors given the combinatorial possibilities with complex mixtures. But they need to do this with relatively few neurons because of constraints on body size. This constraint—lots of odors and relatively few neurons—is the major reason that these odors cannot be processed through a series of dedicated neural pathways.

Synthetic or configural units, as well as the competitive interactions represented by blocking, might arise because of constraint-driven interactions in sensory or more central processing of odors.[89] Some groups of odorants, by virtue of the way their neural codes interact, might give rise to synthetic units more readily than other groups of odorants. Odorants that are processed independently, at least early on in the olfactory system, might be more capable of giving rise to configural (associative) units or competitive interactions. However, it is difficult with behavioral experiments alone to resolve which types of odorants might show which types of interactions. This is the reason we propose that coupling behavioral and physiological analyses might provide a better, more powerful predictive framework for understanding how mixtures are processed. Physiological analysis might identify groups of odors based on an overlap of neural codes such that predictions from behavioral experiments become more precise. This overlap could be in the form of excitatory responses from populations of sensory cells or in terms of inhibitory interactions among glomeruli in the antennal lobe.

12.8.2 MECHANISMS OF OLFACTORY PROCESSING

Both vertebrate and invertebrate olfactory systems face the same constraint, and they have apparently converged on similar means for compressing a multidimensional odor space into a far more limited neural space (Figure 12.10a);[101,102] (see Hildebrand, and Shepherd[26] and Laurent[27] for thorough reviews of neural bases of olfactory processing). In insects, when odorant molecules arrive at the cuticle of a sensory structure, they pass through pores in the cuticle and interact with one or more of a few types of odorant binding proteins located in the lymph of the hair.[103,104] These proteins probably serve to transport the hydrophobic molecule through the lymph to interact with molecular receptors expressed on the dendrites of sensory neurons that project into the sensory structure. Each sensory cell expresses one of 40 to 200 different types of receptor proteins that exist within insect genomes.[105,106] Receptors

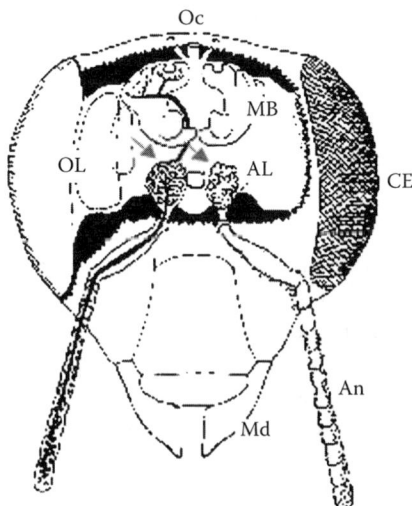

(a)

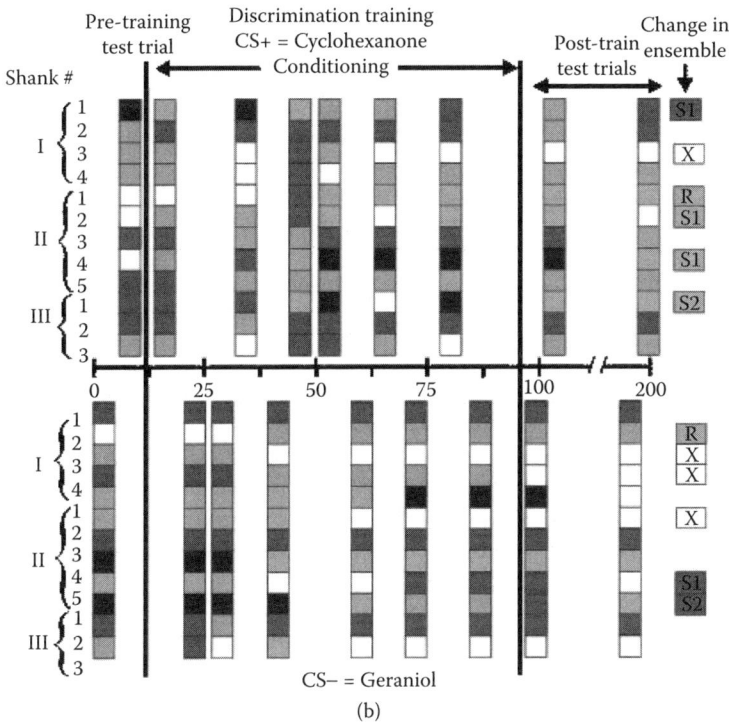

FIGURE 12.10 (a) Schematic of a honeybee worker head with a cutaway view of the brain. AL, antennal lobe; An, antenna; CE, compound eye; MB, mushroom body; Oc, ocellus; OL, optic lobe. (b) (See color insert following page 178.) The trial-by-trial mean changes in neural responses to odor stimulation during differential conditioning (CS+ and CS– odors as described in the text) of the moth, *M. sexta* (from Daly et al.[112]). Unit response data by trial are from a single animal over the course of differential conditioning. Twelve different units (putatively neurons) were identified and recorded on each trial with each odor. Offset of columns indicates that the CS+ and CS– odorants were tested on separate trials. Pre- and post-training trials were conducted without reinforcement. Boxes are color coded at each trial to indicate the statistical change in a unit's response to odor stimulation on those trials: red, excitatory; blue, inhibitory; clear, no change in background firing rate. Overall changes in unit responses over the course of the experiment are summarized on the right: (red "R") an excitatory unit was recruited; (blue "R") an inhibitory unit was recruited; (clear X), an initially responsive unit was dropped; (blue "S1") switched response polarity to one odor and from an excitatory to an inhibitory response; (clear "S2") switched response polarity to both odors (color in figure denotes response at the first post-test). (From Daly, K.C., Christensen, T.A., Lei, H., Smith, B.H., and Hildebrand, J.G., *Proc. Nat. Acad. Sci. USA* 101, 10476–10481, 2004, with permission.)

are tuned to respond to a small range of odorants and are typically most sensitive to a few of those odorants.[69] The tuning broadens with an increase in concentration, because odorants to which the receptor is less sensitive are capable of activating the receptor at higher concentrations. Any odorant will therefore excite several different types of receptor cells, and the pattern of cells excited by two different odorants may partially overlap. Thus the population of approximately 50,000 receptor cells on a honeybee antenna probably set up somewhat broadly tuned, overlapping excitatory sensory representations for odorants and mixtures.[107,108]

Axons from sensory cells that express the same receptor converge to the same glomerulus in the antennal lobe, which is the locus of synaptic interactions with brain interneurons.[109] Thus there is a rough correspondence between the number of receptors and the number of glomeruli, and a glomerulus represents the summarized excitation of at least several hundred sensory cell axons. The dendrites of one or maximally a few projection neurons innervate a glomerulus, and the axons from the projection neuron project the output from the AL to other areas of the brain. Finally, local interneurons interconnect glomeruli.[27] Processes of these interneurons are restricted to the AL. They most likely mediate one or possibly two different types of inhibition.[110]

This network sets up the first-order representations of odors in the brain. It consists of several different types of excitatory and inhibitory interactions among sensory pathways that can change the neural representation for odors.[27] For example, because of the inhibition distributed by the local interneurons, there is not a one-to-one correspondence between sensory input and activation of glomeruli. Sachse and Galizia[72,110] used optical imaging to compare odor-driven glomerular activity elicited by sensory axons with the ensuing activation of projection (output) neurons. The latter activity was characterized by a subset of the more highly activated glomeruli in the sensory pattern. This restriction in activation can be accomplished if highly activated projection neurons effectively suppress relatively weakly activated projection neurons through lateral inhibition. Furthermore, this local network can give rise to complex spatiotemporal patterns that contribute to the code for odor identity.[27,59]

Interactions at the level of the antennal lobe network are made more complex by feedback from other brain neuropils that represents the presence of reinforcement. In the honeybee, a cluster of cells located in the subesophageal ganglion receive input from sucrose-sensitive taste hairs on the mouthparts.[54] They then project their outputs to the antennal lobes, where outputs arborize in most or all of the glomeruli, and to the mushroom bodies, which in the honeybee brain is a multimodal association center. The anatomy and electrophysiological responses of one such cell in this cluster—called VUMmx1—to odors and sucrose have been reasonably well characterized.[54] In naïve honeybees, VUMmx1 responds to sucrose but not to odor. After associative pairing of an odor with sucrose reinforcement, and in a manner that produces robust conditioned responding in behavioral experiments,[36] VUMmx1 responds to odor. Thus VUMmx1 is an important linkage between CS (odor) and US (sucrose) neural pathways in the brain.

Two recent electrophysiological studies of the antennal lobes indicate that neural responses to odor are modified by reinforcement. In the honeybee, glomeruli that are activated by an odor show an increase in responsiveness to that odor after it has been associatively paired with sucrose reinforcement.[111] In the moth, individual units in the antennal lobes show complex changes in response patterns when associated with reinforcement (Figure 12.10b).[112] Excitatory responses in identified cells can increase or decrease. Responses may also switch from excitation to inhibition or vice versa. Differences between honeybees and moths in the modification of responses to odor in the antennal lobe may be due to different measurement techniques.

Nevertheless, both studies reveal significant plasticity in the ensembles in insect antennal lobes, which might be mediated by VUMmx1 or its equivalent in the moth. Disruption of VUM-mediated modulation in the antennal lobe blocks associative conditioning of odor,[113] which further supports the contention that modulation by reinforcement is important for the function of the antennal lobes.

12.8.3 THE ANTENNAL LOBE AS A "MATCHED FILTER"

These data beg the question of why reinforcement is necessary for antennal lobe processing. Answers to this question remain hypothetical. Two computational models of the antennal lobe have suggested that reinforcement tunes the antennal lobe such that it acts as a matched filter for odor detection.[90,91] Coincident activation of odor (sensory) and reinforcement (VUMmx1) pathways changes the response of the neural ensemble in the antennal lobe to that odor. Glomeruli that participate in this reinforced pattern can more effectively inhibit other glomeruli to which they are connected via inhibitory interneurons. Subsequently any pattern of activity in the antennal lobe that is similar to the reinforced pattern will converge to the reinforced pattern. This would solve the variable mixture problem alluded to above (BAXY, CAXZ, DAXT) because the antennal lobe would become tuned to detect AX over the more variable background. And it would solve the blocking problem as well.

12.8.4 INTERACTIONS BETWEEN THE ANTENNAL LOBE AND HIGHER BRAIN CENTERS

Under the matched filter hypothesis, the antennal lobe would be charged with detection of important odors. This information would be passed on to higher centers, such as the mushroom bodies, for integration of that information with contexts from other sensory modalities (e.g., vision). Presumably some of the interactions between odors and visual cues reviewed above would arise first in the mushroom bodies, where multimodal sensory inputs converge.

12.9 CONCLUSION

We have attempted to present a review of behavioral mechanisms that animals use to detect and discriminate natural odors. These mechanisms necessarily involve several different mechanisms of behavioral plasticity, because odors are made relevant, or less relevant, through experience of their association (or lack thereof) with reinforcement. Particularly in regard to learning about natural mixtures, these behavioral mechanisms are complex. It is our feeling that there are not yet enough detailed studies to reach a consensus about which mechanisms are used and in what situations they might come into play.

Nevertheless, a take-home message might be distilled to three basic points, which we have made throughout the discussion. First, and perhaps most obvious, more studies need to be performed that attempt to integrate different behavioral mechanisms and determine their relevance to ecological problems such as, for example, variation in odor composition among flowers. Second, many of the behavioral mechanisms cannot be understood without correlated physiological studies designed to understand how particular odors are processed (independently versus interactively) in the brain. These types of studies may help to clarify which odors might be expected to produce specific behavioral mechanisms in conditioning protocols. Finally, we need to know much more about the ecological problems that learning mechanisms have evolved to solve. In particular, work reviewed above has shown that flower-to-flower variation can have a dramatic impact on what aspects of the mixture honeybees learn. The perceptual properties of the odor memory do not reflect a simple running average of the levels of each of the components. Yet when flower species are chemically characterized, often dozens to hundreds of flowers are extracted and the mean level of each component is reported. Much more information is needed about individual flowers and variations in components among flowers.

ACKNOWLEDGMENTS

This work was supported by grants from the National Institutes of Health (NCRR 9 R01 RR1466 to B.H.S.; NIDCD DC05535 to K.C.D.). Support was provided from a National Science Foundation grant (agreement 0112050) to the Mathematical Biosciences Institute at Ohio State University (to G.A.W.).

REFERENCES

1. Chittka, L., Thomson, J.D., and Waser, N.M., Flower constancy, insect psychology, and plant evolution, *Naturwissenschaften* 86, 361, 1999.
2. Seeley, T.D., *The Wisdom of the Hive*, Harvard University Press, Cambridge, MA, 1996, p. 302.
3. Menzel, R., Das Gedächtnis der Honigbiene für Spektralfarben. I. Kurzzeitiges und langzeitiges Behalten, *Z. Vergl. Physiol.* 60, 82, 1968.
4. Werner, A., Menzel, R., and Wehrhahn, C., Color constancy in the honeybee, *J. Neurosci.* 8, 156, 1988.
5. Backhaus, W., Color opponent coding in the visual system of the honeybee, *Vision Res.* 31, 1381, 1991.
6. Chittka, L., Shmida, A., Troje, N., and Menzel, R., Ultraviolet as a component of flower reflections, and the colour perception of Hymenoptera, *Vision Res.* 34, 1489, 1994.
7. Free, J.B., Effect of flower shapes and nectar guides on the behaviour of foraging honeybees, *Behaviour* 37, 269, 1970.
8. Rodriguez, I., Gumbert, A., Hempel de Ibarra, N., Kunze, J., and Giurfa, M., Symmetry is in the eye of the "beholder": innate preference for bilateral symmetry in flower-naive bumblebees, *Naturwissenschaften* 91, 374, 2004.
9. Kevan, P.G., Texture sensitivity in the life of honeybees, in *Neurobiology and Behavior of Honeybees*, Menzel, R. and Mercer, A., Eds., Springer, Berlin, 1987, p. 96.
10. Scheiner, R., Erber, J., and Page, R.E., Jr., Tactile learning and the individual evaluation of the reward in honey bees (*Apis mellifera* L.), *J. Comp. Physiol. A* 185, 1, 1999.
11. Wright, G.A., Skinner, B.D., and Smith, B.H., Ability of honeybee, *Apis mellifera*, to detect and discriminate odors of varieties of canola (*Brassica rapa* and *Brassica napus*) and snapdragon flowers (*Antirrhinum majus*), *J. Chem. Ecol.* 28, 721, 2002.

12. Raguso, R.A., Flowers as sensory billboards: progress towards an integrated understanding of floral advertisement, *Curr. Opin. Plant Biol.* 7, 434, 2004.

13. von Frisch, K., *The Dance Language & Orientation of Bees*, Harvard University Press, Cambridge, MA, 1967, p. 566.

14. Real, L.A., Animal choice behavior and the evolution of cognitive architecture, *Science* 253, 980, 1991.

15. Stephens, D.W. and Krebs, J.R., *Foraging Theory*, Princeton University Press, Princeton, NJ, 1986, p. 247.

16. Hill, P.S.M., Wells, P.H., and Wells, H., Spontaneous flower constancy and learning in honey bees as a function of colour, *Anim. Behav.* 54, 615, 1997.

17. Menzel, R., Learning, memory and "cognition" in honey bees, in *Neurobiology of Comparative Cognition*, Kesner, R.P. and Olton, D.S., Eds., Erlbaum, Hillsdale, NJ, 1990, p. 237.

18. Michener, C.D., *The Bees of the World*, Johns Hopkins University Press, Baltimore, MD, 2000.

19. Buchmann, S.L. and Nabhan, G.P., *The Forgotten Pollinators*, Island Press, Washington, DC, 1995, p. 292.

20. Pavlov, I.P., *Conditioned Reflexes*, Oxford University Press, New York, 1927.

21. Pearce, J.M., *Animal Learning and Cognition*, 2nd ed., Psychology Press, Sussex, UK, 1997.

22. Bitterman, M.E., Incentive contrast in honey bees, *Science* 192, 380, 1976.

23. Rescorla, R.A. and Holland, P.C., Behavioral studies of associative learning in animals, *Annu. Rev. Psychol.* 33, 265, 1982.

24. Mackintosh, N.J., *The Psychology of Animal Learning*, Academic Press, London, 1974.

25. Rescorla, R.A., Behavioral studies of Pavlovian conditioning, *Annu. Rev. Neurosci.* 11, 329, 1988.

26. Hildebrand, J.G. and Shepherd, G.M., Mechanisms of olfactory discrimination: converging evidence for common principles across phyla, *Annu. Rev. Neurosci.* 20, 595, 1997.

27. Laurent, G., Olfactory network dynamics and the coding of multidimensional signals, *Nat. Rev. Neurosci.* 3, 884, 2002.

28. Bitterman, M.E., Comparative analysis of learning in honeybees, *Anim. Learn. Behav.* 24, 123, 1996.

29. Kuwabara, M., Bilding des bedingten Reflexes von Pavlovs Typus bei der Honigbiene (*Apis mellifica*), *J. Fac. Sci. Hokkaido Univ. Ser. VI Zool.* 13, 458, 1957.

30. Menzel, R. and Bitterman, M.E., Learning by honeybees in an unnatural situation, in *Neuroethology and Behavioral Physiology*, Huber, F. and Markl, H., Eds., Springer-Verlag, New York, 1983, p. 206.

31. Erber, J., Pribbenow, B., Bauer, A., and Kloppenburg, P., Antennal reflexes in the honeybee: tools for studying the nervous system, *Apidologie* 24, 283, 1993.

32. Smith, B.H. and Menzel, R., An analysis of variability in the feeding motor program of the honey bee: the role of learning in releasing a modal action pattern, *Ethology* 82, 68, 1989.

33. Smith, B.H. and Menzel, R., The use of electromyogram recordings to quantify odorant discrimination in the honey bee, *Apis mellifera*, *J. Insect Physiol.* 35, 369, 1989.

34. Smith, B.H., An analysis interaction in binary odorant mixtures, *Physiol. Behav.* 65, 397, 1998.

35. Hellstern, F., Malaka, R., and Hammer, M., Backward inhibitory learning in honeybees: a behavioral analysis of reinforcement processing, *Learn. Mem.* 4, 429, 1998.

36. Bitterman M.E., Menzel, R., Fietz, A., and Schäfer, S., Classical conditioning of proboscis extension in honeybees (*Apis mellifera*), *J. Comp. Physiol.* 97, 107, 1983.

37. Smith, B.H., Abramson, C.I., and Tobin, T.R., Conditional withholding of proboscis extension in honey bees (*Apis mellifera*) during discriminative punishment, *J. Comp. Psychol.* 105, 345, 1992.

38. Mackintosh, N.J., *Conditioning and Associative Learning*, Oxford University Press, New York, 1983.

39. Cardé, R.T. and Minks, A.K., *Insect Pheromone Research*, Chapman & Hall, New York, 1998.

40. Karban, R. and Baldwin, I.T., *Induced Responses to Herbivory*, University of Chicago Press, Chicago, 1997.

41. Murlis, J., Odor plumes and the signals they provide, in *Insect Pheromone Research*, Cardé, R. and Minks, A.K., Eds., Chapman & Hall, New York, 1997, p. 221.

42. Willis, M.A. and Arbas, E.A., Variability in odor-modulated flight by moths, *J. Comp. Physiol. A* 182, 191, 1998.

43. Daly, K.C. and Figueredo, A.J., Habituation of sexual response in *Heliothis* moths, *Physiol. Entomol.* 25, 180, 2000.

44. Smith, B.H., Recognition of female kin by male bees through olfactory signals, *Proc. Natl. Acad. Sci. USA* 80, 4551, 1983.

45. Thompson, R.F. and Spencer, W.A., Habituation: a model phenomenon for the study of substrates of behavior, *Psychol. Rev.* 73, 16, 1966.

46. Peeke, H.V.S., and Petrinovich, L., *Habituation, Sensitization, and Behavior*, Academic Press, Orlando, FL, 1984.

47. Ayasse, M., Schiestl, F.P., Paulus, H.F., Ibarra, F., and Francke, W., Pollinator attraction in a sexually deceptive orchid by means of unconventional chemicals, *Proc. R. Soc. Lond. B Biol. Sci.* 270, 517, 2003.

48. Gemeno, C., Lutfallah, A.F., and Haynes, K.F., Pheromone blend variation and cross-attraction among populations of the black cutworm moth (Lepidoptera: Noctuidae), *Ann. Entomol. Soc. Am.* 93, 1322, 2000.

49. Daly, K.C. and Smith, B.H., Associative olfactory conditioning of the moth, *Manduca sexta*, *J. Exp. Biol.* 203, 2025, 2000.

50. Daly K.C., Chandra, S.B.C., Durtschi, M.L., and Smith, B.H., The generalization of olfactory-based conditioned response reveals unique but overlapping odour representations in the moth, *Manduca sexta*, *J. Exp. Biol.* 204, 3085, 2001.

51. Daly, K.C., Durtschi, M.L., and Smith, B.H., Olfactory-based discrimination learning in the moth, *Manduca sexta*, *J. Insect Physiol.* 47, 375, 2001.

52. Kessler, A. and Baldwin, I.T., Defensive function of herbivore-induced plant volatile emissions in nature, *Science* 291, 2141, 2001.

53. Kessler, A., Halitschke, R., and Baldwin, I.T., Silencing the jasmonate cascade: induced plant defenses and insect populations, *Science* 305, 665, 2004.

54. Hammer, M. and Menzel, R., Learning and memory in the honeybee, *J. Neurosci.* 15, 1617, 1995.

55. Smith, B.H., Merging mechanism and adaptation: learning, generalization, and the control of behavior, in *Insect Learning: Ecological and Evolutionary Perspectives*, Lewis, A.C. and Papaj, D.R., Eds., Chapman & Hall, New York, 1993, p. 126.

56. Shepherd, G.M., Computational structure of the olfactory system, in *Olfaction: A Model System for Computational Neuroscience*, Davis, J.L. and Eichenbaum, H., Eds., MIT Press, Cambridge, MA, 1991, p. 3.

57. Belluscio, L. and Katz, L.C., Symmetry, stereotypy, and topography of odorant representations in mouse olfactory bulbs, *J. Neurosci.* 21, 2113, 2001.

58. Leon, M. and Johnson, B.A., Olfactory coding in the mammalian olfactory bulb, *Brain Res. Rev.* 42, 23, 2003.

59. Daly, K.S., Wright, G.A., and Smith, B.H., Molecular features of odorants systematically influence slow temporal responses across clusters of coordinated antennal lobe units in the moth *Manduca sexta*, *J. Neurophysiol.* 92, 236, 2004.

60. Vareschi, E., Duftunterscheidung bei der Honigbiene. Einzelzell-Ableitungen und Verhaltensreaktionen, *Z. Vergl. Physiol.* 75, 143, 1971.

61. Ditzen, M., Evers, J.F., and Galizia, C.G., Odor similarity does not influence the time needed for odor processing, *Chem. Senses* 28, 781, 2003.

62. Shephard, R.N., Toward a universal law of generalization for psychological science, *Science* 237, 1317, 1987.

63. Wright, G.A. and Smith, B.H., Variation in complex olfactory stimuli and its influence on odour recognition, *Proc. R. Soc. Lond. B* 271, 147, 2004.

64. Wright, G.A. and Smith, B.H., Different thresholds for detection and discrimination of odors in the honeybee (*Apis mellifera*), *Chem. Senses* 29, 127, 2004.

65. Keasar, T., The spatial distribution of nonrewarding artificial flowers affects pollinator attraction, *Anim. Behav.* 60, 639, 2000.

66. Shafir, S., Wiegmann, D.D., Smith, B.H., and Real, L.A., Risk-sensitive foraging: choice behaviour of honeybees in response to variability in volume of reward, *Anim. Behav.* 57, 1055, 1999.

67. Chandra, S.B.C., Hosler, J.S., and Smith, B.H., Heritable variation for latent inhibition and its correlation to reversal learning in the honey bee, *Apis mellifera*, *J. Comp. Psychol.* 114, 86, 2000.

68. Wright, G.A., Lutmerding, A., Dudareva, N., and Smith, B.H., Intensity and the ratios of compounds in the scent of snapdragon flowers affect scent discrimination by honey bees (*Apis mellifera*), *J. Comp. Physiol.* 191, 105, 2005.

69. Shields, V.D. and Hildebrand, J.G., Responses of a population of antennal olfactory receptor cells in the female moth *Manduca sexta* to plant-associated volatile organic compounds, *J. Comp. Physiol. A* 186, 1135, 2001.

70. Atema, J., Borroni, P., Johnson, B., Voigt, R., and Handrich, L., Adaptation and mixture interaction in chemoreceptor cells: mechanisms for diversity and contrast enhancement, in *Perception of Complex Smells and Tastes*, Laing D.G., Ed., Academic Press, Marrickville, NSW, Australia, 1989, p. 83.

71. Stopfer, M., Jayaraman, V., and Laurent, G., Intensity versus identity coding in an olfactory system, *Neuron* 39, 991, 2003.

72. Sachse, S. and Galizia, C.G., The coding of odour-intensity in the honeybee antennal lobe: local computation optimizes odour representation, *Eur. J. Neurosci.* 18, 2119, 2003.

73. Murlis, J. and Jones, C.D., Fine-scale structure of odor plumes in relation to insect orientation to distant pheromone and other attractant sources, *Physiol. Entomol.* 6, 71, 1981.

74. Chandra, S.B.C. and Smith, B.H., An analysis of synthetic processing of odor mixtures in the honey bee (*Apis mellifera*), *J. Exp. Biol.* 201, 3113, 1998.

75. Deisig, N., Lachnit, H., Sandoz, J.C., Lober, K., and Giurfa, M., A modified version of the unique cue theory accounts for olfactory compound processing in honeybees, *Learn. Mem.* 10, 199, 2003.

76. Jinks, A. and Laing, D.G., The analysis of odor mixtures by humans: evidence for a configurational process, *Physiol. Behav.* 72, 51, 2001.

77. Hudson, R., From molecule to mind: the role of experience in shaping olfactory function, *J. Comp. Physiol. A* 185, 297, 1999.

78. Smith, B.H., An analysis of blocking in binary odorant mixtures: an increase but not a decrease in intensity of reinforcement produces unblocking, *Behav. Neurosci.* 111, 57, 1997.

79. Rudy, J.W. and Sutherland, R.J., Configural and elemental associations and the memory coherence problem, *J. Cogn. Neurosci.* 4, 208, 1992.

80. Galizia, C.G. and Menzel, R., The role of glomeruli in the neural representation of odours: results from optical recording studies, *J. Insect Physiol.* 47, 115, 2001.

81. Muller, D., Gerber, B., Hellstern, F., Hammer, M., and Menzel, R., Sensory preconditioning in honeybees, *J. Exp. Biol.* 203, 1351, 2000.

82. Smith, B.H. and Cobey, S., The olfactory memory of honey bee, *Apis mellifera*: II. Blocking between odorants in binary mixtures, *J. Exp. Biol.* 195, 91, 1994.

83. Couvillon, P.A., Arakaki, L., and Bitterman, M.E., Intramodal blocking in honeybees, *Anim. Learn. Behav.* 25, 277, 1997.

84. Couvillon, P.A., Campos, A.C., Bass, T.D., and Bitterman, M.E., Intermodal blocking in honeybees, *Q. J. Exp. Psychol. B* 54, 369, 2001.

85. Hosler, J.S. and Smith, B.H., Blocking and the detection of odor components in blends, *J. Exp. Biol.* 203, 2797, 2000.

86. Blaser, R.E., Couvillon, P.A., and Bitterman, M.E., Backward blocking in honeybees, *Q. J. Exp. Psychol. B* 57, 349, 2004.

87. Thorn, R.S. and Smith, B.H., The olfactory memory of the honeybee, *Apis mellifera* III. Bilateral sensory input is necessary for induction and expression of olfactory blocking, *J. Exp. Biol.* 200, 2045, 1997.

88. Gerber, B. and Ullrich, J., No evidence for olfactory blocking in honeybee classical conditioning, *J. Exp. Biol.* 202, 1839, 1999.

89. Smith, B.H., The role of attention in learning about odorants, *Biol. Bull. MBL* 191, 76, 1996.

90. Linster, C. and Smith, B.H., A computational model of the response of honey bee antennal lobe circuitry to odor blends: overshadowing, blocking and unblocking can arise from lateral inhibition, *Behav. Brain Res.* 87, 1, 1997.

91. Borisyuk, A. and Smith, B.H., Odor interactions and learning in a model of the insect antennal lobe, *Neurocomputing* 1041, 58, 2004.

92. Jakobsen, H.B. and Olsen, C.E., Influence of climatic factors on rhythmic emission of volatiles from *Trifolium repens* L. flowers in situ, *Planta* 192, 365, 1994.

93. Pott, M.B., Pichersky, E., and Piechulla, B., Evening specific oscillations of scent emission, SAMT enzyme activity, and SAMT mRNA in flowers of *Stephanotis floribunda*, *J. Plant Physiol.* 159, 925, 2002.

94. Dudareva, N., Murfitt, L.M., Mann, C.J., Gorenstein, N., Kolosova, N., Kish, C.M., Bonham, C., and Wood, K., Developmental regulation of methyl benzoate biosynthesis and emission in snapdragon flowers, *Plant Cell* 12, 949, 2000.

95. Schiestl, F.P., Ayasse, M., Paulus, H.F., Erdmann, D., and Francke, W., Variation of floral scent emission and postpollination changes in individual flowers of *Ophrys sphegodes* subsp. *sphegodes*, *J. Chem. Ecol.* 23, 2281, 1997.

96. Pham-Delegue, M.H., Bailez, O., Blight, M.M., Masson, C., Picard-Nizou, A.L., and Wadhams, L.J., Behavioral discrimination of oilseed rape volatiles by the honeybee *Apis mellifera* L., *Chem. Senses* 18, 483, 1993.

97. Pham-Delegue, M.H., Blight, M.M., Kerguelen, V., Le Métayer, M., Marion-Poll, F., Sandoz, J.C., and Wadhams, L.J., Discrimination of oilseed rape volatiles by the honeybee: combined chemical and biological approaches, *Entomol. Exp. Appl.* 83, 87, 1997.

98. Andersson, S. and Dobson, H.E., Behavioral forging responses by the butterfly *Heliconius melpomene* to *Lantana camara* floral scent, *J. Chem. Ecol.* 29, 2303, 2003.

99. Reinhard, J., Srinivasan, M.V., Guez, D., and Zhang, S.W., Floral scents induce recall of navigational and visual memories in honeybees, *J. Exp. Biol.* 207, 4371, 2004.

100. Giurfa, M., Zhang, S., Jenett, A., Menzel, R., and Srinivasan, M.V., The concepts of "sameness" and "difference" in an insect, *Nature* 410, 930, 2001.

101. Eisthen H.L., Why are olfactory systems of different animals so similar?, *Brain Behav. Evol.* 59, 273, 2002.

102. Strausfeld, N.J. and Hildebrand, J.G., Olfactory systems: common design, uncommon origins?, *Curr. Opin. Neurobiol.* 9, 634, 1999.

103. Vogt, R.G., Prestwich, G.D., and Lerner, M.R., Odorant-binding-protein subfamilies associate with distinct classes of olfactory receptor neurons in insects, *J. Neurobiol.* 22, 74, 1991.

104. Pelosi, P., Odorant-binding proteins, *Crit. Rev. Biochem. Mol. Biol.* 29, 199, 1994.

105. Clyne, P.J., Warr, C.G., Freeman, M.R., Lessing, D., Kim, J., and Carlson, J.R., A novel family of divergent seven-transmembrane proteins: candidate odorant receptors in *Drosophila*, *Neuron* 22, 327, 1999.

106. Vosshall, L.B., Amrein, H., Morozov, P.S., Rzhetsky, A., and Axel, R., A spatial map of olfactory receptor expression in the *Drosophila* antenna, *Cell* 96, 725, 1999.

107. Getz, W.M. and Akers, R.P., Honeybee olfactory sensilla behave as integrated processing units, *Behav. Neural. Biol.* 61, 191, 1994.

108. Getz, W.M. and Akers, R.P., Partitioning non-linearities in the response of honey bee olfactory receptor neurons to binary odors, *Biosystems* 34, 27, 1995.

109. Vosshall, L.B., Wong, A.M., and Axel, R., An olfactory sensory map in the fly brain, *Cell* 102, 147, 2000.

110. Sachse, S. and Galizia, C.G., Role of inhibition for temporal and spatial odor representation in olfactory output neurons: a calcium imaging study, *J. Neurophysiol.* 87, 1106, 2002.

111. Faber, T., Joerges, J., and Menzel, R., Associative learning modifies neural representations of odors in the insect brain, *Nat. Neurosci.* 2, 74, 1999.

112. Daly, K.C., Christensen, T.A., Lei, H., Smith, B.H., and Hildebrand, J.G., Learning modulates the ensemble representations for odors in primary olfactory networks, *Proc. Natl. Acad. Sci. USA* 101, 10476, 2004.

113. Farooqui, T., Robinson, K., Vaessin, H., and Smith, B.H., Modulation of early olfactory processing by an identified octopaminergic reinforcement pathway in the honeybee, *J. Neurosci.* 23, 5370, 2003.

13 Behavioral Responses to Floral Scent: Experimental Manipulations and the Interplay of Sensory Modalities

Robert A. Raguso

CONTENTS

13.1 INTRODUCTION

With the publication of this book and recent special features in major journals,[1,2] the interdisciplinary field of floral scent research has come of age. This book celebrates an unprecedented decade of advances in our understanding of scent biosynthesis and emission in flowers, and its detection and perception by animals. A major challenge confronting fragrance scientists is how to fully integrate these discoveries into the mainstream of plant reproductive biology, with the goal of reaching a more sophisticated understanding of the selective forces that shape the evolution of scent as a dimension of floral phenotype. This goal is hindered by strong cultural biases among investigators. Despite the general importance of olfaction to insects[3] and other flower-visiting animals, there is a persistent bias among insect visual physiologists and plant reproductive ecologists that pollination is essentially a visually guided phenomenon.[4–6] The traditional omission of scent from experimental studies of pollination and floral evolution may have economic as well as philosophical causes, but recent advances in analytical chemistry provide inexpensive, rigorous methods with which to characterize fragrances (Chapter 1) and incorporate scent chemistry into manipulative experiments.[7] At the opposite extreme, plant biochemists and olfactory physiologists commonly assume that fragrance is necessary for fertilization or outcrossing, and thus is subject to pollinator-mediated natural selection.[8–10] However, not all scented flowers require pollinators, and not all pollinators use scent to find flowers.[11,12] Moreover, selection on floral traits by pollinators may be relatively weak when compared with the impact of herbivores or edaphic factors.[13–15] In such cases, scent chemistry may reflect plant defense, genetic drift, or shared evolutionary history (phylogeny) rather than selection for pollinator attraction or preference.[16–19]

Thus we need to ask when fragrance is important, how it works, and which subsets or blend components elicit behavioral responses from beneficial or detrimental floral visitors. Floral scent rarely functions alone and should be studied in the context of other sensory cues (color, texture, nectar taste) likely to be utilized by flower visitors.[20–22] The seminal reviews by Norris Williams[23] and Heidi Dobson[24] emphasized these issues as keys to understanding the overall contributions of floral scent to plant-pollinator relationships, and substantial progress has been made in the past decade. I revisit these themes here, with a focus on diversity of function at different physical scales and the behavioral context in which odor and other sensory cues are perceived. Complementary behavioral themes are discussed with authority in Chapter 8 (the systematic distribution of fragrance chemistry), Chapter 9 (butterfly feeding), Chapter 10 (sexual deception and gas chromatography-electroantennographic detection [GC-EAD]), and Chapter 12 (odor perception and associative learning).

13.2 WHEN IS SCENT IMPORTANT?

13.2.1 What Constitutes an Olfactory Pattern? The Rationale for Behavioral Assays

A significant barrier to understanding the behavioral relevance of fragrance is our inability, as human observers, to smell and recognize patterns relevant to pollinator olfaction without technological assistance. For example, humans' keen visual acuity, combined with observations dating from Sprengel,[25] support the hypothesis that dark stripes within a pale *Penstemon* corolla visually guide bumblebees to hidden nectar (Figure 13.1A). We recognize that these signals are nested within the larger targets of blue racemes, and expect bees to utilize different visual signals depending upon their distance from the flower.[26,27] Even patterns of ultraviolet (UV) reflectance, undetected by the human eye, make intuitive sense as "targets" when revealed by technological aids.[28,29] Similar contrasts must exist for floral scent, such as the odors of pollen or nectar,[30,31] but our inherent ability to discern such patterns is poor, like the pixilated image of a *Penstemon* flower in Figure 13.1B. Although analytical chemistry provides us with long lists of chemical constituents in fragrances—the pixels of this analogy—we lack the "search images" required to parse relevant olfactory information from complex chemical blends. Thus all identified scent compounds are potentially important until proven otherwise, often a laborious process (Chapters 10 and 12).

The identification of consistent, informative olfactory "patterns" has come slowly, with the accumulation of chemical and behavioral evidence summarized elsewhere[32] (see Chapter 8). In some cases, human sensitivity to the scent of animal waste, carrion, rotting fungi, and fermenting fruit has aided in the identification of the chemical signals that mediate floral mimicry of such substances.[33,34] In other cases, human perception may prevent us from recognizing olfactory patterns when we encounter them. A guild of unrelated Northern Hemisphere plants—*Cornus, Daucus, Heracleum, Sambucus, Sorbus, Viburnum*—produces umbels of small white flowers with highly generalized pollinator spectra. Delpino[34] and Kerner von Marilaum[35] described an unpleasant "aminoid" odor common to these flowers, attributed a century later to the presence of amino acid derivatives such as valine methyl ester, isoleucine-related imines, and 1-pyrroline,[36–38] for which humans have an unusually low perceptual threshold (0.022 ppm for 1-pyrroline[39]). However, these analyses also identified abundant terpenoid and aromatic compounds ubiquitous among angiosperm floral scents, which are likely to attract the broad spectrum of insect visitors that typify such flowers.[40,41] Female tephritid fruit flies are attracted to 1-pyrroline as a component of male fly pheromone,[42] and other insects may also be attracted. However, in the absence of a bioassay, it is unclear why convergent evolution has endowed white umbel flowers with this and similar odors, or whether these compounds are perceived by the plants' broad spectra of beetle, fly, and bee/wasp pollinators. On the other hand, it is possible that these odors or related compounds (e.g., putrescine) defend the exposed nectaries in umbel flowers from exploitation by ants,[43] or alternatively, they might be products of microbial infestation.[31] Our perception of fragrance chemistry may fundamentally differ from that of flower visitors, and our bias toward pollinator attraction may lead us to ignore alternative functions with fitness consequences. Clearly experiments investigating the function of floral scent should include alternative hypotheses in their design.

13.2.2 DECOUPLING FLORAL CUES AT A DISTANCE

Ideally the importance of fragrance variation, like other floral traits, should be measured in the context of pollinator effectiveness, herbivore attack, and reproductive success

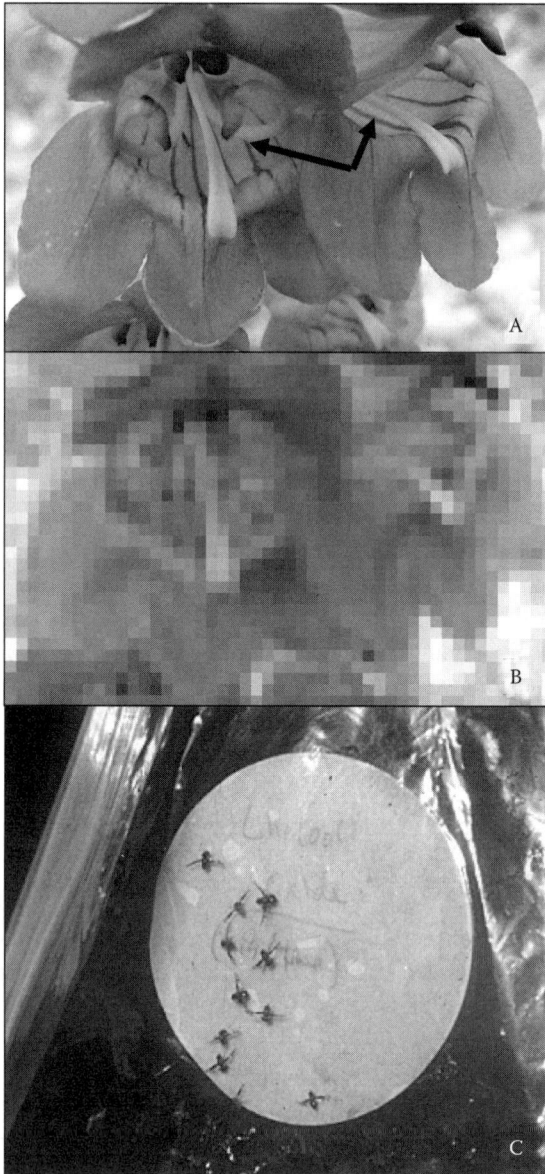

FIGURE 13.1 Identical photographs of *Penstemon* flowers adjusted to (A) adequate (72 dpi) and (B) poor (6 dpi) graphical resolution. In (A), the dark markings (arrow) in the throat contrast visually with the pale corolla and indicate nectar guides for bees. In (B), the grainy resolution is insufficient to distinguish these markings as visual guides. Image (B) represents our current, limited understanding of how the chemical components of floral scent, like visual pixels, are integrated into a complex and functional whole by discerning pollinators. (C) Male *Megalaemyia* (Otitidae) flies were unexpectedly attracted to furanoid linalool oxides used as bait in a euglossine bee trapping experiment at the Maquipucuna field station, Nanegalito, Ecuador. (Photos © Robert A. Raguso.)

(pollen export, fruit or seed set) using quantitative measures of natural selection or path analysis.[44,45] Only recently have such studies begun to include natural or artificial variation in scent,[16,46] and similar experiments are sorely needed on generalized as well as specialized pollination systems. However, there is a long tradition of simple field assays designed to test whether floral scent is attractive to flower visitors, many of which were routinely performed before the advent of modern analytical chemistry.[47,48] For example, the visual cues of scented flowers can be masked by dark gauze or cheesecloth permeable to odor as an intact stimulus.[49,50] Conversely, transparent glass, plastic vessels or bags have been employed to remove fragrance from otherwise visible flowers (Figure 13.2).[51,52] The proximity of such treatments to open control flowers can be used to infer the distance at which scent or visual cues are attractive.[53,54] In this way, *Manduca sexta* moths were shown to use visual or olfactory cues to find flowers from beyond 5 m and a combination of cues to extend the proboscis and feed, but it did not discriminate between scented and scentless flowers bathed in the same odor plume.[55,56] However, distance attraction can be difficult to define and measure due to the differences in body size, foraging range, and habitat selection behaviors of diverse animal pollinators. For animals

FIGURE 13.2 Floral deconstruction. *Oenothera pallida* with petals (A) removed and (B) present; both treatments are wrapped with dyed cheesecloth to control for its presence in other treatments (not shown) in which the entire flower is concealed. Arrow in (B) indicates *Sphecodogastra lusoria*, a crepuscular bee. (C) *Puccinia monoica*-infected *Arabis* leaves placed within transparent plastic vials. The left vial is open, emitting scent, whereas the right vial is capped, providing only visual cues. (D) Outdoor randomized array of *Oenothera caespitosa* flowers with different subsets of floral cues. (1) Controls, (2) flowers concealed within a mesh bag, and (3) scentless paper models decouple odor and visual cues at close range only: the entire array is bathed in fragrance. (Photos © Robert A. Raguso.)

suspected to orient to odor from a distance, it is not uncommon to use sticky traps (or even mist nets) baited with flowers, extracts, or single odorants,[57-59] but such assays are complicated by the potential for secondary attraction to trapped animals. A greater shortcoming of trapping experiments is that only data for the focal species tend to be published, whereas the incidental attraction of unexpected visitors—often herbivores, seed predators, or nectar robbers—indicates potentially important selective agents (Figure 13.1C).[19,60] Laminar flow wind tunnels, Y-tubes, or T-mazes are effective arenas in which to measure locomotory responses to floral scent, while avoiding complex interactions between individuals.[61-63] These assays and others with artificial flowers are discussed at length in a recent treatment by Dobson et al.[64]

13.2.3 FINE-SCALE MANIPULATION: VARIATIONS ON A THEME BY CLEMENTS AND LONG

Once animals arrive at a flower patch, scent may function at much smaller physical scales, usually in conjunction with other sensory signals,[65] for which alternative bioassays are required. The nested nature of floral signals,[66,67] combined with the difficulty of working with some animals in captivity, demands that many plant-pollinator systems must be studied outdoors in natural populations.[68,69] In their groundbreaking book, *Experimental Pollination*, Clements and Long[47] outlined ingenious methods for manipulating the color, posture and symmetry of living flowers *in situ*, providing the tools for modern studies of phenotypic selection.[70-72]

13.2.3.1 Removing Cues from Living Flowers: Floral Deconstruction

At least three categories of floral manipulation have been used to experimentally modify scent and decouple it from other sensory cues. One approach is floral deconstruction, the selective removal or modification of flower parts whose volatile emissions contrast with those of other floral organs (Figure 13.2).[73] The best studied example of olfactory contrast is that between the perianth and androecium in flowers that furnish pollen as a nutritive reward.[30,74,75] Heidi Dobson and Gunnar Bergström pioneered the study of pollen odors and their relevance to specialist and generalist pollen-collecting bees.[76-78] Through the reciprocal transfer of androecia between first- and second-day *Rosa rugosa* flowers, Dobson et al.[79] demonstrated that the olfactory and visual displays of stamen clusters were responsible for discrimination by *Bombus terrestris* bees against second-day flowers. Patt et al.[80] used this approach to control for age-specific changes in spadix and spathe odors in their study of nursery pollination of *Peltandra virginica* (Araceae) by pollen-eating chloropid flies. "Sham inflorescences" of scentless mature bud spadixes (the oviposition substrate) inserted within excised spathes from inflorescences of different ages and genders were used to measure the impact of spathe odor on fly oviposition. The number of eggs decreased with spathe age, total scent production, and changes in emission ratios between the two most abundant compounds, which were new to science.[81] In a related system, Miyake and Yafuso[46] modified *Alocasia odora* (Araceae) spadices to show that the sterile appendix odor attracted drosophilid fly pollinators. Partial or full excision of the appendix significantly reduced fly attraction and fruit set without reducing the visual display (spadices are hidden within spathes). Floral deconstruction clearly holds great promise, but requires the careful use of positive (unmanipulated flowers) and negative controls (for wounding artifacts) and may be impractical for species with small, numerous flowers or delicate floral organs.

13.2.3.2   Adding Odor to Living Flowers: Floral Augmentation

A second approach is floral augmentation, in which extracts, blends, or single odors are added to living flowers, in some cases even from different species,[82] to test a specific behavioral or ecological hypothesis (Figure 13.3). For example, Hossaert-McKey et al.[53] applied pentane extracts from different fig parts or receptive stages to immature figs to identify the source and timing of the most attractive signals, and successfully elicited fig wasp orientation behavior. Dobson et al.[79] found that adding

FIGURE 13.3 Floral augmentation (A, C) and reconstruction (B, D). (A) Microfuge tubes containing floral extracts (in pentane) dissolved in mineral oil were used to augment female strawberry (*F. virginiana*) flowers with the scent of hermaphrodite flowers, anthers, or pentane controls.[83] (Photo © Tia-Lynn Ashman.) (C) Filter papers (arrow and hatched rectangle, right flower) were loaded with methyl benzoate or methyl cinnamate (in ethanol) and inserted between the filaments and corolla of *A. majus* flowers for assays with *Bombus* bees. The closed flower lip (left) concealed the filter paper, but augmented odor was detected in floral headspace.[86] (B) Flower dummies used to study attraction to Arabis leaves infected with *Puccinia* rust fungus. Positive control (center) is an infected "pseudoflower." Yellow (right) and white (left) paper squares were pinned to terry cloth wicks, to which odor extracts or solvent controls were added.[54] A *Dialictus* bee (arrow) landed on the yellow, scented flower dummy. (D) Artificial *Fragaria* flowers with paper corollas and odor emitters with thread anthers and plastic-headed pins as carpels, placed among foliage of living plants. Arrow indicates a fly that pollinates authentic *Fragaria* flowers. (All photos except (A) © Robert A. Raguso.)

eugenol or tetradecyl acetate (pollen odors) to emasculated *Rosa* flowers significantly increased bumblebee landing rates over negative controls, whereas other
components of pollen odor had no effect on visitation. No treatment led to as many
landings as positive controls, underscoring the contribution of visual display to the
bees' search image. More recently, Ashman et al.[83] used floral augmentation to
investigate pollinator discrimination against female flowers in gynodioecious
Fragaria virginiana (Rosaceae). Female flowers augmented with pentane extracts
of hermaphroditic flowers or anthers (but not petals or pentane controls) were
approached at levels on a par with hermaphroditic flowers, but again the stamens'
visual stimuli were required for landing. Cunningham et al.[84] used flower augmentation to test whether *Helicoverpa armigera* moths' conditioned responses to phenylacetaldehyde or α-pinene could be transferred to *Nicotiana tabacum* (Solanaceae)
flowers spiked with those odors. An important implication of this study is that moths
may use subtle differences between odor blends to discriminate between flowers of
the same species (see Chapter 12).

However, the results of floral augmentation can be unpredictable. Manning[85]
added novel odors (lavender and rose oil) to plants of *Cynoglossum officinale*
(Boraginaceae) with and without flowers, respectively, to test their impact on bumblebee foraging decisions. These odors had no effect on bee approaches, whether
flowers were present or not, but reduced landings by half when flowers were present,
either masking or disrupting the combined odor-color stimulus learned by the bees.
Odell et al.[86] suspected that a polymorphism for methyl benzoate and methyl cinnamate contributed to significant bumblebee preference for yellow over white Sonnet
snapdragon (*Antirrhinum majus*: Scrophulariaceae) flowers. Flowers of each color
morph were augmented with the missing odors at both relevant and supernormal
concentrations and set within a randomized array.[87,88] However, the bees' preference
for yellow flowers was not altered, and scent variation appears to have been selectively neutral to *Bombus appositus* in this system. Finally, Baldwin et al.[89] artificially
enhanced the emissions of benzyl acetone, the most abundant volatile in *Nicotiana
attenuata* (Solanaceae) flowers, at the whole-plant level. Seed set did not significantly increase, but seed predation by *Cormelina* bugs did, either because floral odor
was stronger or because it continued during the day, whereas *N. attenuata* flowers
normally emit scent on a nocturnal rhythm.[90] These vignettes highlight the importance
of odor emission rates in experimental pollination studies, and the potential for odors
to have different functions or consequences at different concentrations.[79,91]

13.2.3.3 Building Floral Phenotypes from the Ground up: Floral Reconstruction

A third experimental approach is floral reconstruction, the preparation of artificial
flower models or dummies to which natural odors or rewards are added (Figure
13.3).[68] The simplest cases involve single sensory cues, such as when bees and wasps
feed from colored cardboard squares,[92] and when male euglossine bees or tephritid
flies collect fragrances applied to blotter paper or silica gel TLC plates.[23,93]
A common misinterpretation of single-modality assays is that the missing cues are
unnecessary for pollinator attraction, when in fact additional treatments may reveal

that other floral stimuli are redundant or complementary signals eliciting attraction, landing, or probing.[27,94] More commonly, the full range of pollinator activity at a flower requires interplay between several cues, as illustrated by the classic studies of pseudocopulation in *Ophrys* orchids, in which odors were added to velvet flower dummies to imitate the texture of the orchids' labella (Chapter 10).[95] Gibernau and Hossaert-McKey[96] added pentane extracts of receptive *Ficus carica* figs to paraffin dummy figs and learned that distance attraction of *Blastophaga psenes* wasps is guided by odor, whereas close range orientation and ostiole entry is mediated by contact chemoreception. Andersson[97] designed flower models in which the odors and colors of flowers versus vegetation could be matched in all possible combinations to test for innate and learned preferences among three butterfly species. These assays revealed consistent preferences for floral over vegetative odors as feeding stimulants, especially when combined with pink color, and innate bias for some floral scents over others (Chapter 9).

When successful, floral reconstruction distills complex suites of floral traits into fundamental units of attraction, akin to the "sign-stimulus" experiments of Tinbergen.[98] Schiestl and Peakall's[99,100] elegant studies of sexual deception in Australian orchids reconstituted the stimuli eliciting natural levels of male thynnine wasp attraction with black plastic beads and a single odorant, and provide rare measurements of the magnitude and direction of pollinator-mediated selection on *Chiloglottis* labellum size and odor concentration. However, such assays can be frustrating when flower models are avoided, for reasons ranging from density dependence[101,102] to inappropriate visual reflectance.[103] Also, negative results in these assays are meaningless without positive controls to verify that pollinators were attracted to natural flowers.[56] An additional pitfall in floral reconstruction is the unintended omission of more than one critical cue. Thom et al.[104] recently performed binary choice assays with *M. sexta* moths, using conical paper flowers scented with essential oil. The moths showed an innate preference for flower models that emitted carbon dioxide (CO_2) at levels measured by Guerenstein et al.[105] from flowers of *Datura wrightii* (Solanaceae) over those with ambient levels of CO_2. Differences in the UV reflectance or texture of artificial flowers may also create problems in floral attraction or handling time.[106,107]

One hazard common to floral augmentation and reconstruction is the bias toward attraction; investigators do not frequently test for repellence. Ômura et al.[108] reported deterrence of nectar feeding in *Pieris rapae* butterflies by floral scent and extracts from *Osmanthus fragrans* (Oleaceae). Binary combinations of 2-phenylethanol (2PE), a proven feeding stimulant, plus other single components of *O. fragrans* scent were added to artificial flowers. Blends of 2PE plus γ-decalactone or α-ionone were inferred to be repellent when they significantly reduced feeding in comparison to control flowers scented with 2PE alone. Other measures of repellence involve reduced attraction to a flower model with a single compound, in comparison to the negative control.[79,91] Unexpected interaction effects also could occur in floral augmentation or reconstruction, as was documented for simple odor blends from male euglossine bee-pollinated orchids in Central America. Methyl cinnamate attracts four genera of euglossines, but when combined with one to three additional compounds from a given species of *Gongora* or *Stanhopea*, only that orchid's pollinator is attracted.[109] Furthermore, generalized pollinator responses to single scent components do not always indicate that they are the sole functional attractants. Roy and Raguso[54] observed generalized attraction of halictid bees to several yellow paper models charged with different single components of *Puccinia* rust fungus odor (methyl benzoate, phenylacetaldehyde, 2PE) as well as reconstituted blends (Figure 13.3), whereas anthomyiid flies only responded to the intact rust-infected *Arabis* controls. On the other hand, generalized odor attraction appears to be precisely the mechanism that drives competition among spring flowers for naïve foraging bees in both rewarding and food-deceptive flowers.[110,111] Given the subtle, complementary differences between classes of floral manipulation and the information gleaned from each approach, experiments that incorporate multiple methods at different spatial scales[84] are most likely to lead to a balanced understanding of the roles of fragrance and other sensory cues in their model system.

13.3   HOW DOES SCENT WORK?

13.3.1   Sensory Interaction and Context in Floral Advertisement

The preceding examples highlight a degree of flexibility inherent to the information content of floral scent compounds and the neural organization of flower-visiting animals. Methyl cinnamate attracts neotropical euglossine bees to orchids, but is ignored by Rocky Mountain bumblebees visiting snapdragon flowers, whereas 2PE is a critical pollen odor for the pollinators of strawberry flowers, a proboscis extension trigger for nectar-feeding butterflies, and a general attractant for noctuid moths and halictid bees. These patterns, in parallel with those of floral colors,[112] indicate that few individual floral scent compounds possess universal significance for flower-visiting animals; rather it is their limitless combinatorial potential with other odors, shapes, colors, floral geometry, and other contextual cues that determines the species-specific signals that promote constancy[113,114] (Chapter 12) and putatively guild-specific patterns of "pollination syndromes" that have generated so much recent debate.[6,115–117] Exceptions to this trend might be expected in cases of sexual deception or pollination by animals with unusual sensory capabilities. An examination of three such cases—sexually deceptive orchids, brood deceptive aroids, and New World bat-pollinated flowers—revealed the information content of floral odors to be highly context specific.[118] The odor produced by *Chiloglottis* orchid labella matches the cuticular pheromone of female *Neozeleboria* wasps,[99] but is irrelevant to males when presented more than 25 cm above the ground, the height at which female wasps perch when calling males.[119] *Helicodiceros muscivorus* (Araceae), a ground-dwelling plant native to Mediterranean islands, produces inflorescences that are remarkable carrion mimics in all respects, with dimethyl oligosulfide odors characteristic of rotting flesh.[120] Floral augmentation with oligosulfides and a heating coil revealed that female blowflies are attracted to these brood-site mimics by the appropriate odor signal,[121] but are more likely to land on the appendix and enter the floral chamber when it is heated.[122] The same sulfurous odors are present in a guild of unrelated bat-pollinated flowers in Central America, but with a completely distinct combination of flower shape, color, and height above the ground, and a nectar-based pollination system; they are by no means carrion mimics.[123] Both laboratory-raised and wild bats show significant preferences for flower models with sulfurous odors, but are versatile enough to use visual cues or echolocation alone to find and feed from flowers.[124,125]

13.3.2 Synergism between Odor and other Floral Cues

These examples illustrate the ubiquity of synergism between different sensory modalities in pollinator attraction, whether food-based or deceptive, and the futility of attempting to categorize floral phenotype space in one sensory dimension. Olle Pellmyr's pre-*Yucca* research provides several examples of behavioral synergism among floral signals, as revealed through the floral manipulations profiled above. Deconstruction of *Lysichiton americanum* (Araceae) inflorescences showed that its staphylinid beetle pollinators were strongly attracted to yellow objects (such as the spathe) only in the presence of the spadix's odor.[126] Pellmyr[127] pioneered floral augmentation to test whether odor differences among altitudinal races of *Cimicifuga simplex* (Ranunculaceae) contributed to divergence in their pollinator spectra. When isoeugenol and methyl anthranilate, odors specific to butterfly-pollinated *C. simplex*, were applied individually to bumblebee-pollinated flowers, they had no effect, but a binary blend of these compounds induced *Argynnis* butterflies to land and probe longer on flowers of the latter race. Finally, floral reconstruction of *Zygogynum viellardii* (Winteraceae) flowers with yellow paper models revealed that *Sabatinca* moths approach and huddle at flower models only when they are scented with the four-component odor blend.[128] In sum, synergistic interactions between visual and olfactory floral cues are prevalent across the spectrum of plant-pollinator interactions, from alpine tundra to tropical rainforests. Even when hummingbirds are primary pollinators and ignore floral volatiles, mites carried in the birds' nostrils are attracted to the nectar's odor, and their activities impact nectar standing crops, hummingbird movement, and pollen flow between plants.[129,130] These observations suggest that biologists who model pollination dynamics without accounting for both visual and olfactory cues may draw inaccurate conclusions about the selective forces in their system.

13.4 WHICH SUBSETS OF SCENTS ARE BEHAVIORALLY ACTIVE?

13.4.1 Signal, Noise, and Electroantennograms

Once floral scent is determined to be important through the methods detailed above, how does one identify the fractions or individual blend components that show behavioral activity, and what is the biological significance of scent components that lack such activity? Ideally, complex fragrances could be partitioned into subsets that either have or do not have measurable consequences for plant reproduction. The former—"signals"—should include pollinator attractants, landing or probing cues, repellents of floral predators, herbivores, or thieves, compounds that prime or synergize the preceding functions, and odors that enhance floral constancy.[131] The latter—"noise"—includes selectively neutral by-products of signal biosynthesis, genetic drift, or phylogenetic artifacts.[18,132,133] Once signal components are identified, information concerning their sites of biosynthesis or emission could be incorporated into selection gradient studies on correlated odor morphometric traits.[134,135] An increasingly common approach to screening complex blends for active constituents is to combine gas chromatography and electroantennographic detection (GC-EAD) with conditioned proboscis extension reflex (PER) or another behavioral assay.[136,137] It is now possible, with chiral GC columns, to test the stimulation of antennae or single sensory neurons by specific enantiomers using GC-EAD.[138] The relative merits and efficacy of antennal screening, particularly in studies of sexual deception and brood site mimicry, are discussed in detail elsewhere (Chapter 10).[12,118,139]

13.4.2 PHYLOGENETIC CONSTRAINT AS A SOURCE OF NOISE: DO LINEAGES HAVE SPECIFIC ODORS?

In the remainder of this section, I address one source of "noise" in fragrance blends by reviewing the available evidence for phylogenetic constraint in scent composition across different plant lineages. This issue arose in the heyday of chemical taxonomy, when Adams, Mabry, Rodman, and others[140–142] established that specific classes of plant secondary metabolites could be considered shared derived traits diagnostic for their lineages. Several researchers have examined floral headspace composition in diverse angiosperm lineages and have found evidence suggestive of species-specific odor patterns,[143,144] particularly when beetle pollination is involved.[145,146] Other trends to emerge from these studies were odor similarities associated with biogeography,[147,148] divergence among related species due to pollinator shifts,[149] and the presence of unusual compounds shared by pairs of closely related species.[150] Scent complexity was often found to be reduced, if not absent, in species that are autogamous or pollinated by hummingbirds,[143,151,152] and many, but not all, biosynthetic classes of odors were present in most lineages. Similar patterns were observed in two studies in which scent data from single genera (*Narcissus* [Amaryllidaceae] and *Clusia* [Guttiferae]) were analyzed using cluster analysis, with the resulting dendrograms examined for consistency with taxonomic hypotheses.[82,153] Barkman[133,154] systematically compared the taxonomic signal of *Cypripedium* orchid scents using phenetic versus cladistic approaches, and argued that scent compounds should not be coded as independent characters because they are produced in suites of highly correlated biosynthetic pathway products.

The general consensus among such studies is that scent chemistry is so evolutionarily labile (homoplaseous) that it is poorly suited to phylogenetic reconstruction.[155] Instead, scent chemistry might profitably be mapped onto phylogenetic trees generated by nonfloral or molecular data. When this was done for the Magnoliaceae (30 species, 3 genera)[156] and Nyctaginaceae (20 species, 2 genera),[18,152] whole classes of compounds were revealed as shared ancestral (uninformative) traits, such as monoterpenes in *Magnolia* and *cis*-jasmone (a defense-signaling compound) in *Mirabilis*, but other compounds (e.g., lactones) showed patterns consistent with sister taxon relationships based on molecular data, even across continents. Species of

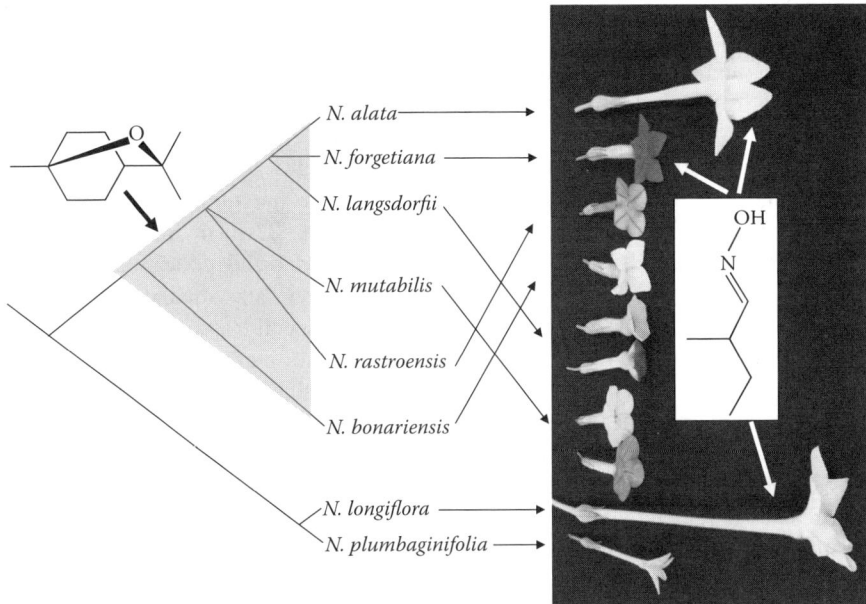

FIGURE 13.4 Phylogenetic mapping of scent compounds in *Nicotiana* section *Alatae*. Arrows link species names with photographs of their flowers. 1,8-cineole (upper left) and small amounts of monoterpenes are emitted by flowers of all species in *Alatae* s.s., regardless of floral morphology or pollinator affinities. Nitrogenous aldoximes (right panel) universally associated with hawkmoth pollination are present only in sphingophilous species (*N. longiflora, N. alata*) and in hummingbird-pollinated *N. forgetiana*, the closest relative of *N. alata*. Thus fragrance chemistry may reflect both the phylogenetic relationship and pollination strategy. Species topology is a cartoon of relationships in *Nicotiana* section *Alatae*, as inferred from References 158–159.

Mirabilis had substantially fewer compounds on average (24) than did *Acleisanthes* (53), with less biosynthetic diversity, but plants in both lineages effectively attract the same species of hawkmoth.[18,152] A subsequent study on *Nicotiana* (Solanaceae) section *Alatae*[157] revealed similar patterns; the independent evolution of nitrogenous aldoximes in both ingroup and outgroup hawkmoth-pollinated species, and the presence of 1,8-cineole and other monoterpenes, emitted from the corolla limb on a nocturnal rhythm, in all species within section *Alatae* regardless of pollinator affinities (Figure 13.4). Taken as a whole, this body of work indicates that the chemical composition of floral scent blends reflects both the evolution of signals which mediate pollinator behavior and shared history among related species, and that some lineages are constrained in the number and kinds of volatile compounds available to their floral physiology.

13.5 CONCLUSION

I have described several complementary experimental approaches used to measure animals' behavioral responses to floral scent at different physical scales. The obvious potential to test such responses using transgenic plants or interspecific hybrids with modified floral phenotypes has not yet begun to be realized. The evidence presented here and throughout this book indicates that fragrance serves numerous, sometimes unanticipated roles in plant reproductive ecology, eliciting both innate and learned responses from flower-visiting animals, often in the context of nonvolatile floral cues. It is also apparent that not all scent components contribute to behavioral activity, and that not all fragrances have reproductive functions. The pioneers of fragrance research—Stefan Vogel, Bertil Kullenberg, Bastiaan Meeuse, Gunnar Bergström, Calaway Dodson, and their many students and colleagues—studied plants whose powerful or unusual fragrances behaviorally mediate pollination through sexual deception, brood-site mimicry, or sex-specific rewards. These exotic reproductive strategies and floral phenotypes depart so markedly from the nectar- and pollen-based reward systems of "typical" plant-pollinator interactions, that floral scent may not have seemed relevant to previous generations of pollination biologists. Here we have argued that fragrance is broadly relevant to plant reproductive biology, even when pollinators ignore odor or flowers are highly generalized. The technological and theoretical tools are now in place to broadly test these ideas by integrating the study of fragrance variation and its behavioral consequences into the mainstream of experimental pollination biology.

ACKNOWLEDGMENTS

I am grateful to Natalia Dudareva and Eran Pichersky for the invitation to contribute to this book, and to Tia-Lynn Ashman, Kristina Jones, Almut Kelber, Bitty Roy, Mark Willis, and Michelle Zjhra for shared adventures in floral manipulation. Special thanks to Gunnar Bergström, Heidi Dobson, Jette Knudsen, and John Law for bringing us together in Oxford and Ventura. The preparation of this manuscript was supported by a National Institutes of Health (NIH) grant (RR-P20 RR 016461), from the BRIN Program of the U.S. National Center for Research Resources, and a U.S. National Science Foundation (NSF) grant (DEB-0317217). The contents are solely the responsibility of the author and do not necessarily represent the official views of the NIH or NSF.

REFERENCES

1. *Plant Species Biology* 14(2), 1999.
2. Plant Physiology 135(4), 2004.
3. Hansson, B.S., *Insect Olfaction*, Springer-Verlag, Berlin, 1999.
4. Kevan, P.G., Floral colours through the insect eye: what they are and what they mean, in *Handbook of Experimental Pollination Biology*, Jones, C.E. and Little, R.J., Eds., Van Nostrand-Reinhold, New York, 1983, p. 3.
5. Menzel, R. and Shmida, A., The ecology of flower colours and the natural colour vision of insect pollinators: the Israeli flora as a case study, *Biol. Rev.* 68, 81, 1993.
6. Waser, N.M., Chittka, L., Price, M.V., Williams, N.M., and Ollerton, J., Generalization in pollination systems and why it matters, *Ecology* 77, 1043, 1996.

7. Raguso, R.A., Why do flowers smell? The chemical ecology of fragrance-driven pollination, in *Advances in Insect Chemical Ecology*, Cardé, R.T. and Millar, J.G., Eds., Cambridge University Press, Cambridge, 2004, p. 151.

8. Metcalf, R.L., Plant volatiles as insect attractants, *CRC Crit. Rev. Plant Sci.* 5, 251, 1987.

9. Matile, P. and Altenburger, R., Rhythms of fragrance emission in flowers, *Planta* 174, 242, 1988.

10. Barkman, T.J., Evidence for positive selection on the floral scent gene isoeugenol-O-methyltransferase, *Mol. Biol. Evol.* 20, 168, 2003.

11. Motten, A.F. and Stone, J.L., Heritability of stigma position and the effect of stigma-anther separation on outcrossing in a predominantly self-fertilizing weed, *Datura stramonium* (Solanaceae), *Am. J. Bot.* 87, 339, 2000.

12. Raguso, R.A., Floral scent, olfaction and scent-driven foraging behavior, in *Cognitive Ecology of Pollination: Animal Behavior and Floral Evolution*, Chittka, L. and Thomson, J.D., Eds., Cambridge University Press, Cambridge, 2001, p. 83.

13. Herrera, C.M., Selection on floral morphology and environmental determinants of fecundity in a hawk moth-pollinated violet, *Ecol. Monogr.* 63, 251, 1993.

14. Fineblum, W.L. and Rausher, M.D., Do floral pigmentation genes also influence resistance to enemies? The W locus in *Ipomoea purpurea*, *Ecology* 78, 1446, 1997.

15. Armbruster, W.S., Can indirect selection and genetic context contribute to trait diversification? A transition-probability study of blossom colour evolution in two genera, *J. Evol. Biol.* 15, 468, 2002.

16. Ackerman, J.D., Melendez-Ackerman, E.J., and Salguero-Faria J. Variation in pollinator abundance and selection on fragrance phenotypes in an epiphytic orchid, *Am. J. Bot.* 84, 1383, 1997.

17. Pichersky, E. and Gershenzon, J., The formation and function of plant volatiles: perfumes for pollinator attraction and defense, *Curr. Opin. Plant Biol.* 5, 237, 2002.

18. Levin, R.A., McDade, L.A., and Raguso, R.A., The systematic utility of floral and vegetative fragrance in two genera of Nyctaginaceae, *Syst. Biol.* 52, 334, 2003.

19. Theis, N. and Lerdau, M., The evolution of function in plant secondary metabolites, *Int. J. Plant Sci.* 164, S93, 2003.

20. Bergström, G., Role of volatile chemicals in *Ophrys*-pollinator interactions, *Biochemical Aspects of Plant and Animal Coevolution*, Harborne, G., Ed., Academic Press, New York, 1978, p. 207.

21. Kunze, J. and Gumbert, A., The combined effect of color and odor on flower choice behavior of bumble bees in flower mimicry systems, *Behav. Ecol.* 12, 447, 2001.

22. Gardener, M.C. and Gillman, M.P., The taste of nectar—a neglected area of pollination ecology, *Oikos* 98, 552, 2002.

23. Williams, N.H., Floral fragrances as cues in animal behavior, in *Handbook of Experimental Pollination Biology*, Jones, C.E. and Little, R.J., Eds., Van Nostrand-Reinhold, New York, 1983, p. 51.

24. Dobson, H.E.M., Floral volatiles in insect biology, in *Insect-Plant Interactions*, vol. 5, Bernays, E., Ed., CRC Press, Boca Raton, FL, 1994, p. 47.

25. Sprengel, F.C., *Das entdeckte Geheimnis der Natur im Bau und in der Befruchtung der Blumen*, Englemann, Leipzig, 1894, p. 1793.

26. Weiss, M.R., Floral colour changes as cues for pollinators, *Nature* 354, 227, 1991.

27. Lunau, K., Innate recognition of flowers by bumblebees: orientation of antennae to visual stamen signals, *Can. J. Zool.* 70, 2139, 1992.

28. Dafni, A. and Kevan, P.G., Floral symmetry and nectar guides: ontogenetic constraints from floral development, colour pattern rules and functional significance, *Bot. J. Linn. Soc.* 120, 371, 1996.

29. Johnson, S.D. and Andersson, S., A simple field method for manipulating ultraviolet reflectance of flowers, *Can. J. Bot.* 80, 1325, 2002.

30. Dobson, H.E.M., Groth, I., and Bergström, G., Pollen advertisement: chemical contrasts between whole-flower and pollen odors, *Am. J. Bot.* 83, 877, 1996.

31. Raguso. R.A., Why are some floral nectars scented?, *Ecology* 85, 1486, 2004.

32. Kaiser, R., *The Scent of Orchids*, Elsevier Science, Amsterdam, 1993.

33. Kite, G.C. et al. Inflorescence odours and pollinators of *Arum* and *Amorphophallus* (Araceae), in *Reproductive Biology*, Owens, S.J. and Rudall, P.J., Eds., Royal Botanic Gardens, Kew, UK, 1998, p. 295.

34. Delpino, F., Ulteriori osservazioni e considerazioni sulla dicogamia nel regno vegetale. 2 (IV). Delle piante zoidifile, *Atti Soc. Ital. Sci. Nat.* 16, 151, 1874.

35. Kerner von Marilaum, A., *The Natural History of Plants; Their Forms, Growth, Reproduction and Distribution*, Blackie and Son, London, 1895.

36. Harper, R., Bate-Smith, E.C., and Land, D.G., *Odour Description and Odour Classification*, American Elsevier, New York, 1968.

37. Joulain, D., The composition of the headspace from fragrant flowers: further results, *Flavour Fragr. J.* 2, 149, 1987.

38. Kaiser, R., Trapping, investigation and reconstitution of flower scents, in *Perfumes: Art, Science and Technology*, Müller, P.M. and Lamparsky, D., Eds., Elsevier Applied Science, London, 1991, p. 213.

39. Amoore, J.E., Forrester, L.J., and Buttery, R.G., Specific anosmia to 1-pyrroline: the spermous primary odor, *J. Chem. Ecol.* 1, 299, 1975.

40. Tollsten, L., Knudsen, J.T., and Bergström, G., Floral scent in generalistic *Angelica* (Apiaceae)—an adaptive character?, *Biochem. Syst. Ecol.* 22, 161, 1994.

41. Borg-Karlson, A.-K., Valterová, I., and Nilsson, L.A., Volatile compounds from flowers of six species in the family Apiaceae: bouquets for different pollinators?, *Phytochemistry* 35, 111, 1994.

42. Robacker, D.C., Roles of putrescine and 1-pyrroline in attractiveness of technical-grade putrescine to the Mexican fruit fly (Diptera: Tephritidae), *Fla. Entomol.* 84, 679, 2001.

43. Galen, C., Flowers and enemies: predation by nectar thieving ants in relation to variation in floral form of an alpine wildflower, *Polemonium viscosum*, *Oikos* 85, 426, 1999.

44. Campbell, D.R., Measurements of selection in a hermaphroditic plant: variation in male and female pollination success, *Evolution* 43, 318, 1989.

45. Conner, J.K., Understanding natural selection: an approach integrating selection gradients, multiplicative fitness components and path analysis, *Ethol. Ecol. Evol.* 8, 387, 1996.

46. Miyake, T. and Yafuso, M., Floral scents affect reproductive success in fly-pollinated *Alocasia odora* (Araceae), *Am. J. Bot.* 90, 370, 2003.

47. Clements, F.E. and Long, F.L., *Experimental Pollination: An Outline of the Ecology of Flowers and Insects*, Carnegie Institute, Washington, DC, 1923.

48. Kullenberg, B., Field experiments with chemical sexual attractants on aculeate hymenopteran males I, *Zool. Bidrag Uppsala* 31, 253, 1956.

49. Nilsson, L.A. et al. Ixoroid secondary pollen presentation and pollination by small moths in the Malagasy treelet *Ixora platythyrsa* (Rubiaceae), *Plant Syst. Evol.* 170, 161, 1990.

50. Knudsen, J.T., Andersson, S., and Bergman, P., Floral scent attraction in *Geonoma macrostachys*, an understorey palm of the Amazonian rain forest, *Oikos* 85, 409, 1999.

51. Knoll, F., Insekten und Blumen IV. Die *Arum*-Blütenstände und ihre Besucher, *Abhandlungen der Kaiserlich-Königlichen Zoologisch-botanischen Gesallschaft in Wien* 12, 383, 1926.

52. Gottsberger, G. and Silberbauer-Gottsberger, I., Olfactory and visual attraction of *Erioscelis emarginata* (Cyclocephalini, Dynastinae) to the inflorescences of *Philodendron selloum* (Araceae), *Biotropica* 23, 23, 1991.

53. Hossaert-McKey, M., Gibernau, M., and Frey, J.E., Chemosensory attraction of fig wasps to substances produced by receptive figs, *Entomol. Exp. Appl.* 70, 185, 1994.

54. Roy, B.A. and Raguso, R.A., Olfactory vs. visual cues in a floral mimicry system, *Oecologia* 109, 414, 1997.

55. Raguso, R.A. and Willis, M.A., Synergy between visual and olfactory cues in nectar feeding by naïve hawkmoths, *Manduca sexta*, *Anim. Behav.* 63, 685, 2002.

56. Raguso, R.A. and Willis, M.A., Synergy between visual and olfactory cues in nectar feeding by wild hawkmoths, *Manduca sexta*, *Anim. Behav.* 65, 407, 2004.

57. Ware, A.B. and Compton, S.G., Breakdown of pollinator specificity in an African fig tree, *Biotropica* 24, 544, 1992.

58. Meagher, R.L., Jr., Trapping noctuid moths with synthetic floral volatile lures, *Entomol. Exp. Appl.* 103, 219, 2002.

59. Mikich, S.B., Bianconi, G.V., Maia, B.H., and Teixeira, S.D., Attraction of the fruit eating bat *Carollia perspicillata* to *Piper gaudichaudianum* essential oil, *J. Chem. Ecol.* 29, 2379, 2003.

60. Irwin, R.E., Brody, A.K., and Waser, N.M., The impact of floral larceny on individuals, populations and communities, *Oecologia* 129, 161, 2001.

61. Plepys, D., Ibarra, F., Francke, W., and Lofstedt, C., Odour-mediated nectar foraging in the silver Y moth, *Autographa gamma* (Lepidoptera: Noctuidae): behavioural and electrophysiological responses to floral volatiles, *Oikos* 99, 75, 2002.

62. Cook, S.M., Bartlet, E., Murray, D.A., and Williams, I.H., The role of pollen odour in the attraction of pollen beetles to oilseed rape flowers, *Entomol. Exp. Appl.* 104, 43, 2002.

63. Dufaÿ, M., Hossaert-McKey, M., and Anstett, M.C., When leaves act like flowers: how dwarf palms attract their pollinators, *Ecol. Lett.* 6, 28, 2003.

64. Dobson, H.E.M. et al. Scent as an attractant, in *Practical Pollination Biology*, Dafni, A. and Kevan, P.G., Eds., Enviroquest, Cambridge, Ontario, 2005.

65. Barth, F.G., *Insects and Flowers; The Biology of a Partnership*, Princeton University Press, Princeton, NJ, 1991.

66. Brantjes, N.B.M., Senses involved in the visiting of flowers by *Cucullia umbratica* (Noctuidae: Lepidoptera), *Entomol. Exp. Appl.* 20, 1, 1976.

67. Marden, J.H., Remote perception of floral nectar by bumblebees, *Oecologia* 64, 232, 1984.

68. Kearns, C.A. and Inouye, D.W., Techniques for pollination biologists, University Press of Colorado, Niwot, CO, 1993.

69. Dafni, A., *Pollination Ecology; A Practical Approach*, Oxford University Press, Oxford, 1993.

70. Waser, N.M. and Price, M.V., Pollinator behaviour and natural selection for flower colour in *Delphinium nelsonii*, *Nature* 302, 422, 1983.

71. Wilson, P., Selection for pollination success and the mechanical fit of *Impatiens* flowers around bumblebee bodies, *Biol. J. Linn. Soc.* 55, 355, 1995.

72. Johnson, S.D. and Steiner, K.E., Long-tongued fly pollination and evolution of floral spur length in the *Disa draconis* complex (Orchidaceae), *Evolution* 51, 45, 1997.

73. Johnson, S.D. and Midgley, J.J., Fly pollination of *Gorteria diffusa* (Asteraceae) and a possible mimetic function for dark spots on the capitulum, *Am. J. Bot.* 84, 429, 1997.

74. Vogel, S., Evolutionary shifts from reward to deception in pollen flowers, in *The Pollination of Flowers by Insects*, Richards, A.J., Ed., Academic Press, London, 1978, p. 9.

75. D'Arcy, W.G., D'Arcy, N.S., and Keating, R.C., Scented anthers in the Solanaceae, *Rhodora* 92, 50, 1990.

76. Dobson, H.E.M., Bergström, G., and Groth, I., Differences in fragrance chemistry between flower parts of *Rosa rugosa* Thunb. (Rosaceae), *Isr. J. Bot.* 39, 143, 1990.

77. Dobson, H.E.M. and Bergström, G., The ecology and evolution of pollen odors, *Plant Syst. Evol.* 222, 63, 2000.

78. Bergström, G., Dobson, H.E.M., and Groth, I., Spatial fragrance patterns within the flowers of *Ranunculus acris* (Ranunculaceae), *Plant Syst. Evol.* 195, 221, 1995.

79. Dobson, H.E.M., Danielson, E.M., and van Wesep, I.D., Pollen odor chemicals as modulators of bumble bee foraging on *Rosa rugosa* Thunb. (Rosaceae), *Plant Species Biol.* 14, 153, 1999.

80. Patt, J.M., French, J.C., Schal, C., Lech, J., and Hartman, T.G., The pollination biology of Tuckahoe, *Peltandra virginica* (Araceae), *Am. J. Bot.* 82, 1230, 1995.

81. Patt, J.M., Hartman, T.G., Creekmore, W., Elliott, J., Schal, C., Leck, J., and Rosen, R.T., The floral odor of *Peltandra virginica* contains novel trimethyl-2,5-dioxabicyclo[3.2.1.]nonanes, *Phytochemistry* 31, 487, 1992.

82. de L. Nogueira, P.C., Bittrich, V., Shepherd, G.J., Lopes, A.V., and Marsaioli, A.J., The ecological and taxonomic importance of flower volatiles of *Clusia* species (Guttiferae), *Phytochemistry* 56, 443, 2001.

83. Ashman, T.L., Cole, D.H., Bradburn, M., Blaney, B., and Raguso, R.A., The scent of a male: the role of floral volatiles in pollination of a gender dimorphic plant, *Ecology* 86, 2099, 2005.

84. Cunningham, J.P., Moore, C.J., Zalucki, M.P., and West, S.A., Learning, odour preference and flower foraging in moths, *J. Exp. Biol.* 207, 87, 2004.

85. Manning, A., Some aspects of the foraging behaviour of bumble-bees, *Behaviour* 9, 164, 1956.

86. Odell, E., Raguso, R.A., and Jones, K.N., Bumblebee foraging responses to variation in floral scent and color in snapdragons (*Antirrhinum*: Scrophulariaceae), *Am. Midl. Nat.* 142, 257, 1999.

87. Galen, C. and Kevan, P.G., Bumblebee foraging and floral scent dimorphism: *Bombus kirbyellus* Curtis (Hymenoptera: Apidae) and *Polemonium viscosum* Nutt. (Polemoniaceae), *Can. J. Zool.* 61, 1207, 1983.

88. Jones, K.N., Analysis of pollinator foraging: tests for non-random behaviour, *Funct. Ecol.* 11, 255, 1997.

89. Baldwin, I.T., Preston, C.A., Euler, M.A., and Gorham, D., Patterns and consequences of benzyl acetone floral emissions from *Nicotiana attenuata* plants, *J. Chem. Ecol.* 23, 2327, 1997.

90. Euler, M.A. and Baldwin, I.T., The chemistry of defense and apparency in the corollas of *Nicotiana attenuata*, *Oecologia* 107, 102, 1996.

91. Henning, J.A., Peng, Y.S., Montague, M.A., and Teuber, L.R., Honey bee (Hymenoptera: Apidae) behavioral responses to primary alfalfa (Rosales: Fabaceae) floral volatiles, *J. Econ. Entomol.* 85, 233, 1992.

92. Real, L.A., Uncertainty and pollinator-plant interactions: the foraging behavior of bees and wasps on artificial flowers, *Ecology* 62, 20, 1981.

93. Nishida, R., Shelly, T.E., and Kaneshiro, K., Acquisition of female-attracting fragrance by males of oriental fruit fly from a Hawaiian lei flower, *Fagraea berteriana*, *J. Chem. Ecol.* 23, 2275, 1997.

94. Winter, Y. and von Helversen, O., Bats as pollinators: foraging energetics and floral adaptations, in *Cognitive Ecology of Pollination: Animal Behavior and Floral Evolution*, Chittka, L. and Thomson, J.D., Eds., Cambridge University Press, Cambridge, 2001, p. 148.

95. Borg-Karlson, A.-K., Chemical and ethological studies of pollination in the genus *Ophrys* (Orchidaceae), *Phytochemistry* 29, 1359, 1990.

96. Gibernau, M. and Hossaert-McKey, M., Are olfactory signals sufficient to attract fig pollinators?, *Ecoscience* 5, 306, 1998.

97. Andersson, S., Foraging responses in the butterfly *Inachis io, Aglais urticae* (Nymphalidae) and *Gonepteryx rhamni* (Pieridae) to floral scents, *Chemoecology* 13, 1, 2003.

98. Tinbergen, N., Social releasers and the experimental method required for their study, *Wilson Bull.* 60, 5, 1948.

99. Schiestl, F.P., Peakall, R., Mant, J.G., Ibarra, F., Schulz, C., Franke, S., and Francke, W., The chemistry of sexual deception in an orchid-wasp pollination system, *Science* 302, 437, 2003.

100. Schiestl, F.P., Floral evolution and pollinator mate choice in a sexually deceptive orchid, *J. Evol. Biol.* 17, 67, 2004.

101. Harder, L.D. and Barrett, S.C.H., Mating cost of large floral displays in hermaphrodite plants, *Nature* 373, 512, 1995.

102. Ohashi, K. and Yahara, T. Behavioral responses of pollinators to variation in floral display size and their influences on the evolution of floral traits, in *Cognitive Ecology of Pollination: Animal Behavior and Floral Evolution*, Chittka, L. and Thomson, J.D., Eds., Cambridge University Press, Cambridge, 2001, p. 274.

103. White, R.H., Stevenson, R.D., Bennett, R.R., Cutler, D.E., and Haber, W.A., Wavelength discrimination and the role of ultraviolet vision in the feeding behavior of hawkmoths, *Biotropica* 26, 427, 1994.

104. Thom, C., Guerenstein, P.G., Mechaber, W.L., and Hildebrand, J.G., Floral CO_2 reveals flower profitability to moths, *J. Chem. Ecol.* 30, 1285, 2004.

105. Guerenstein, P.G., Yepez, A., Van Haren, J., Williams, D.G., and Hildebrand, J.G., Floral CO_2 emission may indicate food abundance to nectar feeding moths, *Naturwissenschaften* 91, 329, 2004.

106. Kevan, P.G. and Lane, M.A., Flower petal microtexture is a tactile cue for bees, *Proc. Natl. Acad. Sci. USA* 82, 4750, 1985.

107. Kevan, P.G., Giurfa, M., and Chittka, L., Why are there so many and so few white flowers?, *Trends Plant Sci.* 1, 280, 1996.

108. Ômura, H., Honda, K., and Hayashi, N., Floral scent of *Osmanthus fragrans* discourages foraging behavior of cabbage butterfly, *Pieris rapae, J. Chem. Ecol.* 26, 655, 2000.

109. Dodson, C.H., Dressler, R.L., Hills, H.G., Adams, R.M., and Williams, N.H., Biologically active compounds in orchid fragrances, *Science* 164, 1243, 1969.

110. Bergström, G., Birgersson, G., Groth, I., and Nilsson, A., Floral fragrance disparity between three taxa of lady's slipper, *Cypripedium calceolus* (Orchidaceae), *Phytochemistry* 31, 2315, 1992.

111. Borg-Karlson, A.-K., Unelius, C.R., Valterova, I., and Nilsson, L.A., Floral fragrance chemistry in the early flowering shrub, *Daphne mezereum* (Thymelaeaceae), *Phytochemistry* 41, 1477, 1996.

112. Chittka, L. and Menzel, R., The evolutionary adaptation of flower colours and the insect pollinator's colour vision, *J. Comp. Physiol. A* 170, 171, 1992.

113. Gegear, R.J. and Laverty, T.M., The effect of variation among floral traits on the flower constancy of pollinators, in *Cognitive Ecology of Pollination*, Chittka, L. and Thomson, J.D., Eds., Cambridge University Press, Cambridge, 2001, p. 1.
114. Wright, G.A., Skinner, B.D., and Smith, B.H., Ability of honeybee, *Apis mellifera*, to detect and discriminate odors of varieties of canola (*Brassica rapa* and *Brassica napus*) and snapdragon flowers (*Antirrhinum majus*), *J. Chem. Ecol.* 28, 721, 2002.
115. Ollerton, J., Reconciling ecological processes with phylogenetic patterns: the apparent paradox of plant-pollinator systems, *J. Ecol.* 84, 767, 1996.
116. Johnson, S.D. and Steiner, K.E., Generalization versus specialization in plant pollination systems, *Trends Ecol. Evol.* 15, 140, 2000.
117. Fenster, C.B., Armbruster, W.S., Wilson, P., Dudash, M.R., and Thomson, J.D., Pollination syndromes and floral specialization, *Annu. Rev. Ecol. Evol. Syst.* 35, 375, 2004.
118. Raguso, R.A., Flowers as sensory billboards: progress towards an integrated understanding of floral advertisement, *Curr. Opin. Plant Biol.* 7, 434, 2004.
119. Handel, S.N. and Peakall, R., Thynnine wasps discriminate among heights when seeking mates: tests with a sexually deceptive orchid, *Oecologia* 95, 241, 1993.
120. Kite, G.C., Inflorescence odour of the foul-smelling aroid, *Helicodiceros muscivorus*, *Kew Bull.* 55, 237, 2000.
121. Stensmyr, M.C., Urru, I., Collu, I., Celander, M., Hansson, B.S., and Angioy, A.-M., Rotting smell of dead horse arum florets, *Nature* 420, 625, 2002.
122. Angioy, A.-M., Stensmyr, M.C., Urru, I., Puliafito, M., Collu, I., and Hansson, B.S., Function of the heater: the dead horse arum revisited, *Proc. R. Soc. Lond. B* 271(suppl. 3), S13, 2003.
123. von Helversen, O., Winkler, L., and Bestmann, H.J., Sulphur-containing "perfumes" attract flower-visiting bats, *J. Comp. Physiol. A* 186, 143, 2000.
124. Winter, Y., López, J., and von Helversen, O., Ultraviolet vision in a bat, *Nature* 425, 612, 2003.
125. von Helversen, D. and von Helversen, O., Object recognition by echolocation: a nectar-feeding bat exploiting the flowers of a rain forest vine, *J. Comp. Physiol. A* 189, 327, 2003.
126. Pellmyr, O. and Patt, J.M., Function of olfactory and visual stimuli in pollination of *Lysichiton americanum* (Araceae) by a staphylinid beetle, *Madroño* 33, 47, 1986.
127. Pellmyr, O., Three pollination morphs in *Cimicifuga simplex*: incipient speciation due to inferiority in competition, *Oecologia* 68, 304, 1986.
128. Pellmyr, O., Thien, L.B., Bergström, G., and Groth, I., Pollination of New Caledonian Winteraceae: opportunistic shifts or parallel radiation with their pollinators?, *Plant Syst. Evol.* 173, 143, 1990.
129. Heyneman, A.J., Colwell, R.K., Naeem, S., Dobkin, D.S., and Hallet, B., Host plant discrimination: experiments with hummingbird flower mites, in *Plant-Animal Interactions: Evolutionary Ecology in Tropical and Temperate Regions*, Price, P.W., Lewinsohn, T.M., Fernandes, G.W., and Benson, W.W., Eds., John Wiley & Sons, New York, 1991, p. 455.
130. Lara, C. and Ornelas, J.F., Effects of nectar theft by flower mites on hummingbird behavior and the reproductive success of their host plant, *Moussonia deppeana* (Gesneriaceae), *Oikos* 96, 470, 2002.
131. Raguso. R.A., Olfactory landscapes and deceptive pollination: signal, noise and convergent evolution in floral scent, in *Insect Pheromone Biochemistry and Molecular Biology*, Blomquist, G.J. and Vogt, R., Eds., Academic Press, New York, 2003, p. 631.

132. Steele, C.L., Crock, J., Bohlmann, J., and Croteau, R., Sesquiterpene synthases from grand fir (*Abies grandis*): comparison of constitutive and wound-induced activities, and cDNA isolation, characterization and bacterial expression of δ-selinene synthase and γ-humulene synthase, *J. Biol. Chem.* 273, 2078, 1998.
133. Barkman, T.J., Character coding of secondary chemical variation for use in phylogenetic analyses, *Biochem. Syst. Ecol.* 29, 1, 2001.
134. Armbruster, W.S., Estimating and testing adaptive surfaces: the morphology and pollination of *Dalechampia* blossoms, *Am. Nat.* 135, 14, 1990.
135. Cresswell, J.E. and Galen, C., Frequency-dependent selection and adaptive surfaces for floral character combinations: the pollination of *Polemonium viscosum*, *Am. Nat.* 138, 1342, 1991.
136. Wadhams, L.J., Blight, M.M., Kerguelen, V., Le Métayer, M., Marion-Poll, F., Masson, C., Pham-Delègue, M.-H., Woodcock, C.M., Discrimination of oilseed rape volatiles by honey bee: novel combined gas chromatographic-electrophysiological behavioral assay, *J. Chem. Ecol.* 20, 3221, 1994.
137. Schiestl, F.P. and Marion-Poll, F., Detection of physiologically active flower volatiles using gas chromatography coupled with electroantennography, in *Molecular Methods of Plant Analysis*, vol. 21, *Analysis of Taste and Aroma*, Jackson, J.F, Linskens, H.F., and Inman, R., Eds., Springer, Berlin, 2002, p. 173.
138. Stranden, M., Borg-Karlson, A.-K., and Mustaparta, H., Receptor neuron discrimination of the germacrene D enantiomers in the moth *Helicoverpa armigera*, *Chem. Senses* 27, 143, 2002.
139. Honda, K., Ômura, H., and Hayashi, N., Identification of floral volatiles from *Ligustrum japonicum* that stimulate flower-visiting by cabbage butterfly, *Pieris rapae*, *J. Chem. Ecol.* 24, 2167, 1998.
140. Adams, R.P., von Rudloff, E., and Hogge, L., Chemosystematic studies of the western North American junipers based on their volatile oils, *Biochem. Syst. Ecol.* 11, 189, 1983.
141. Clement, J.S. and Mabry, T.J., Pigment evolution in the Caryophyllales: a systematic overview, *Bot. Acta* 109, 360, 1996.
142. Rodman, J.E., Divergence, convergence and parallelism in phytochemical characters: the glucosinolate-myrosinase system, in *Phytochemistry and Angiosperm Phylogeny*, Young, D.A. and Seigler, D.S., Eds., Praeger Scientific, New York, 1981, p. 43.
143. Dahl, Å.E., Wassgren, A.-B., and Bergström, G., Floral scents in *Hypecoum* sect. *Hypecoum* (Papaveraceae): chemical composition and relevance to taxonomy and mating system, *Biochem. Syst. Ecol.* 18, 157, 1990.
144. Knudsen, J.T. and Mori, S.A., Floral scents and pollination in neotropical Lecythidaceae, *Biotropica* 28, 42, 1996.
145. Thien, L.B., Heimermann, W.H., and Holman, R.T., Floral odors and quantitative taxonomy of *Magnolia* and *Liriodendron*, *Taxon* 24, 557, 1975.
146. Jürgens, A., Webber, A.C., and Gottsberger, G., Floral scent compounds of Amazonian Annonaceae species pollinated by small beetles and thrips, *Phytochemistry* 55, 551, 2000.
147. Gerlach, G. and Schill, R., Fragrance analysis, an aid to taxonomic relationships of the genus *Coryanthes* (Orchidaceae), *Plant Syst. Evol.* 168, 159, 1989.
148. Jürgens, A., Floral scent compounds in nocturnal *Conophytum* species: chemical composition and its relevance to taxonomy and pollination biology, in *Dumpling and His Wife—New Views of the Genus Conophytum*, Hammer, S. and Barnhill, C., Eds., EAE Creative Color, Norwich, UK, 2002, p. 322.

149. Jürgens, A., Witt, T., and Gottsberger, G., Flower scent composition in *Dianthus* and *Saponaria* species (Caryophyllaceae) and its relevance for pollination biology and taxonomy, *Biochem. Syst. Ecol.* 31, 345, 2003.

150. Kite, G.C. and Hetterschieid, W.L.A., Inflorescence odours of *Amorphophallus* and *Pseudodracontium* (Araceae), *Phytochemistry* 46, 71, 1997.

151. Lindberg, A.B., Knudsen, J.T., and Olesen, J.M., Independence of floral morphology and scent chemistry as trait groups in a set of *Passiflora* species, *Det Norske Videnskaps-Akademi. I. Matematisk Naturvidenskapelige Klasse, Skrifter, Ny Ser.* 39, 91, 2000.

152. Levin, R.A., Raguso, R.A., and McDade, L.A., Fragrance chemistry and pollinator affinities in Nyctaginaceae, *Phytochemistry* 58, 429, 2001.

153. Dobson, H.E.M., Arroyo, J., Bergström, G, and Groth, I., Interspecific variation in floral fragrances within the genus *Narcissus* (Amaryllidaceae), *Biochem. Syst. Ecol.* 25, 685, 1997.

154. Barkman, T.J., Beaman, J.H., and Gage, D.A., Floral fragrance variation in *Cypripedium*: implications for evolutionary and ecological studies, *Phytochemistry* 44, 875, 1997.

155. Williams, W.M. and Whitten, N.H., Molecular phylogeny and floral fragrances of male euglossine bee-pollinated orchids: a study of *Stanhopea* (Orchidaceae), *Plant Species Biol.* 14, 129, 1999.

156. Azuma, H., Thien, L.B., and Kawano, S., Molecular phylogeny of *Magnolia* (Magnoliaceae) inferred from cpDNA sequences and evolutionary divergence of the floral scents, *J. Plant Res.* 112, 291, 1999.

157. Raguso, R.A., Levin, R.A., Foose, S.E., Holmberg, M.W., and McDade, L.A., Fragrance chemistry, nocturnal rhythms and pollination "syndromes" in *Nicotiana*, *Phytochemistry* 63, 265, 2003.

158. Buckler, E.S., IV, Ippolito, A., Holtsford, T.P., The evolution of ribosomal DNA: Divergent paralogues and phylogenetic implications, *Genetics* 145, 821, 1997.

159. Chase, M.W., Knapp, S., Cox, A.V., Clarkson, J.J., Butsko, Y., Joseph, J., Savolainen, V., Parokonny, A.S., Molecular systematics, GISH and the origin of hybrid taxa in *Nicotiana* (Solanaceae), *Annals of Botany* 92, 107, 2003.

Section V

Commercial Aspects of Floral Scent

14 Molecular Engineering of Floral Scent

Joost Lücker, Harrie A. Verhoeven, Linus H.W. van der Plas, and Harro J. Bouwmeester

CONTENTS

14.1 INTRODUCTION

Although humans do not play a major role in the pollination of wild plants, floral scent has always been highly appreciated by humans for its sensual pleasure. As a result of this admiration, floral fragrance compounds have been heavily commercialized. In Figure 14.1 some examples of floral scent compounds are shown. Synthetically produced, but also natural floral volatiles are important constituents of perfumes and cosmetics, and are also used as additives in foods and beverages.[1]

Despite the appreciation by humans, the floral scent of many commercial cut flowers seems to have been lost during plant breeding over the decades, possibly as a result of excluding scent as a selection criterion.[2] Today there is growing interest in the scent of flowers, so breeders have put the scent back on their list of selection criteria. One option to breed for scent is to use wild species to cross scent back into commercial cut flowers.[3] However, this would require many years of back-crossing and selection to create a variety having all the desired characteristics. Hence it may be more time efficient to introduce scent into commercial cut flowers by metabolic engineering.[2,4,5] In addition, genetically modified flowers emitting new volatiles may turn out to be more resistant to insect pests and pathogens, which could result in a decrease in the use of environmentally damaging crop protection agents. Tobacco plants with elevated levels of the diterpene cembratrienol in their trichomes were

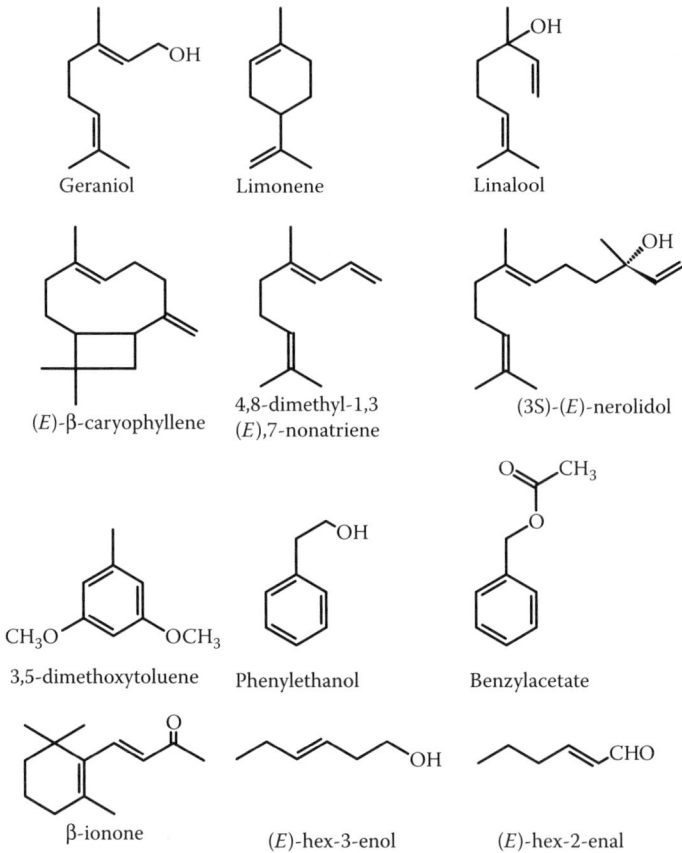

FIGURE 14.1 Examples of volatiles that contribute to the floral scent of plants.

obtained by antisense suppression of the P450 hydroxylase responsible for the conversion of cembatrienol to cembatriendiol.[6,7] These transgenic plants showed an increased resistance to aphids when tested in the field. When flowers emit elevated and novel scents, they may, if the right scents are chosen, possibly deter certain herbivores from the plants. Volatiles may also protect plants against plant pathogens. For example, some monoterpenes and sesquiterpenes have strong antimicrobial properties.[8–10]

In the past decade, a number of research groups have used metabolic engineering approaches to alter volatile formation in a range of plant species. These studies have uncovered unexpected caveats determining the success of pathway engineering. Crucial factors for successful engineering were, for example, precursor availability, enzyme localization, pH of the targeted cell compartment, and the presence of endogenous, modifying enzymes that metabolized the newly produced compound into unexpected products with a different functionality.[11–17] These studies, the possible pitfalls, and the exciting possibilities that lie ahead are discussed in this chapter.

14.2   METABOLIC ENGINEERING OF FLORAL SCENT

There are a number of metabolic pathways that can be targeted to engineer floral scent. In Chapter 4, the metabolic pathways that are involved in floral scent and their regulation are described. Research on the enzymatic regulation of these pathways provides knowledge about the strategies that can be employed to generate plants with novel scents.[4] However, floral scent can also be altered unexpectedly, as was shown in transgenic carnation with antisense flavonoid-3-hydroxylase. The gene encoding this enzyme was knocked out with the goal of altering the color of the flowers.[18] Flowers of the plants with the highest reduction in flavonoid-3-hydroxylase activity showed a strong increase in the emission of the volatile ester methyl benzoate, which significantly altered the overall scent of the flowers.[18] As an explanation for this increase in fragrance production, a shift of the flux of metabolites from anthocyanin biosynthesis to benzoic acid production was postulated. Both originate from the phenylpropanoid pathway, and the production of methyl benzoate is regulated by the level of benzoic acid.[4,18] In *Petunia*, introduction of a strawberry alcohol acyltransferase resulted in the emission of new esters, but only upon substrate feeding of detached flowering stems with isoamyl alcohol, suggesting that ester formation is substrate regulated.[12]

Since terpenoids are among the major components of floral scent, most research on the genetic engineering of flower fragrance has focused on this class of compounds, and thus it is the focus of this chapter. During the past decade, various authors have speculated about the possibilities of genetic or metabolic engineering of terpene biosynthesis.[4,19–27] Apart from commercial applications, such as improved floral scent of ornamentals, these reviews also describe the possibility of modification of essential oil crops or scientific opportunities for ecological studies such as altered herbivore, pollinator, and tritrophic interactions. In a number of these reviews it was also discussed that the cellular fate of monoterpenes produced in tissues not specifically adapted for secretion and storage of such volatile hydrophobic metabolites was unknown.[22–25] It was suggested that transgenic production of monoterpenes in nonadapted tissues could lead to deposition of oil droplets. Alternatively, the relatively high vapor pressure of the monoterpenes could promote evaporation from the producing cells to the atmosphere, which of course is highly desirable for scent production. Another possibility is that the volatiles may be retained in the waxy cuticle of the epidermal cell layer (see Chapter 7).[22]

14.2.1   METABOLIC ENGINEERING OF MONOTERPENE BIOSYNTHESIS

In order to explain the different possible approaches to metabolic engineering of monoterpene biosynthesis, Figure 14.2 shows a schematic outline of terpene metabolism in plants. For monoterpene biosynthesis, geranyl diphosphate (GPP) synthase and monoterpene synthases are considered to be effective points of regulation, but also various steps of the methylerythritol 4-phosphate (MEP) pathway provide multiple control points to direct precursor flux into monoterpene biosynthesis.[22,26] The MEP pathway is controlled by tight feedback regulation of its first step, deoxyxylulose-5-phosphate (DOX) synthase.[28]

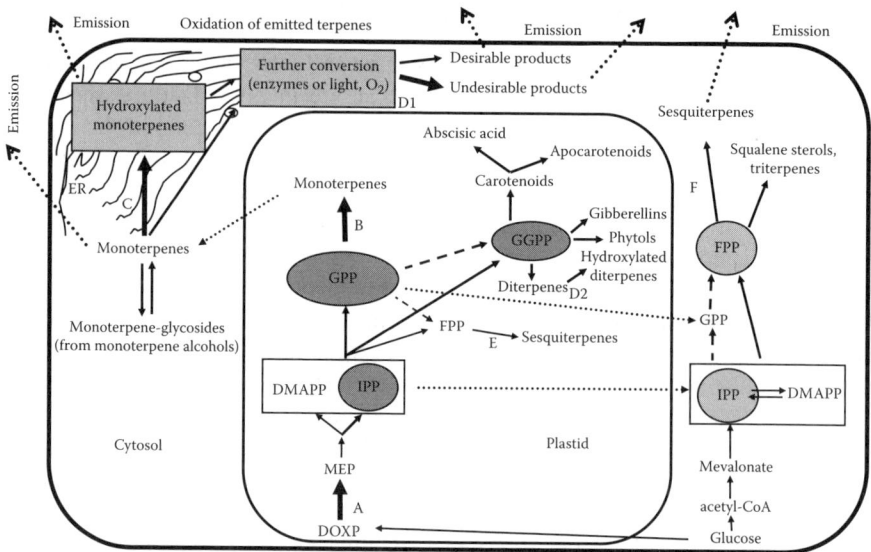

FIGURE 14.2 Schematic outline of terpene metabolism in plants. Steps of terpene biosynthesis that have been modified by metabolic engineering are indicated with solid arrows and letters. Dotted arrows indicate (possible) transport of either precursors or terpenes. Dashed arrows indicate steps or locations that are uncertain. The letters indicate pathway engineering by (A) up-regulation of the MEP pathway,[32] (B) introduction or up-regulation of single or multiple monoterpene synthases,[13–15,17,34,35,41] (C) introduction/up-regulation of monoterpene hydroxylase cytochrome P450 enzymes,[16] (D1–2) down-regulation of further conversions leading to undesirable products in the pathway,[6,7,32] (E) introduction of a sesquiterpene synthase into the plastids,[11] and (F) introduction of a sesquiterpene synthase into the cytosol.[11,58,59]

Monoterpene pathway engineering was initially attempted in essential oil-containing plants, as these plants have specialized compartments for monoterpene production and are capable of producing large amounts of monoterpenes.[29–31] One approach was to increase the total yield of the monoterpenes produced. An increase of 50% was achieved in an experiment with peppermint, where a reductoisomerase of the MEP pathway was overexpressed (Figure 14.2A).[32] The same paper also reported on an attempt to change the composition of the essential oil. To achieve this, the production of an undesired metabolite in peppermint essential oil, menthofuran, was down-regulated by the introduction of an antisense construct of the cytochrome P450 enzyme menthofuran synthase (Figure 14.2D1).[32,33] The level of menthofuran could be reduced by 50% compared to the amount present in the wild-type plants.[32] Another attempt to modify the composition or quality of the essential oil was carried out by transformation of a (−)-limonene synthase (Figure 14.2B) into peppermint. Introduction of this enzyme resulted in small qualitative and quantitative variations in the composition of the essential oil.[34]

Tobacco has proven to be a suitable model for monoterpene metabolic engineering. Transgenic plants expressing the *Perilla frutescens* limonene synthase produced limonene in the leaves (Figure 14.2B).[35] Production of limonene in plastid-targeted limonene synthase-transformed plants was much higher than when the introduced limonene synthase was targeted to the cytosol. Plants with limonene targeted to the endoplasmic reticulum did not produce any limonene.[35] Flowers of these plants were not analyzed for limonene emission.

Introduction of three lemon monoterpene synthases, all regulated by CaMV-d35S promoters, in tobacco was achieved by successive crossing of primary transgenic lines with single-gene inserts that were obtained by three individual transformation experiments (Figure 14.2B).[17] This approach was used previously for the engineering of a polyhydroxybutyrate pathway into plants, where three genes performing subsequent steps of the polyhydroxybutyrate pathway, all regulated by a CaMV-35S promoter, were shown to be active in one plant after subsequent crossings.[36] In the experiments of Lücker et al., for the first time three foreign genes that were competing for the same substrate were introduced into one plant.[37] By the introduction of three monoterpene synthases, the transgenic tobacco plants emitted significant amounts of three new products, γ-terpinene, (+)-limonene, and (−)-β-pinene (Figure 14.3).[17,38] In addition to the main products, a large number of minor compounds were detected: α-thujene, (+/−)-α-pinene, (+)-β-pinene, myrcene, (+/−)-sabinene, α-terpinene, (−)-limonene, p-cymene, and terpinolene (Figure 14.3).[17,38] All of the major and minor compounds were, by heterologous characterization,

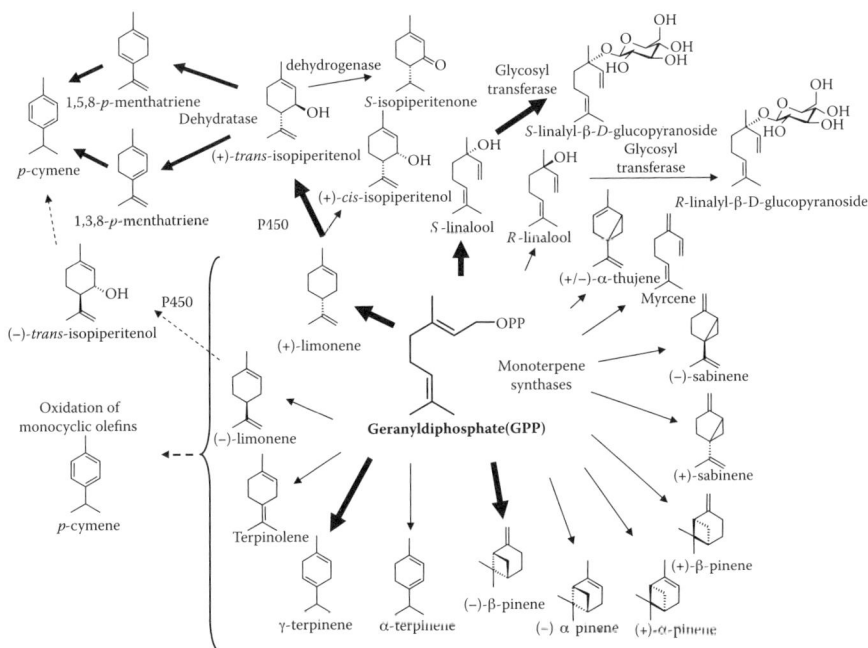

FIGURE 14.3 Monoterpenes and derivatives formed from GPP after metabolic engineering of tobacco and petunia by Lücker et al.[15–17] The main products are indicated with bold arrows. The formation of minor products is indicated by normal arrows. Dashed arrows indicate putative conversions. P450 is the *M. spicata* "Crispa" limonene hydroxylase (cytochrome P450) that was used for tobacco transformation.[16]

shown to be products of the three introduced enzymes.[38] Finally, the volatile products were found to be emitted by the leaves as well as the flowers of the transgenic plants. A first attempt at sensory analysis of these plants by a human panel resulted in the detection of a significant difference between the transgenic and the control plants.[39] However, in a series of sensory attributes, no significant differences could be discriminated, probably because of insufficient panel training.[39]

The introduction of monoterpene formation by metabolic engineering sometimes proves to be less straightforward than expected. Several studies have shown that endogenous enzymes of the transgenic host plant modified the monoterpenoids introduced by metabolic engineering. Such further modification occurred when the linalool synthase from *Clarkia breweri*[40] was introduced in petunia, tomato, and carnation (Figure 14.2B).[13–15] In tomato, linalool production was restricted to the fruit by using the fruit-specific *E8* promoter, and this resulted in high levels of linalool and 8-hydroxy-linalool, probably as a consequence of the presence of an endogenous P450 enzyme that can hydroxylate linalool.[14] Carnation normally does not produce any monoterpenes, but in transgenic *C. breweri* linalool synthase-expressing carnation, emission of linalool was detected from flowers and leaves, albeit at low levels. Most of the introduced linalool in carnation was oxidized to linalool oxides, which were detected at higher levels in the plant tissue than linalool itself.[13] Introduction of the *C. breweri* (*S*)-linalool synthase into *Petunia hybrida* W115, another plant with no detectable monoterpenes, led to the production of high levels of (*S*)-linalool (Figure 14.2B).[15] However, the linalool was not detectable in the headspace, as it was directly converted by a highly efficient endogenous glycosyltransferase to nonvolatile (*S*)-linalyl-β-D-glucopyranoside (Figure 14.3).[15] The same binary vector containing the (*S*)-linalool synthase complementary DNA (cDNA) was used to transform microtom tomato and tobacco,[41] resulting in accumulation of a glycosylated form of linalool. However, in contrast to *Petunia*, in tobacco the glycosylation was incomplete, since substantial amounts of free linalool were detected in the headspace.[41] When a linalool/nerolidol synthase from strawberry was introduced into *Arabidopsis*, another plant species lacking special storage organs for monoterpenes, part of the produced linalool was converted to a series of further oxidized or glycosylated metabolites.[11] Nevertheless, a substantial amount of free linalool was also produced and emitted by *Arabidopsis* leaves and flowers. In conclusion, the results obtained by several groups with the transformation of linalool synthases to various plant species show that monoterpene alcohols, produced by metabolic engineering, in tissues not adapted to these compounds, can be partially converted by oxidation and glycosylation into less reactive or less phytotoxic products (Figure 14.2). Monoterpenes such as linalool, at high concentrations, are known to be detrimental to biological tissues.[42–44] While conjugated linalool can be found in large amounts, indicating that there is a lot of available GPP and linalool synthase activity is probably very high, free linalool is never detected in large amounts in nonadapted transgenic plant tissues.[11,15]

An even more challenging metabolic engineering approach was the introduction of two consecutive steps, spanning two cell compartments, of a monoterpene biosynthetic pathway in one plant (Figure 14.2B and C).[16] A transgenic tobacco line already expressing three monoterpene synthases[17] was subsequently transformed with a cDNA encoding a cytochrome P450 limonene hydroxylase from *Mentha spicata* "Crispa" (Figure 14.2C). In the obtained transgeniclines expressing the P450 cDNA this resulted in the detection of a new major product in the headspace of the flowers and the leaves, which was identified as (+)-*trans*-isopiperitenol.[16] In addition to (+)-*trans*-isopiperitenol, a number of side products could be detected such as isopiperitenone and *p*-cymene (Figure 14.3).[16] Isopiperitenone was probably formed from isopiperitenol by an endogenous alcohol dehydrogenase of tobacco. Indeed, it was shown that monoterpene alcohols, such as carveol, can be converted to the corresponding ketone by cultured suspension cells of tobacco, indicating the presence of endogenous alcohol dehydrogenase activity.[45] The increase in the level of *p*-cymene that was detected can be explained by assuming an endogenous dehydratase activity.[16] Two other products, tentatively identified as 1,3,8-*p*-menthatriene and 1,5,8-*p*-menthatriene could be putative intermediates in such dehydratase-initiated conversion of isopiperitenol to *p*-cymene (Figure 14.3). In this way a variety of monoterpenoids were created in transgenic tobacco plants and the volatile profile extensively altered. Products and possible conversions are indicated in Figure 14.3. The engineered cytochrome P450 was most likely localized in the endoplasmic reticulum, where it functions as a complex with an endogenous NADPH-cytochrome P450 reductase (Figure 14.2). As the monoterpene synthases catalyze the production of monoterpenes in the plastids, this implies that a transport mechanism between the plastids and the cytosol, possibly via lipophilic oil bodies, also operates on monoterpenes introduced through metabolic engineering. Such a transport mechanism was suggested to be present in plants containing essential oil, but is apparently also present in plants, such as tobacco, not explicitly adapted to the production of large amounts of monoterpenes.[25,46,47]

The results of these metabolic engineering experiments sometimes shed new light on hypotheses about regulation of terpene biosynthesis. For example, it is assumed that there is no isopentenyl diphosphate (IPP) isomerase present in the MEP pathway, which would result in a constant ratio of dimethylallyl diphosphate (DMAPP) to IPP of 1:5 in the plastids (Figure 14.2).[48] For the formation of GPP, only one molecule of IPP and one molecule of DMAPP are required. For the formation of geranylgeranyl diphosphate (GGPP), only three molecules of IPP and one molecule of DMAPP are required. Therefore there is probably an excess of IPP in the plastids that might be transferred to the cytosol (Figure 14.2). Reports show that plastidic IPP indeed was incorporated into sesquiterpenes in chamomile and peppermint.[49,50] In experiments using specific pathway inhibitors for the MEP or the mevalonate pathway, indirect evidence was found for an export system of isoprenoid precursors from the plastids to the cytosol in *Arabidopsis*.[51] Others have shown that isolated membrane fractions (including chloroplast membranes) from three different plant species transported IPP and GPP across the membranes, suggesting the presence of a unidirectional proton symport system for the transport of specific isoprenoid intermediates across plastid membranes (Figure 14.2).[52] Due to the up-regulated formation of monoterpenoids in the transgenic tobacco plants described above, there may be less isoprenoid precursors transported to the cytosol for the production of sesquiterpenes. Indeed, the emission level of the sesquiterpene β-caryophyllene from flowers of transgenic tobacco plants expressing the three lemon monoterpene synthases was two- to threefold lower in the transgenic flowers than in control flowers, except for the flower bud stage.[17]

Interestingly, the emission levels of linalool remained unaltered. In a recent paper, where the flux of MEP pathway-derived IPP and DMAPP to the sesquiterpene nerolidol was determined by deuterated precursor feeding experiments in snapdragon, it was shown that sesquiterpene biosynthesis in snapdragon exclusively relies on plastid-derived precursors produced via the diurnally regulated MEP pathway.[53] Therefore it is likely that the level of β-caryophyllene production in tobacco is severely reduced by an increase in the flux of IPP and DMAPP to GPP, triggered by the activity of overexpressed lemon monoterpene synthase enzymes in the transgenic plants.[17]

14.2.2 METABOLIC ENGINEERING OF OTHER TERPENOID PATHWAYS

Another way to elevate the volatile emissions from flowers would be to enzymatically release glycosidically bound volatiles from the site of storage by introduction of a glycosidase. From intact leaves of tobacco, where a fungal β-glycosidase was introduced by genetic engineering targeted to specific cell compartments, the emission of β-caryophyllene, cembrene, and 2-ethyl-hexanol was elevated, although the emitted sesquiterpene and diterpene were not stored as glycosides in the plant.[54] In addition, elevated emission of linalool, nerol, furanoid cis-linalool oxide, 4-methyl-1-pentanol, 6-methyl-hept-5-en-2-ol and 2-ethylhexanol, and 3-hydroxyl-β-ionone was detected compared to the controls after the leaves had been crushed.[54]

With the recent discovery that carotenoid cleavage dioxygenases (CCDs) are responsible for the formation of the highly fragrant apocarotenoids (e.g., β-ionone, which is a component of the flower scents of many plants), new ways to alter flower scent have become feasible.[55,56] Although there are no publications yet on the upregulation of these enzymes in plants by genetic engineering, in tomato a CCD enzyme has been expressed in antisense orientation, resulting in a significant decrease of the flavor volatiles β-ionone, pseudoionone, and geranylacetone.[57]

Attempts to increase the production of sesquiterpenes by metabolic engineering have so far resulted in minor increases in the desired product, likely as a result of the limited precursor availability and tight regulation of the cytosolic mevalonate pathway.[21] When a fungal trichodiene synthase, a sesquiterpene synthase, was transformed to tobacco plants, only traces of trichodiene were formed, although enzyme activity could be easily detected (Figure 14.2F).[58] Transformation of a chicory germacrene A synthase to *Arabidopsis* and an *Artemisia annua* amorpha-4,11-diene synthase to tobacco resulted in the production of only trace amounts of the expected products (Figure 14.2F).[11,59] Apparently, for a significant increase of sesquiterpene formation through metabolic engineering, precursor availability has to be improved.[27] Sesquiterpene biosynthesis occurs in the cytosol and depends on the supply of farnesyl diphosphate (FPP) produced by FPP synthase. FPP is produced from precursors IPP and DMAPP originating from the mevalonate pathway, which is tightly regulated and limited by hydroxymethyl glutaryl-CoA reductase (HMGR), but can use IPP and DMAPP derived from the plastidial MEP pathway.[52,53] A new possible approach using metabolic engineering would be to introduce the production of FPP and of sesquiterpenes from this FPP into the plastids. Indeed, in transgenic *Arabidopsis*, where a strawberry dual function linalool/nerolidol synthase was introduced into the plastids, also the sesquiterpene nerolidol was formed in addition to the expected linalool (Figure 14.2E). This indicates that there is probably already some FPP available in the plastids that is accessible for an introduced sesquiterpene synthase.[11]

An alternative approach for the modification of floral scent profiles would be the use of transcription factors. Transcription factors, in contrast to structural genes, can control multiple pathway steps, so they could possibly be useful for the alteration of complex metabolic pathways.[60,61] Up-regulation of certain transcription factors could also lead to unexpected up-regulation of a series of other, perhaps undesired, genes.[62,63] It is likely that certain transcription factors may enhance the transport and deposition of metabolites in addition to increasing the pathway flux. This could represent a potential advantage of this approach for metabolic engineering, in contrast to engineering multiple steps of a pathway, as one of the overexpressed steps might limit the accumulation process of the end product in the pathway.[61] For terpenoid biosynthetic pathways, no transcription factors have been described to date, except for an ORCA3 transcription factor that regulates the terpenoid indole alkaloid pathway in *Catharanthus roseus*.[64] However, the genes affected by this transcription factor were all involved in the production of the precursors for the indole part of the pathway. The genes involved in the production of the terpenoid precursor were not affected.[64]

14.3 FUNCTIONAL IMPLICATIONS OF METABOLIC ENGINEERING

14.3.1 ECOLOGICAL CONSEQUENCES

Modification of volatile formation in plants by metabolic engineering can also have effects on how the plant interacts with its environment. Pollinator attraction can be altered or the deterrence of insects can be lost or induced.[65–70] Even attracting predators of herbivores in tritrophic interactions should be considered.[19,71] *In vitro* experiments showed that monoterpenes, sesquiterpenes, and phenylpropanoids can have fungicidal or antimicrobial effects.[72–75] Therefore it is probably advisable that plants emitting novel volatiles be compared with their wild type with respect to their ecological interactions before commercialization. Petunia plants expressing a linalool synthase showed a delayed and less severe natural infection by mildew than the nontransformed plants under standard greenhouse conditions.[41] Microtom tomato fruits, transformed with the same gene, were much more resistant to postharvest pathogens than the nontransgenic controls.[41] Also, effects of transgenic, volatile-producing plants on insects have been reported. In dual-choice assays, *Arabidopsis* plants transformed with the dual function linalool/nerolidol synthase, emitting higher linalool levels than the control plants, significantly repelled *Myzus persicea*.[11] Transgenic tobacco plants transformed with lemon monoterpene synthases, mainly emitting limonene, β-pinene, and γ-terpinene were much less visited by herbivorous insects (e.g., white flies), but more by fruit flies, than wild-type tobacco plants in the same greenhouse compartment.[41] Fruit flies have been reported to be attracted to limonene.[69] Diterpenoid cembratrienols that were increased in trichome exudates of transgenic tobacco plants were found to result in greater resistance to aphids, even in field tests.[6,7]

An even more complicated interaction between a number of organisms is present in multitrophic interactions.[71] In this case, plants start de novo biosynthesis of volatiles (mostly terpenoids) in order to attract carnivores (predators and parasitoids) of the herbivores that are foraging on the plant.[71] Such interactions have been shown to occur as a resistance mechanism in cotton, potato, cucumber, *Nicotiana attenuata*, lima bean, maize, and *Arabidopsis*.[19,76–79] Among the novel volatiles that are emitted by plants as a response to herbivory are some monoterpenes such as linalool and ocimene. Experiments with transgenic potato, emitting linalool, have already demonstrated that a change in volatile emission can affect tritrophic interactions. Predatory mites were significantly more attracted to uninfested, transgenic, linalool-producing potato plants than to uninfested wild-type plants.[19]

14.3.2 PHYSIOLOGICAL EFFECTS

The introduction of a new flux toward the production of high levels of monoterpenes may also alter the natural flux of precursors to other related pathways.[17] The decrease in the precursor flux to the sesquiterpenoid pathway in transgenic monoterpene-producing tobacco, which led to a lower emission of β-caryophyllene, was described in Section 14.2.1. There, the introduced monoterpene synthases competed for the plastidic pool of IPP, thus lowering the flux of IPP to the cytosol, and thus sesquiterpene biosynthesis. Changes in the flux to higher terpenoids, such as the carotenoids, gibberellins, or abscisic acid, could have dramatic physiological effects in transgenic plants (Figure 14.2). Indeed, overexpression of a phytoene synthase in tomato resulted in a dwarf phenotype that was explained by severe inhibition of gibberellin biosynthesis by competition for the common substrate GGPP (Figure 14.2).[80] Transgenic *Arabidopsis* lines with a high expression of strawberry linalool/nerolidol synthase also showed growth retardation compared with control plants.[11] Tobacco plants that were homozygous for single monoterpene synthases showed a reduction in the length of the shoot of about 20% compared with the control plants (Figure 14.4).[41] The decrease in length resulted from a decrease in internodal length, which was most severe on the internodes just below the inflorescence of the mature tobacco plants.[41] If the endogenous pool of plastidic isoprenoid precursors is depleted by the introduced monoterpene synthases, fewer precursors are available for the synthesis of GGPP and subsequently derived plant growth regulators, such as gibberellins and abscisic acid. Increased gibberellin biosynthesis in transgenic trees has been shown to stimulate growth.[81]

The growth reduction in the transgenic *Arabidopsis* and tobacco plants[11,41] may also have been caused by a phytotoxic effect of the monoterpenes themselves produced in large amounts in a nonadapted organ.[11,43,82,83] Seeds of the transgenic tobacco plants did not show any dormancy after long-time storage at room temperature, while the control seeds did.[41] A lower level of abscisic acid, which is also produced from plastidic precursors, could perhaps explain this decrease in dormancy, as maternal abscisic acid is responsible for the onset of dormancy in developing seeds.[84] Also, in some of the linalool synthase-expressing microtom tomato lines, many seeds showed

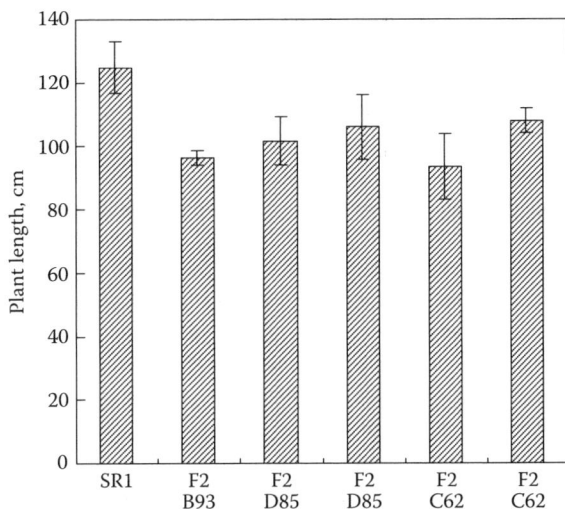

FIGURE 14.4 Total plant length (sum of all internodes) of control (SR1) and five homozygous transgenic lines, expressing γ-terpinene synthase (B93, one plant line), β-pinene synthase (D85, two plants lines), and limonene synthase (C62, two plants lines) (average of six plants ± standard deviation).

precocious germination,[41] suggesting a deficiency in abscisic acid during seed maturation. This is supported by the fact that the fruits of transgenic plants were orange instead of red, probably as a result of a lower carotenoid (lycopene) level. As abscisic acid is synthesized from carotenoid precursors, a concomitant decrease of lycopene and abscisic acid would not be unlikely.[85] It is clear that overexpression of monoterpene synthases can lead to phytotoxic effects and a decrease in precursor flux to other terpenoid pathways (Figure 14.2).[41]

14.4 COMMERCIAL ASPECTS AND GOALS FOR THE FUTURE

The possibility of the construction of a functional two-step or two-compartment pathway for monoterpene biosynthesis[16] opens up the possibility of producing any desired flavor or fragrance compound in plants. With combinations of monoterpene synthases and hydroxylating enzymes and different levels of enzyme activity, a whole range of changes in scent profiles of a plant can be accomplished, as displayed in Figure 14.3. When this approach is used for ornamental plants, they could be genetically engineered to produce fragrant monoterpenes and monoterpene derivatives according to the wishes of the consumer. However, care has to be taken to carefully examine the resulting plants, since endogenous conversions could result in the production of unexpected metabolites. On the other hand, better metabolite profiling techniques are becoming available that will aid in the evaluation of metabolically engineered plants.[86] The different monoterpenes produced in our experiments gave a different scent to the flowers, depending upon which monoterpene synthase was introduced. The combination of three monoterpene synthases in one plant resulted in the observation of a fruity smell in tobacco flowers and leaves. But most striking was the menthol-like fragrance of the plants obtained after transformation with the gene for cytochrome P450, in which (+)-*trans*-isopiperitenol was emitted in addition to all the other monoterpenoids.[41] Chemically synthesized isopiperitenol is described as having a pleasant musky-menthol odor and could become a base material in the perfume industry.[87]

Several ecological effects have been observed on transgenic plants producing novel terpenoids, which apparently modify the degree of resistance to pests and diseases.[6,7,11,41] Therefore resistance to certain pests and diseases could be improved by modification of the terpenoid profile in transgenic plants. However, before possible effective compounds can be implemented for plant resistance, more ecological, molecular, and biochemical research is necessary, since the biosynthesis of such compounds may need to be specifically targeted and regulated. It has been shown that it is feasible to obtain an extensive elevation and alteration of the monoterpene fragrance profile of a plant without the need to produce the desired volatiles targeted to specialized secretory structures using rather straightforward approaches of metabolic engineering.

Now that proof of the concept has been given, metabolic engineering can be applied commercially in the near future, for example, to produce flower-specific attractants to improve pollination of seed crops or to produce volatiles attractive to humans in commercial cut flowers. Scientific challenges will be the engineering of sesquiterpene formation and the fine-tuning of metabolic engineering by using specific promoters to target gene expression to where and when it is needed. Once specific transcription factors have been identified that alter the expression of fragrance metabolite-producing pathways, these will also be applied to modify floral scent. In addition, the instrument of metabolic engineering will create opportunities to study the importance of volatile metabolites in interactions between plants and other organisms, and will increase our understanding of the regulation of volatile metabolite biosynthesis.

REFERENCES

1. Burdock, G.A., *Fenaroli's Handbook of Flavor Ingredients*, 4th ed., CRC Press, Boca Raton, FL, 2002.
2. Zuker, A., Tzfira, T., and Vainstein, A., Genetic engineering for cut-flower improvement, *Biotechnol. Adv.* 16, 33, 1998.
3. Verhoeven, H.A., Blaas, J., and Brandenburg. W.A., Fragrance profiles of wild and cultivated roses, in *Encyclopedia of Rose Science*, Roberts, A.V. and Debener, T., Eds., Academic Press, Amsterdam, 2003, p. 240.
4. Dudareva, N. and Pichersky, E., Biochemical and molecular genetic aspects of floral scents, *Plant Physiol.* 122, 627, 2000.
5. Vainstein, A., Lewinsohn, E., Pichersky, E., and Weiss, D., Floral fragrance: new inroads into an old commodity, *Plant Physiol.* 127, 1383, 2001.

6. Wang, E.M., Hall, J.T., and Wagner, J.G., Transgenic *Nicotiana tabacum* L. with enhanced trichome exudate cembratrieneols has reduced aphid infestation in the field, *Mol. Breed.* 13, 49, 2004.
7. Wang, E.M., Wang, R., DeParasis, J., Loughrin, J.H., Gan, S., and Wagner, G.J., Suppression of a P450 hydroxylase gene in plant trichome glands enhances natural-product-based aphid resistance, *Nat. Biotechnol.* 19, 371, 2001.
8. Griffin, S.G., Wyllie, S.G., Markham, J.L., and Leach, D.N., The role of structure and molecular properties of terpenoids in determining their antimicrobial activity, *Flavour Fragr. J.* 14, 322, 1999.
9. Beckstrom-Sternberg, S.M. and Duke, J.A., *Handbook of Medicinal Mints (Aromathematics): Phytochemicals and Biological Activities*, CRC Press, Boca Raton, FL, 1996.
10. Neirotti, E., Moscatelli, M., and Tiscornia, S., Antimicrobial activity of the limonene, *Arq. Biol. Tecnol.* 39, 233, 1996.
11. Aharoni, A., Giri, A.P., Deuerlein, S., Griepink, F., de Kogel, W.J., Verstappen, F.W., Schwab, W., and Bouwmeester, H.J., Terpenoid metabolism in wild-type and transgenic *Arabidopsis* plants, *Plant Cell* 15, 2866, 2003.
12. Beekwilder, J., Alvarez-Huerta, M., Neef, E., Verstappen, F.W., Bouwmeester, H.J., and Aharoni, A., Functional characterization of enzymes forming volatile esters from strawberry and banana, *Plant Physiol.* 135, 1865, 2004.
13. Lavy, M., Zuker, A., Lewinsohn, E., Larkov, O., Ravid, U., Vainstein, A., and Weiss, D., Linalool and linalool oxide production in transgenic carnation flowers expressing the *Clarkia breweri* linalool synthase gene, *Mol. Breed.* 9, 103, 2002.
14. Lewinsohn, E., Schalechet, F., Wilkinson, J., Matsui, K., Tadmor, Y., Nam, K.-H., Amar, O., Lastochkin, E., Larkov, O., Ravid, U., Hiatt, W., Gepstein, S., and Pichersky, E., Enhanced levels of the aroma and flavor compound *S*-linalool by metabolic engineering of the terpenoid pathway in tomato fruits, *Plant Physiol.* 127, 1256, 2001.
15. Lücker, J., Bouwmeester, H.J., Schwab, W., Blaas, J., van der Plas, L.H.W., and Verhoeven, H.A., Expression of *Clarkia S*-linalool synthase in transgenic petunia plants results in the accumulation of *S*-linalyl-β-D-glucopyranoside, *Plant J.* 27, 315, 2001.
16. Lücker, J., Schwab, W., Franssen, M.C., van der Plas, L.H.W., Bouwmeester, H.J., and Verhoeven, H.A., Metabolic engineering of monoterpene biosynthesis: two-step production of (+)-*trans*-isopiperitenol by tobacco, *Plant J.* 39, 135, 2004.
17. Lücker, J., Schwab, W., van Hautum, B., Blaas, J., van der Plas, L.H.W., Bouwmeester, H.J., and Verhoeven, H.A., Increased and altered fragrance of tobacco plants after metabolic engineering using three monoterpene synthases from lemon, *Plant Physiol.* 134, 510, 2004.
18. Zuker, A., Tzfira, T., Ben-Meir, H., Ovadis, M., Shklarman, E., Itzhaki, H., Forkmann, G., Martens, S., Neta-Sharir, I., Weiss, D., and Vainstein, A., Modification of flower color and fragrance by antisense suppression of the flavanone 3-hydroxylase gene, *Mol. Breed.* 9, 33, 2002.
19. Bouwmeester, H.J., Kappers, I.F., Verstappen, F.W.A., Aharoni, A., Luckerhoff, L.L.P., Lücker, J., Jongsma, M.A., and Dicke, M., Exploring multi-trophic plant-herbivore interactions for new crop protection methods, in *The International Congress Crop Science and Technology*, British Crop Protection Council, Alton, UK, Glasgow, 2003, p. 1123.
20. Bohlmann, J., Meyer-Gauen, G., and Croteau, R., Plant terpenoid synthases: molecular biology and phylogenetic analysis, *Proc. Natl. Acad. Sci. USA* 95, 4126, 1998.

21. Chappell, J., The genetics and molecular genetics of terpene and sterol origami, *Curr. Opin. Plant Biol.* 5, 151, 2002.
22. Haudenschild, C.D. and Croteau, R.B., Molecular engineering of monoterpene production, *Genet. Eng.* 20, 267, 1998.
23. McCaskill, D. and Croteau, R.B., Prospects for the bioengineering of isoprenoid biosynthesis, *Adv. Biochem. Eng. Biotechnol.* 55, 107, 1997.
24. McCaskill, D. and Croteau, R., Some caveats for bioengineering terpenoid metabolism in plants, *Trends Biotechnol.* 16, 349, 1998.
25. Little, D.B., and Croteau, R.B., Biochemistry of essential oil terpenes: a thirty year overview, in *Flavor Chemistry: 30 Years of Progress*, Teranishi, R. and Wick, E.L., Eds., Kluwer Academic/Plenum Publishers, New York, 1999, p. 239.
26. Mahmoud, S.S. and Croteau, R.B., Strategies for transgenic manipulation of monoterpene biosynthesis in plants, *Trends Plant Sci.* 7, 366, 2002.
27. Chappell, J., Valencene synthase: a biochemical magician and harbinger of transgenic aromas, *Trends Plant Sci.* 9, 266, 2004.
28. Wolfertz, M., Sharkey, T.D., Boland, W., and Kuhnemann, F., Rapid regulation of the methylerythritol 4-phosphate pathway during isoprene synthesis, *Plant Physiol.* 135, 1939, 2004.
29. Fahn, A., *Secretory Tissues in Plants*, Academic Press, London, 1979.
30. Gershenzon, J., Maffei, M., and Croteau, R., Biochemical and histochemical localization of monoterpene biosynthesis in the glandular trichomes of spearmint (*Mentha spicata*), *Plant Physiol.* 89, 1351, 1989.
31. Gershenzon, J., McConkey, M.E., and Croteau, R.B., Regulation of monoterpene accumulation in leaves of peppermint, *Plant Physiol.* 122, 205, 2000.
32. Mahmoud, S.S. and Croteau, R.B, Metabolic engineering of essential oil yield and composition in mint by altering expression of deoxyxylulose phosphate reductoisomerase and menthofuran synthase, *Proc. Natl. Acad. Sci. USA* 98, 8915, 2001.
33. Bertea, C.M., Schalk, M., Karp, F., Maffei, M., and Croteau, R., Demonstration that menthofuran synthase of mint (*Mentha*) is a cytochrome P450 monooxygenase: cloning, functional expression, and characterization of the responsible gene, *Arch. Biochem. Biophys.* 390, 279, 2001.
34. Krasnyanski, S., May, R.A., Loskutov, A., Ball, T.M., and Sink, K.C., Transformation of the limonene synthase gene into peppermint (*Mentha x piperita* L.) and preliminary studies on the essential oil profiles of single transgenic plants, *Theor. Appl. Genet.* 99, 676, 1999.
35. Ohara, K., Ujihara, T., Endo, T., Sato, F., and Yazaki, K., Limonene production in tobacco with *Perilla* limonene synthase cDNA, *J. Exp. Bot.* 54, 2635, 2003.
36. Nawrath, C., Poirier, Y., and Somerville, C., Targeting of the polyhydroxybutyrate biosynthetic pathway to the plastids of *Arabidopsis thaliana* results in high levels of polymer accumulation, *Proc. Natl. Acad. Sci. USA* 91, 12760, 1994.
37. Capell, T. and Christou, P., Progress in plant metabolic engineering, *Curr. Opin. Biotech.* 15, 148, 2004.
38. Lücker, J., El Tamer, M.K., Schwab, W., Verstappen, F.W., van der Plas, L.H.W., Bouwmeester, H.J., and Verhoeven, H.A., Monoterpene biosynthesis in lemon (*Citrus limon*): cDNA isolation and functional analysis of four monoterpene synthases, *Eur. J. Biochem.* 269, 3160, 2002.
39. El Tamer, M.K., Smeets, M., Holthuysen, N., Lücker, J., Tang, A., Roozen, J., Bouwmeester, H.J., and Voragen, A.G.J., The influence of monoterpene synthase transformation on the odour of tobacco, *J. Biotechnol.* 106, 15, 2003.

40. Dudareva, N., Cseke, L., Blanc, V.M., and Pichersky, E., Evolution of floral scent in *Clarkia*: novel patterns of S-linalool synthase gene expression in the *C. breweri* flower, *Plant Cell* 8, 1137, 1996.

41. Lücker, J., Metabolic engineering of monoterpene biosynthesis in plants, Ph.D. dissertation, Wageningen University, 2002.

42. Izumi, S., Takashima, O., and Hirata, T., Geraniol is a potent inducer of apoptosis-like cell death in the cultured shoot primordia of *Matricaria chamomilla*, *Biochem. Biophys. Res. Commun.* 259, 519, 1999.

43. Vaughn, S.F. and Spencer, G.F., Volatile monoterpenes inhibit potato tuber sprouting, *Am. Potato J.* 68, 821, 1991.

44. Weidenhamer, J.D., Macias, F.A., Fischer, N.H., and Williamson, G.B., Just how insoluble are monoterpenes?, *J. Chem. Ecol.* 19, 1799, 1993.

45. Suga, T. and Hirata, T., Biotransformation of exogenous substrates by plant cell cultures. *Phytochemistry* 29, 2393, 1990.

46. Bosabalidis, A.M., Ontogenesis, ultrastructure and morphometry of the petiole oil ducts of celery (*Apium graveolens* L.), *Flavour Fragr. J.* 11, 269, 1996.

47. Bouwmeester, H.J., Gershenzon, J., Konings, M.C.J.M., and Croteau, R., Biosynthesis of the monoterpenes limonene and carvone in the fruit of caraway. I. Demonstration of enzyme activities and their changes with development, *Plant Physiol.* 117, 901, 1998.

48. Rohdich, F., Hecht, S., Gartner, K., Adam, P., Krieger, C., Amslinger, S., Arigoni, D., Bacher, A., and Eisenreich, W., Studies on the nonmevalonate terpene biosynthetic pathway: metabolic role of IspH (LytB) protein, *Proc. Natl. Acad. Sci. USA* 99, 1158, 2002.

49. Adam, K.P. and Zapp, J., Biosynthesis of the isoprene units of chamomile sesquiterpenes, *Phytochemistry* 48, 953, 1998.

50. McCaskill, D. and Croteau, R., Monoterpene and sesquiterpene biosynthesis in glandular trichomes of peppermint (*Mentha x piperita*) rely exclusively on plastid-derived isopentenyl diphosphate, *Planta* 197, 49, 1995.

51. Laule, O., Führholz, A., Chang, H.S., Zhu, T., Wang, X., Heifetz, P.B., Gruissem, W., and Lange, B.M., Crosstalk between cytosolic and plastidial pathways of isoprenoid biosynthesis in *Arabidopsis thaliana*, *Proc. Natl. Acad. Sci.* 100, 6866, 2003.

52. Bick, J.A. and Lange, B.M., Metabolic cross talk between cytosolic and plastidial pathways of isoprenoid biosynthesis: unidirectional transport of intermediates across the chloroplast envelope membrane, *Arch. Biochem. Biophys.* 415, 146, 2003.

53. Dudareva, N., Andersson, S., Orlova, I., Gatto, N., Reichelt, M., Rhodes, D., Boland, W., and Gershenzon, J., The nonmevalonate pathway supports both monoterpene and sesquiterpene formation in snapdragon flowers, *Proc. Natl. Acad. Sci. USA* 102, 933, 2005.

54. Wei, S., Marton, I., Dekel, M., Shalitin, D., Lewinsohn, E., Bravdo, B.A., and Shoseyov, O., Manipulating volatile emission in tobacco leaves by expressing *Aspergillus niger* β-glucosidase in different subcellular compartments, *Plant Biotech. J.* 2, 341, 2004.

55. Bouvier, F., Suire, C., Mutterer, J., and Camara, B., Oxidative remodeling of chromoplast carotenoids: Identification of the carotenoid dioxygenase CsCCD and CsZCD genes involved in crocus secondary metabolite biogenesis, *Plant Cell* 15, 47, 2003.

56. Simkin, A.J., Underwood, B.A., Auldridge, M., Loucas, H.M., Shibuya, K., Schmelz, E., Clark, D.G., and Klee, H.J., Circadian regulation of the PhCCD1 carotenoid cleavage dioxygenase controls emission of beta-ionone, a fragrance volatile of petunia flowers, *Plant Physiol.* 136, 3504, 2004.

57. Simkin, A.J., Schwartz, S.H., Auldridge, M., Taylor, M.G., and Klee, H.J., The tomato carotenoid cleavage dioxygenase 1 genes contribute to the formation of the flavor volatiles beta-ionone, pseudoionone, and geranylacetone, *Plant J.* 40, 882, 2004.

58. Hohn, T.M. and Ohlrogge, J.B., Expression of a fungal sesquiterpene cyclase gene in transgenic tobacco, *Plant Physiol.* 97, 460, 1991.

59. Wallaart, T.E., Bouwmeester, H.J., Hille, J., Poppinga, L., and Maijers, N.C.A., Amorpha-4,11-diene synthase: cloning and functional expression of a key enzyme in the biosynthetic pathway of the novel antimalarial drug artemisinin, *Planta* 212, 460, 2001.

60. Gantet, P. and Memelink, J., Transcription factors: tools to engineer the production of pharmacologically active plant metabolites, *Trends Pharmacol. Sci.* 23, 563, 2002.

61. Broun, P., Transcription factors as tools for metabolic engineering in plants, *Curr. Opin. Plant Biol.* 7, 202, 2004.

62. Bruce, W., Folkerts, O., Garnaat, C., Crasta, O., Roth, B., and Bowen, B., Expression profiling of the maize flavonoid pathway genes controlled by estradiol-inducible transcription factors CRC and P, *Plant Cell* 12, 65, 2000.

63. Grotewold, E., Chamberlin, M., Snook, M., Siame, B., Butler, L., Swenson, J., Maddock, S., and Bowen, B., Engineering secondary metabolism in maize cells by ectopic expression of transcription factors, *Plant Cell* 10, 721, 1998.

64. van der Fits, L. and Memelink, J., ORCA3, a jasmonate-responsive transcriptional regulator of plant primary and secondary metabolism, *Science* 289, 295, 2000.

65. Raguso, R.A. and Pichersky, E., A day in the life of a linalool molecule: chemical communication in a plant-pollinator system. Part 1: linalool biosynthesis in flowering plants, *Plant Species Biol.* 14, 95, 1999.

66. Pichersky, E. and Gershenzon, J., The formation and function of plant volatiles: perfumes for pollinator attraction and defense, *Curr. Opin. Plant Biol.* 5, 237, 2002.

67. Hori, M., Repellency of rosemary oil against *Myzus persicae* in a laboratory and in a screenhouse, *J. Chem. Ecol.* 24, 1425, 1998.

68. Hori, M. and Harada, H., Screening plants resistant to green peach aphid, *Myzus persicae* (Sulzer) (Homoptera: Aphididae), *Appl. Entomol. Zool.* 30, 246, 1995.

69. Jacobson, M., Plants, insects, and man—their interrelationships, *Econ. Bot.* 36, 346, 1982.

70. De Moraes, C.M., Mescher, M.C., and Tumlinson, J.H., Caterpillar-induced nocturnal plant volatiles repel conspecific females, *Nature* 29, 577, 2001.

71. Dicke, M. and van Loon, J.J.A., Multitrophic effects of herbivore-induced plant volatiles in an evolutionary context, *Entomol. Exp. Appl.* 97, 237, 2000.

72. Belaiche, T., Tantaoui, E., and Ibrahimy, A., Application of a two levels factorial design to the study of the antimicrobial activity of three terpenes, *Sci. Aliment.* 15, 571, 1995.

73. Caccioni, D.R.L., Guizzardi, M., Biondi, D.M., Renda, A., and Ruberto, G., Relationship between volatile components of citrus fruit essential oils and antimicrobial action on *Penicillium digitatum* and *Penicillium italicum*, *Int. J. Food Microbiol.* 43, 73, 1998.

74. Tsao, R. and Zhou, T., Antifungal activity of monoterpenoids against postharvest pathogens *Botrytis cinerea* and *Monilinia fructicola*, *J. Essent. Oil Res.* 12, 113, 2000.

75. Agarwal, K.K., Khanuja, S.P.S., Ahmad, A., Santha Kumar, T.R., Gupta, V.K., Kumar, S., Antimicrobial activity profiles of the two enantiomers of limonene and carvone isolated from the oils of *Mentha spicata* and *Anethum sowa*, *Flavour Fragr. J.* 17, 59, 2002.

76. Baldwin, I.T., An ecologically motivated analysis of plant-herbivore interactions in native tobacco, *Plant Physiol.* 127, 1449, 2001.

77. Bouwmeester, H.J., Verstappen, F.W., Posthumus, M.A., and Dicke, M., Spider mite-induced (3*S*)-(E)-nerolidol synthase activity in cucumber and lima bean. The first dedicated step in acyclic C11-homoterpene biosynthesis, *Plant Physiol.* 121, 173, 1999.

78. van Poecke, R.A., Posthumus, M.A., and Dicke, M., Herbivore-induced volatile production by *Arabidopsis thaliana* leads to attraction of the parasitoid *Cotesia rubecula*: chemical, behavioral, and gene-expression analysis, *J. Chem. Ecol.* 27, 1911, 2001.

79. Weissbecker, B., Schütz, S., Klein, A., and Hummel, H.E., Analysis of volatiles emitted by potato plants by means of a Colorado beetle electroantennographic detector, *Talanta* 44, 2217, 1997.

80. Fray, R.G., Wallace, A., Fraser, P.D., Valero, D., Hedden, P., Bramley, P.M., and Grierson, D., Constitutive expression of a fruit phytoene synthase gene in transgenic tomatoes causes dwarfism by redirecting metabolites from the gibberellin pathway, *Plant J.* 8, 693, 1995.

81. Eriksson, M.E., Israelsson, M., Olsson, O., and Moritz, T., Increased gibberellin biosynthesis in transgenic trees promotes growth, biomass production and xylem fiber length, *Nat. Biotechnol.* 18, 784, 2000.

82. Romagni, J.G., Allen, S.N., and Dayan, F.E., Allelopathic effects of volatile cineoles on two weedy plant species, *J. Chem. Ecol.* 26, 303, 2000.

83. Vaughn, S.F. and Spencer, G.F., Synthesis and herbicidal activity of modified monoterpenes structurally similar to cimmethylin, *Weed Sci.* 44, 7, 1996.

84. Bewley, J.D., Seed germination and dormancy, *Plant Cell* 9, 1055, 1997.

85. Chernys, J.T. and Zeevaart, J.A., Characterization of the 9-cis-epoxycarotenoid dioxygenase gene family and the regulation of abscisic acid biosynthesis in avocado, *Plant Physiol.* 124, 343, 2000.

86. Trethewey, R.N., Metabolite profiling as an aid to metabolic engineering in plants, *Curr. Opin. Plant Biol.* 7, 196, 2004.

87. Guillon, J., Rioult, J.-P., and Robba, M., New synthesis of isopiperitenol, previously isolated from species of *Cymbopogon*, *Flavour Fragr. J.* 15, 223, 2000.

Index

A

Abelia grandiflora, 173
Abies grandis, 57
Acacia, 43
Achillea millefolium, 206, 210
Acnistus arborescens, 95
Agrostemma githago, 201
Alocasia odora, 160, 302
Amorphophallus eichleri, 159
Amorphophallus rivieri, 158
Anacamptis pyramidalis, 201, 204–205
Angraecum sesquipedale, 176
Annonaceae, 153
Antirrhinum majus, 8–9, 56–57, 61, 63–65, 115, 117, 119, 126, 128, 134, 138, 303–304
Arabidopsis, 57, 80
Arabidopsis lyrata, 86
Arabidopsis thaliana, 4, 5, 8, 40, 61, 70, 79, 113, 130
    attracting insect pollinators, 86
    as emitting little scent, 80
    inflorescences and, 82
    "reverse genetics" and, 80
Araujia sericofera, 168
Aristolochia gigantea, 158–159
Arum creticum, 155
Arum italicum, 106
Arum maculatum, 159
Arum palaestinum, 160
Asarum proboscideum, 160
Asimina triloba, 160
Aster pilosus, 200
AtTPS genes; *see also* Genes
    expression of, 82
    identifying their function, 84–85
    localizing their expression, 85–86

B

Bats as pollinators
    chiropterophily, 178–179
    flower traits attracting bats, 178
    sulfur-containing volatiles and, 178
Beetles
    bruchid, 155

    carrion, 158
    hopliine, 152
    Scarabaeidae, tropical dynastine, 152–153
    in temperate regions, 154
    tropical nondynastine, 153
Benzenoids, 35, 59–60, 128, 152, 204
Benzoic acid carboxyl methyltransferase (BAMT), 60, 65, 119, 128
Berberis aquifolium, 173
Biochemical pathways; *see also* Biosynthetic pathways
    acyltransferases, 60–62
    CoA-independent non-β--oxidative pathway, 59
    CoA-dependent β-oxidative pathway, 59
    dimethylallyl diphosphate (DMAPP), 56
    geranylgeranyl diphosphate synthase (GGPPS), 56–57
    isopentenyl diphosphate (IPP), 56
    lignin pathway, 59
    lipoxygenase pathway, 57
    methylerythritol phosphate (MEP) pathway, 56
    methyltransferases, 62
    mevalonic acid (MVA) pathway, 56
    orcinol OMTs, 60
    P450 enzyme, 57
    shikimate pathway, 65
Biosynthetic pathways; *see also* Biochemical pathways
    BEAT enzyme, 63
    IEMT enzyme, 63
    LIS enzyme, 63
    spatial and temporal regulation, 62–65
Birds as pollinators
    flower traits attracting birds, 177–178
    role of floral and inflorescence morphology, 177
Browneopsis disepala, 180
Bruguiera gymnorrhiza, 177
Brush or bilabiate flower forms, 153–154
Buddleja davidii, 7, 170, 201, 207, 209–210
Butterflies as pollinators; *see also* Lepidoptera as pollinators
    antennae and olfactory receptors, 200
    adult butterflies, foods of, 200
    butterfly- and moth-pollinated plants, 202, 203–205

339